T5-CVG-053

Statistical Inference

Theory and Practice

THEORY AND DECISION LIBRARY

General Editors: W. Leinfellner and G. Eberlein

SERIES B: MATHEMATICAL AND STATISTICAL METHODS

Volume 17

Scope

The series focuses on the application of methods and ideas of logic, mathematics and statistics to the social sciences. In particular, formal treatment of social phenomena, the analysis of decision making, information theory and problems of inference will be central themes of this part of the library. Besides theoretical results, empirical investigations and the testing of theoretical models of real world problems will be subjects of interest. In addition to emphasizing interdisciplinary communication, the series will seek to support the rapid dissemination of recent results.

The titles published in this series are listed at the end of this volume.

Statistical Inference

Theory and Practice

Edited by

Tadeusz Bromek and Elżbieta Pleszczyńska

Institute of Computer Science of the Polish Academy of Sciences

KLUWER ACADEMIC PUBLISHERS

DORDRECHT/BOSTON/LONDON

PWN—POLISH SCIENTIFIC PUBLISHERS

WARSAW

Library of Congress Cataloging-in-Publication Data

Teoria i praktyka wnioskowania statystycznego. English.
Statistical inference: theory and practice/edited by Tadeusz Bromek and Elżbieta Pleszczyńska.
p. cm.—(Theory and decision library. Series B)
Translation of: Teoria i praktyka wnioskowania statystycznego.
Includes bibliographical references.
ISBN 0-7923-0718-6
1. Mathematical statistics. 2. Probabilities. I. Bromek, Tadeusz. II. Pleszczyńska, Elżbieta. III. Title. IV. Series: Theory and decision library. Series B. Mathematical and statistical methods.
QA276 T4413 1990 90-4293
519.5—dc20
ISBN 0-7923-0718-6

Translated by: Jerzy Bachrach (Ch. 10), Ewa Bednarczuk (Chs. 1, 2, 4, 9, 11, Introduction, Closing remarks and Appendix), Jan Koniarek (Chs. 6 and 7), Maria Radziwiłł (Ch. 8)

This translation has been made from *Teoria i praktyka wnioskowania statystycznego*, published by Państwowe Wydawnictwo Naukowe, Warszawa 1988

English Edition published by Polish Scientific Publishers, Warszawa, Poland, in co-publication with Kluwer Academic Publishers, P.O. Box 17, 3300 AA Dordrecht, The Netherlands

Distributors for Albania, Bulgaria, Cuba, Czechoslovakia, Hungary, Korean People's Democratic Republic, Mongolia, People's Republic of China, Poland, Romania, the U.S.S.R., Vietnam, and Yugoslavia,
ARS POLONA
Krakowskie Przedmieście 7, 00-068 Warszawa, Poland

Distributors for the U.S.A. and Canada
Kluwer Academic Publishers,
101 Philip Drive, Norwell, MA 02061, U.S.A.

Distributors for all remaining countries
Kluwer Academic Publishers,
P. O. Box 322, 3300 AH Dordrecht, The Netherlands

Printed in Poland by D.N.T.

Preface

Use and misuse of statistics seems to be the *signum temporis* of past decades. But nowadays this practice seems slowly to be wearing away, and common sense and responsibility recapturing their position.

It is our contention that little by little statistics should return to its starting point, i.e., to formalizing and analyzing empirical phenomena. This requires the reevaluation of many traditions and the rejection of many myths.

We hope that our book would go some way towards this aim. We show the sharp conflict between what is needed and what is feasible. Moreover, we show how slender are the links between theory and practice in statistical inference, links which are sometimes no more than mutual inspiration.

In Part One we present the consecutive stages of formalization of statistical problems, i.e., the description of the experiment, the presentation of the aim of the investigation, and of the constraints put upon the decision rules. We stress the fact that at each of these stages there is room for arbitrariness. We prove that the links between the real problem and its formal counterpart are often so weak that the solution of the formal problem may have no rational interpretation at the practical level. We give a considerable amount of thought to the reduction of statistical problems.

In Part Two we discuss selected theoretical topics, namely discriminant analysis, screening, and measuring stochastic dependence, and we emphasize their interrelations. But it is only the practical problems discussed in Part Three that make it plain to the reader what are the essential difficulties which have to be tackled by statistics. The comparison of Parts Two and Three proves that ready—made theoretical schemes are of limited value in practice. This becomes evident in a particularly drastic manner when the reader compares Chapter 7 with Chapter 3. Although formally father recognition is a special case of two-class discriminant analysis, in a practical approach to this problem we make very little use of general theory.

This book has been written for a wide circle of readers. Users of statistical methods will probably be particularly interested in the practical problems contained in Part Three. We hope that our book would enhance their efficiency in solving their own statistical problems. The scheme of inference presented in Part One and illustrated in Parts Two and Three is meant for all readers interested in statistical methodology. Moreover, the book may be treated as a source of unconventional examples which can be used in teaching and in studying statistics.

A certain amount of experience and familiarity with statistics will considerably facilitate the reading of this book but broadly speaking the only prerequisite is a knowledge of the basic notions of the theory of probability and of mathematical analysis. The few passages which refer to less elementary notions may be omitted without breaking the continuity of reading. Moreover, the proofs of the theorems given in the book are dispensed with.

Nevertheless, the reader should be warned that the study of the book may prove somewhat difficult because of the subtle controvertiality inherent in statistical inference and the difficulties involved. What is more, the unified system of inference which we have introduced aims at unmasking some traditional inconsistencies in concepts, terminology and notation. This effect is strengthened by the comparison of various domains of applications.

The reader is not compelled to read the book from cover to cover since the chapters are more or less independent, as presented in Fig. 1. Moreover, readers having no formal training in statistics may get an insight into the contents of the book by reading through the Introduction to the book, the introductions to the chapters of Parts Two and Three, and the Closing Remarks.

We are very grateful to all those who have helped us to write this book. Special thanks are due to the reviewers, Tomas Havranek and Jerzy Kucharczyk, and to Jan Oderfeld, Robert Bartoszyński, Janusz Roguski, Jadwiga Lewkowicz, Jacek Koronacki, Stanisław Gnot, Henryk Kotlarski, Maria Moszyńska, Joanna Tyrcha, Ignacy Wald, Anna Bittner, and Ryszard Solich.

Contents

Contributors:

Tadeusz Bromek
Jan Ćwik
Wiera Gáfriková
Małgorzata Gołembiewska-Skoczylas
Kazimierz Grygiel
Teresa Kowalczyk
Adam Kowalski
Jan Mielniczuk
Magdalena Niewiadomska-Bugaj
Elżbieta Pleszczyńska
Wiesław Szczesny
Hubert Szczotka
Zofia Szczotkowa

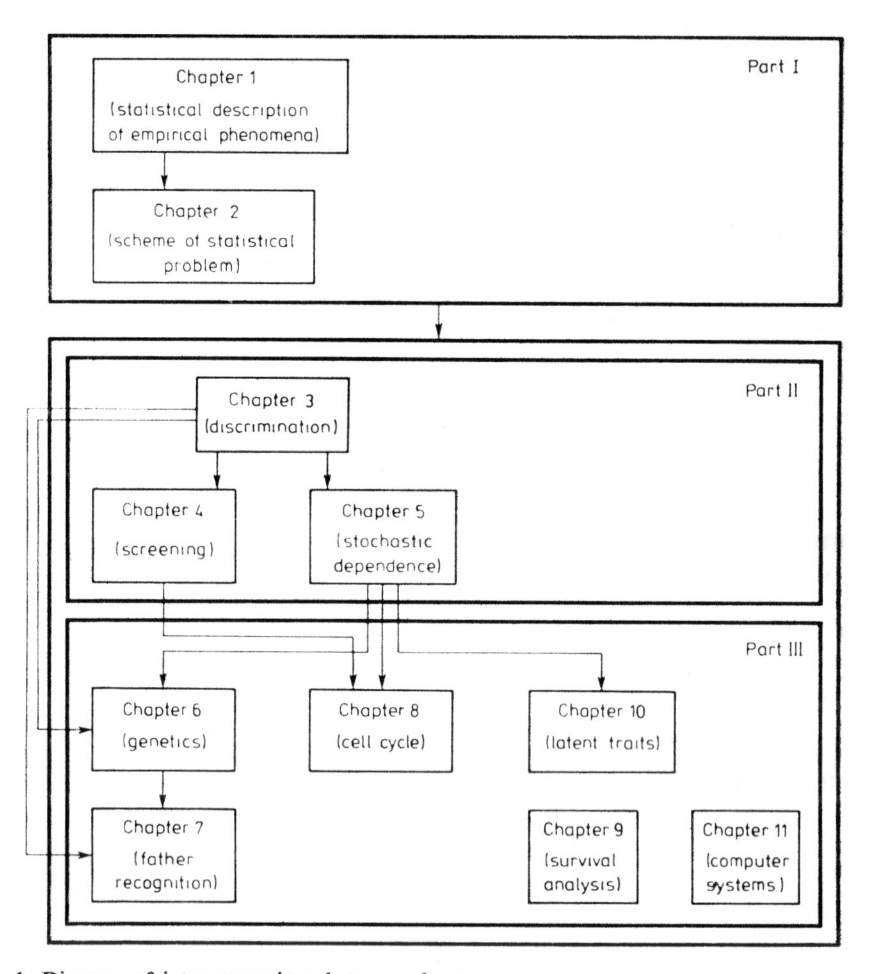

FIG. 1. Diagram of interconnections between chapters.

Introduction

Representing a real-life problem as a formally stated statistical problem is an essential stage of the statistician's work. It is perhaps also the most difficult stage for there exist no formal rules on how to proceed. Handbooks on statistics deal mostly with typical problems while in practice we are faced with a variety of nontypical situations.

Our aim is to discuss the difficulties arising when one tries to formalize practical problems. Whenever possible we shall point out the diversity of approaches existing within the same general framework.

We being with a presentation of several strongly simplified and nontypical examples of statistical problem constructing. These examples concern a special group of problems, namely the so called proof-reading processes. These problems have little in common with real book proof-reading but are of significant illustrative value. Besides, it should be stressed here that all real-life problems analyzed in this volume can be formalized in several different ways and it is unavoidable that each conceivable formalization is more or less controversial, each has its own advantages and disadvantages.

Consider a proof-reading process in which a book is checked by two proof-readers working independently with identical copies of the text. Let n_1 and n_2 denote the number of errors detected by the first and the second proof-reader respectively, and let m denote the number of those errors which were detected by *both* the first and the second proof-reader. Assume that these three numbers are available after completing a proof-reading process. Knowing these numbers, the editor should answer the question whether the number of undetected errors n_0 is small enough to justify submitting the book for printing. To answer this question some information is needed about the total number of errors n. Clearly, we have the equality

$$n_0 = n - n_1 - n_2 + m.$$

The editor is interested in estimating the number of errors n by exploiting the numbers n_1, n_2, m.

Suppose the successive errors appearing in the book are indexed by natural numbers, i.e., the error appearing as the i-th in the book is indexed by i, $i = 1, \ldots, n$. Thus, a realization (though not observable!) of the proof-reading process can be described by a sequence

$$\omega = (n, x_1, y_1, \ldots, x_n, y_n), \tag{1}$$

where x_i $(i = 1, \ldots, n)$ takes on the value 1 if the i-th error is detected by the first proof-reader, and the value 0 otherwise, y_i $(i = 1, \ldots, n)$ takes on the value 1 if the i-th error is detected by the second proof-reader, and the value 0 otherwise.

Each time a particular course of the proof-reading process is the result of many interfering factors. However, when we intend to estimate the total number of errors it seems to be justified to make a simplifying assumption that the realization of the process is generated by a random mechanism.

Moreover, let us assume a unification of errors and a stationary character of the proof-readers' work. These assumptions enables us to assign to each proof-reader a constant probability of error detection; a probability which is independent of the index of an error, and independent of the number, importance and frequency of errors already detected. Denote by p_1 and p_2 the probability of error detection by the first and the second proof-reader, respectively, and by $\pi(n)$ the probability that the total number of errors in the book is equal to n. According to our assumptions, we now use the numbers p_1, p_2, and the distribution π of the total number of errors to define the random mechanism of the process, i.e., the distribution on the set of realizations of the process. If a book under the proof-reading process is regarded as being randomly selected from a population of books being edited, then $\pi(n)$, $n = 1, 2, \ldots, n$, corresponds to those books in which the total number of errors is equal to n. But if we restrict our considerations to one particular book with a fixed, though unknown, number of errors n, then the whole mass of π is concentrated in n, i.e., $\pi(n) = 1$.

The probability of x_i is equal to p_1 if $x_i = 1$ and is equal to $1-p_1$ if $x_i = 0$. Hence, this probability can be written as $p_1^{x_i}(1-p_1)^{1-x_i}$. In the same way we can write down the probability of y_i. Denote by P_{π, p_1, p_2} the probability distribution according to which a realization of the proof-reading process is randomly chosen (i.e., at the first stage a book is randomly selected, and then the proof-reading process is simulated by random detection of successive errors). In view of the independence of the actions of both proof-readers, we have

$$P_{\pi, p_1, p_2}(n, x_1, y_1, \ldots, x_n, y_n) = \pi(n) p_1^{n_1}(1-p_1)^{n-n_1} p_2^{n_2}(1-p_2)^{n-n_2} \tag{2}$$

where $n_1 = x_1 + \ldots + x_n$, $n_2 = y_1 + \ldots + y_n$.

Denote

$$\theta = (\pi, p_1, p_2).$$

If no a priori information about π, p_1, p_2 is available, then θ is an arbitrary element of the Cartesian product of the set of all probability distributions defined on the set of natural numbers and of two intervals $(0, 1)$. The set of all possible values of θ is denoted by $\boldsymbol{\Theta}$. Thus, we conclude that the realizations of the proof-reading process are randomly generated according to a certain distribution from the family

$$\mathscr{P} = (P_\theta,\ \theta \in \boldsymbol{\Theta}).$$

Some assumptions concerning the family $\mathscr{P}$ express explicitly certain known facts about the proof-reading process, e.g., the assumption that both proof-readers work independently of each other. Other assumptions are subjective; they are simply believed to be satisfied, e.g., the assumption that random detections of errors are mutually independent.

A realization ω is accessible only through the observed data n_1, n_2, m. Using these data we wish to estimate the total number of errors, n. To this aim we need a function, called an *estimator of the number of errors*, which assigns to the triples (n_1, n_2, m) a certain number regarded as an approximate n. We can proceed in two ways: either try to determine an estimator with the required properties, or choose an estimator, investigate its properties, and, depending on the results, accept or discard it. We follow the second way and consider the estimator defined by Polya (1976).

According to our assumptions, for $n > 0$, the numbers of errors n_1, n_2 have binomial distributions with the probability of success p_1, p_2, respectively:

$$\text{for } n > 0 \quad (n_i \,|\, n) \sim B(n, p_i), \quad i = 1, 2,$$

and the number of errors m has a binomial distribution with the probability of success $p_1 p_2$:

$$\text{for } n > 0 \quad (m \,|\, n) \sim B(n, p_1 p_2).$$

Hence, when $n > 0$, the expectations of the number of errors n_1, n_2, m are equal to np_1, np_2, np_1p_2, respectively. Representing n in the form

$$n = \frac{(np_1)\,(np_2)}{np_1 p_2}$$

and replacing np_1, np_2, np_1p_2 by n_1, n_2, m, respectively, we obtain the expression $n_1 n_2/m$ which can serve, for $m > 0$, as an estimate of the number n of errors. For $m = 0$, the number of errors is equal at least to $n_1 + n_2$. The expression

$$\hat{n} = \begin{cases} \dfrac{n_1 n_2}{m} & \text{if} \quad m > 0, \\ n_1 + n_2 & \text{if} \quad m = 0, \end{cases}$$

will be called the *Polya estimator of the number of errors n*. Intuitively, the Polya estimator provides a good approximation of the unknown number n of errors at least in the case of large n.

For fixed n, the distribution of the estimator $\hat{n}$ is determined by p_1 and p_2. It is not easy to find an explicit formulae for this distribution and its parameters, e.g. for the expected value $E_{n,p_1,p_2}(\hat{n})$ and the standard deviation $\sigma_{n,p_1,p_2}(\hat{n})$. However, we can investigate the asymptotic properties (when $n \to \infty$) of these parameters.

Denote

$$q = \left(\frac{1}{p_1} - 1\right)\left(\frac{1}{p_2} - 1\right).$$

We find that

$$\lim_{n\to\infty} \left(E_{n,p_1,p_2}(\hat{n}) - n\right) = q, \tag{3}$$

$$\lim_{n\to\infty} \left(\sigma^2_{n,p_1,p_2}(\hat{n}) - nq\right) = 2q^2 + \frac{5}{p_1 p_2} + \frac{1}{p_1^2} + \frac{1}{p_2^2} - \frac{1}{p_1^2 p_2^2} - 1. \tag{4}$$

The difference $n - E_{n,p_1,p_2}(\hat{n})$ is called the *bias of the estimator* $\hat{n}$, and the number q is called the *asymptotic bias of* $\hat{n}$. The estimator $\hat{n}$ is asymptotically biased, i.e., $q > 0$, for $p_1, p_2 \in (0, 1)$. This is an undesirable property. However, this bias can be eliminated by substracting the expression $\hat{q}$ from $\hat{n}$, where $\hat{q}$ is obtained from q by representing p_1 and p_2 in the forms $(np_1p_2)/(np_2)$ and $(np_1p_2)/(np_1)$, respectively, and next replacing np_1, np_2, np_1p_2 by n_1, n_2, m, respectively. In this way we get a modified Polya estimator in the form

$$\tilde{n} = \begin{cases} \hat{n} - \dfrac{n_1 n_2}{m^2} + \dfrac{n_1}{m} + \dfrac{n_2}{m} - 1 & \text{if} \quad m > 0, \\ n_1 + n_2 & \text{if} \quad m = 0. \end{cases} \tag{5}$$

The estimator $\tilde{n}$ is asymptotically unbiased, i.e.,

$$\lim_{n\to\infty} \left(E_{n,p_1,p_2}(n) - n\right) = 0,$$

and the sequence of the standard deviations $\sigma_{n,p_1,p_2}(\tilde{n})$ has the following property

$$\lim_{n\to\infty} \left(\sigma^2_{n,p_1,p_2}(\tilde{n}) - nq\right) = q^2 + \frac{q}{p_1 p_2}. \tag{6}$$

The definition of the estimator $\tilde{n}$ was given in Mielniczuk (1986), where the asymptotic properties of sequences of biases and standard deviations for both estimators were investigated.

The estimates of the exact values of standard deviations, obtained from (4) and (6) for a fixed n, will be denoted by $\sigma^{AS}_{n,p_1,p_2}(\hat{n})$ and $\sigma^{AS}_{n,p_1,p_2}(\tilde{n})$. For several pairs (p_1, p_2) with $p_1 = p_2$ and $n = 50$, the corresponding estimates and the asymptotic bias can be found in Table 1.

Now we analyse the above estimates and regard them, for a while, as equal to the exact values. As we see, the bias q and the standard deviations increase very fast with the decrease of the probabilities p_1 and p_2. For $p_1 = p_2 = 0.3$ and $n = 50$, large values of standard deviations eliminate both estimators as being inadequate for our purposes; in addition, the estimator $\hat{n}$ is considerably biased. If p_1 and p_2 are greater than 0.5, the bias q of the estimator $\hat{n}$ is less than 1, its standard deviation is less than 8 and only slightly exceeds the standard deviation of the estimator $\tilde{n}$. Hence, though the estimator $\tilde{n}$ is better than $\hat{n}$, this is of no importance in practice; more significant differences between properties of these estimators for $n = 50$

TABLE 1.

Asymptotic estimates (AS) and Monte-Carlo estimates (MC) of the bias of the estimator $\hat{n}$ and the standard deviations of the estimators $\hat{n}$ and $\tilde{n}$, for $n = 50$.

		$p_1 = p_2 = 0.3$	$p_1 = p_2 = 0.5$	$p_1 = p_2 = 0.7$	$p_1 = p_2 = 0.9$
AS	q	5.44	1	0.18	0.01
	$\sigma^{AS}_{50,p_1,p_2}(\hat{n})$	19.04	7.94	3.17	0.80
	$\sigma^{AS}_{50,p_1,p_2}(\tilde{n})$	16.91	7.42	3.10	0.80
MC	q^{MC}		1.24	0.20	0.02
	$\sigma^{MC}_{50,p_1,p_2}(\hat{n})$		8.15	3.10	0.79
	$\sigma^{MC}_{50,p_1,p_2}(\tilde{n})$		6.44	2.82	0.75

can be observed only beyond the domain $p_i \geqslant 0.5$, $i = 1, 2$. But then both estimators are useless.

Is it, however, of any practical value to analyze situations in which the proof-reader detects on an average less than half of the errors? It seems that such situations should be eliminated from our considerations by taking $p_i \geqslant 0.5$, $i = 1, 2$. We also restrict the admissible distributions π by taking only those for which the probability of drawing a book with less than 50 errors is sufficiently small. Bearing in mind these modifications of the set $\boldsymbol{\Theta}$, we can conclude that if the numbers contained in the upper part of Table 1 were exact values of bias and standard deviations, each of the estimators $\hat{n}$ and $\tilde{n}$ could be accepted as an adequate tool for estimating the number n of errors.

Now it remains to investigate the accuracy of the asymptotic estimates for $n \geqslant 50$. In Mielniczuk (1986) this is done by using a Monte-Carlo method which simulates the proof-reading process as follows. Suppose we can build up a certain random mechanism which, for each pair (p_1, p_2) and a fixed number N, performs N independent draws of realizations of the proof-reading process according to the distribution P_{50,p_1,p_2}. For each of these N randomly selected realizations we calculate the values of the estimators $\hat{n}$ and $\tilde{n}$. Next, by using these values, we calculate the expected value and the standard deviation for both estimators. When the number of drawings is sufficiently large (e.g. $N = 1000$) the resulting Monte-Carlo estimates (MC) approximate the exact values of the parameters of our estimators and can be used for verifying the accuracy of the asymptotic estimates (AS).

The MC estimates obtained for $N = 1000$ and $p_i \geqslant 0.5$ are listed in the lower part of Table 1. They nearly coincide with the corresponding AS estimates. This allows us to recognize the AS estimates as sufficiently accurate for $n \geqslant 50$.

The term "Monte-Carlo method" refers to the roulette wheel, the repeated use of which provides us with realizations of a random variable uniformly distributed over (0, 1). In our investigations the role of the roulette wheel is played by computer programmes, namely by computer generators of pseudorandom numbers with uniform distributions. Using a sequence of $2n$ numbers generated by such a programme, we determine the values of indicators x_i and y_i, $i = 1, \ldots, n$, by putting $x_i = 1$ if the $(2i-1)$-th number exceeds p_1, and $y_i = 1$ if the $(2i)$-th number exceeds p_2.

It is a natural question to ask whether it is worth determining the AS estimates while the MC estimates can be made as accurate as we wish by an appropriate choice of the number of random selections N. In our opinion the answer is clearly positive. First of all, it is expensive and time consuming to get the MC estimates, and hence the triples (n, p_1, p_2) for which the simulation process is performed should be carefully chosen. On the other hand, the AS estimates enable to perform a general analysis of the properties of the estimators and to choose these triples (n, p_1, p_2) for which the MC estimates are particularly needed. Thus, an analysis of the properties based jointly on AS and MC estimates is recommended.

*

Formal investigations of the proof-reading process can be modified in different ways. In particular, we can impose some requirements on the error generation mechanism and investigate quantities characterizing the process. In the simplest (though somewhat unreal) situation we can assume that for any book from a certain population the expected number of errors on each page is constant. Denote this number by λ and assume that the number of errors on arbitrary j pages has a Poisson distribution with parameter $j\lambda$.

Let us now partition the whole population of books into separate subpopulations by including into the same subpopulation all books with the same number of pages s. The book to be investigated (with a known number of pages) is regarded as randomly selected from the corresponding subpopulation. The distribution of the realizations of the proof-reading process in the subpopulation of books with s pages will be denoted by $P^{(s)}_{\lambda, p_1, p_2}$. Hence,

$$P^{(s)}_{\lambda, p_1, p_2}(n, x_1, y_1, \ldots, x_n, y_n)$$

$$= \left[\frac{(\lambda s)^n}{n} \exp(-\lambda s) \right] p_1^{n_1}(1-p_1)^{n-n_1} p_2^{n_2}(1-p_2)^{n-n_2}. \tag{7}$$

The parameter λ which characterizes the error generation mechanism can be estimated, e.g. by the estimator

$$\tilde{\lambda}^{(s)} = \frac{\tilde{n}}{s}.$$

It is shown in Mielniczuk (1986) that the expected value of the estimator $\tilde{\lambda}^{(s)}$ tends to λ with the increase of the number of pages, i.e.,

$$\lim_{s \to \infty} E^{(s)}_{\lambda, p_1, p_2}(\tilde{\lambda}^{(s)}) = \lambda,$$

and that

$$\lim_{s \to \infty} \operatorname{var}^{(s)}_{\lambda, p_1, p_2} s(\tilde{\lambda}^{(s)} - \lambda) = q^2 + \frac{q}{p_1 p_2},$$

where $E^{(s)}_{\lambda,p_1,p_2}$ and $\mathrm{var}^{(s)}_{\lambda,p_1,p_2}$ denote the expected value and the variance, respectively, of the distribution $P^{(s)}_{\lambda,p_1,p_2}$.

Another problem related to the proof–reading process is the evaluation of the proof-readers' performance. The probability of error detection can serve as a measure of this performance. Hence, we want to estimate the probabilities p_1, p_2. The natural estimators of these probabilities are

$$\hat{p}_1 = \begin{cases} \dfrac{m}{n_2} & \text{if} \quad n_2 > 0, \\ 0 & \text{if} \quad n_2 = 0, \end{cases}$$
$$\hat{p}_2 = \begin{cases} \dfrac{m}{n_1} & \text{if} \quad n_1 > 0, \\ 0 & \text{if} \quad n_1 = 0. \end{cases} \tag{8}$$

These are biased estimators. For instance, for $\hat{p}_1$ we have

$$E^{(s)}_{\lambda,p_1,p_2}(\hat{p}_1) = p_1\big(1-\exp(-p_2\,\lambda s)\big).$$

If $p_2 > 0$, the bias of the estimator $\hat{p}_1$ tends to zero with s tending to infinity, and the greater is p_2 the faster is the convergence. For small values of p_2 and $s\lambda$, the estimator $\hat{p}_1$ is not satisfactory.

It is also of interest to compare the performance of proof-readers. To do this we choose a number $\varepsilon > 0$ by means of which we want to distinguish the negligible differences between p_1 and p_2 from the ones held as significant, and we formulate the hypothesis

$$|p_1 - p_2| < \varepsilon. \tag{9}$$

Our goal is to verify whether the hypothesis (9) is true or false. A natural way to do this is to compare the value of the expression $|\hat{p}_1 - \hat{p}_2| - \varepsilon$ with a prescribed threshold value. If the value of this expression exceeds the threshold value, we can conclude that the hypothesis is false and we can regard that proof-reader as being better for whom the estimate of the probability of error detection has a greater value. The choice of the threshold value depends on conditions which one imposes upon the consequences of this procedure.

In the former examples the process of proof-reading can be substituted by analogous processes of statistical quality control in which a homogeneous product e.g., a roll of a textile, is examined under the assumption that the number of defects in a unit of the product has the Poisson distribution. Instead of a homogeneous product with defects we can also consider any signal detection process in which signals are detected randomly and for which, say, the total number of signals in a fixed period of time, has to be estimated.

In statistical quality control as well as in signal detection we admit different schemes of control. The above mentioned system of two independent controllers (proof-readers) is an example of such a scheme.

It is clear that problems arising in different domains of life can be formalized in the same way. On the other hand, a formal description of a particular problem depends on the adopted simplifying assumptions, which are often made quite arbitrarily (e.g., the assumption that the number of defects has the Poisson distribution, or the assumption that the probability of error detection is the same for each error). The same observation refers to the formalization of the research goal and of the requirements imposed on inference procedures.

*

The examples analyzed above illustrate the variety of approaches to a formal description of the phenomenon itself, of the purposes for which a particular study is set, and the variety of requirements imposed on the inference procedures (or, more strictly, on the consequences of their application). In addition, those examples also reveal the existence of a unified framework within which we can formalize a statistical problem. Roughly speaking, investigation of empirical phenomena consists in finding a way of estimating the information we seek by exploiting the information which is available. In such investigations we can distinguish three elements:

— description of the available information;

— description of the information sought for;

— description of the requirements imposed on the performance of inference procedures (which transform the available information into estimates of the information sought for).

Thus, we being with the description of the phenomenon, next specify the goals of a particular study, and eventually we formulate the requirements on inference procedures. The same stages can be distinguished in formal descriptions of problems as well as in intuitive and informal ones.

Note that in our approach the problem of investigating an empirical phenomenon does not consist in estimating the information sought for from observable data in the realization currently obtained; in fact, it consists in finding suitable method of estimating the required information, suitable for any realization which might occur. At this stage we consider the whole available information: about the realization and about the mechanism of generating realizations. Once a suitable estimator is found, we can apply it to observable data in any particular realization. For instance, the first problem formulated for the proof-reading process was not to estimate the number of undetected errors in a particular book but to find a method of estimating the number of undetected errors in an arbitrary book.

The three elements of investigation listed above can be distinguished in any inference about empirical phenomena. But the mathematical tools used for the formalization of a random course of the phenomena have specific traits which definitely differentiate statistical problems from other problems of applied mathematics.

The first part of the book deals with the formalization of statistical problems. We investigate the formal description of available and sought-for information, and various specifications of the requirements to be satisfied by the inference procedures.

*

We assume that the reader is familiar with the elementary terminology of probability theory. It is only sporadically that we use such less known terms as: stochastically ordered random variables, transition probability function, subdistribution function (i.e., an increasing function F: $\boldsymbol{R} \to [0, 1]$ which is continuous on the right and $F(-\infty) = 0$), stochastic (or random) process, renewal process, birth-and-death process, limit distribution. Moreover, we use some simplifying conventions. In particular we say "probability distribution on the set A" instead of "probability distribution on a given σ-field of subsets of A", we say "probability density with respect to a measure μ" assuming tacitly that μ is σ-finite.

Statistical notions are defined in the text. However, this does not apply to some terms used in the final remarks at the end of chapters.

*

SYMBOLS

Probability distributions:

$B(n, p)$	binomial distribution with parameters n, p;
$M_k(\boldsymbol{n}, \boldsymbol{p})$	multinomial distribution with parameters $\boldsymbol{n} = (n_1, \ldots, n_k)$, $\boldsymbol{p} = (p_1, \ldots, p_k)$;
$P(\lambda)$	exponential distribution with parameter λ;
$\Gamma(\lambda)$	gamma distribution with parameter λ (i.e., distribution with the density $f(x) = x^{\lambda-1}e^{-x}(\Gamma(\lambda), x > 0)$;
$N(\mu, \sigma^2)$	normal distribution with parameters μ, σ^2;
$N_2(\mu_1, \mu_2, \sigma_1^2, \sigma_2^2, \varrho)$	bivariate normal distribution with parameters $\mu_1, \mu_2, \sigma_1^2, \sigma_2^2, \varrho$;
$N_k(\boldsymbol{\mu}, \boldsymbol{\Sigma})$	k-variate normal distribution on $\boldsymbol{R}^k$ with parameters $\boldsymbol{\mu}$, $\boldsymbol{\Sigma}$;
$U(a, b)$	uniform distribution on $[a, b]$;
$L(\mu, \sigma^2)$	logistic distribution with parameters μ, σ^2 (with density $f(x) = \dfrac{\pi\exp[-\pi(x-\mu)/(\sigma\sqrt{3})]}{\sigma\sqrt{3}\{1+\exp[-\pi(x-\mu)/(\sigma\sqrt{3})]^2\}}$);
$\Gamma_2(\alpha, \mu_1, \mu_2, \sigma_1^2, \sigma_2^2, \varrho)$	bivariate gamma distribution with parameters $\alpha, \mu_1, \mu_2, \sigma_1^2, \sigma_2^2, \varrho$ (cf. Sections 4.4 and 5.3).

For parameters of probability distributions we use interchangeably the following symbols

$\varkappa(P)$ parameter of distribution P;

$\varkappa_P(X)$ parameter of random variable X defined on $(\boldsymbol{\Omega}, \mathscr{A}, P)$, or shortly: parameter of random variable X;

$\varkappa(X)$ parameter of random variable X when distribution P of X is clear from the context.

In particular:

$E(P)$, $E_P(X)$, $E(X)$	expected value,
$\mathrm{var}(P)$, $\mathrm{var}_P(X)$, $\mathrm{var}(X)$	variance,
$\mathrm{cov}_P(X, Y)$, $\mathrm{cov}(X, Y)$	covariance,
$\mathrm{cor}(P)$, $\mathrm{cor}_P(X, Y)$, $\mathrm{cor}(X, Y)$	correlation coefficient,
$\mathrm{cov}(\boldsymbol{X})$	covariance matrix,
$E(X \mid Y = y)$	value of regression function of random variable X on random variable Y at point y,
$q_P(\alpha)$	α-th quantile of distribution P,
F_P, F_X	distribution functions of distribution P and of random variable X, respectively.

Others symbols:

$X \sim P$	random variable X has distribution P,
$X \sim Y$	random variables X and Y have the same distributions,
$X \sim \mathscr{P}$, $X \in \mathscr{P}$	random variable X has a distribution belonging to family $\mathscr{P}$,
$X \leqslant_{\mathrm{st}} Y$	random variable X is stochastically smaller than random variable Y (i.e., $F_X(a) \geqslant F_Y(a)$, $a \in \boldsymbol{R}$),
$\xrightarrow{\mathscr{L}}$	weak convergence of distributions,
$\xrightarrow{P}$	convergence in probability,
$\to$ a.e. P	convergence almost everywhere (almost surely) in distribution P,
$\boldsymbol{R}$	set of real numbers,
$\boldsymbol{R}^+$	set of nonnegative real numbers,
$\boldsymbol{R}^k$	k-dimensional real space,
$\# A$	cardinality of finite set A,
1_A	identity on sets A,
χ_A	characteristic function of set A,
$f \circ g$	composition of functions f and g,
$\boldsymbol{A}^T$	transpose of matrix $\boldsymbol{A}$,
$\{\ \}$	set; we distinguish between set $\{x_i, i = 1, \dots, n\}$ and sequence $(x_i, i = 1, \dots, n)$; the latter may contain equal elements. Consequently, set $\{P_\theta, \theta \in \boldsymbol{\Theta}\}$ and indexed family $(P_\theta, \theta \in \boldsymbol{\Theta})$ are different objects.

Part One

FORMALIZATION OF STATISTICAL PROBLEMS

Chapter 1

Statistical description of empirical phenomena

Key words: *population, feature, simple random sample, identifiability, empirical distribution, likelihood function, observability, maximal observable statistic.*

1.1. STATISTICAL SPACE

The first step towards a formal description of an empirical phenomenon consists in collecting relevant information about the possible practical course of the phenomenon. In doing this, we want to retain only that information which is of value for the specific study we have in mind. Not always are we able to decide which part of the available information have to be taken into account. However, in practice there usually exists some natural description of the realization which is convenient for the next stages of formalization.

Formally expressed information about a possible course of a phenomenon will be called *realization of that phenomenon.* The set of all realizations will be denoted by $\boldsymbol{\Omega}$. In the proof-reading process considered in the Introduction a realization of the process is formally represented as a sequence ω,

$$\omega = (n, x_1, y_1, \ldots, x_n y_n),$$

and $\boldsymbol{\Omega}$ is the set of all sequences ω, for $n = 0, 1, 2, \ldots$, and $x_i, y_i \in \{0, 1\}$.

The next step of the formalization process is to describe what we know about the realization generation mechanism. Usually we are faced with situations justifying the assumption that a realization is selected by chance from the set $\boldsymbol{\Omega}$ according to a certain probability distribution. In some cases such a randomness is an inherent feature of the phenomenon. In other cases the phenomenon is, in principle, deterministic but its realizations are generated by a mechanism which is very complex or not known. In these cases its substitution by a random mechanism may simplify the description and at the same time may be allowable from the point of view of the research to be carried on. Such a substitution has been made in investigating the proof-reading process, where, in fact, detection of an error depends on many factors whose mutual relations are not known.

The distribution according to which we draw a realization is usually not known. The available information allows us only to say that this distribution belongs to a certain set $\mathscr{P}$ of distributions defined on $\boldsymbol{\Omega}$. More precisely, for technical reasons we distinguish a σ-field $\mathscr{A}$ of subsets of $\boldsymbol{\Omega}$ such that all distributions belonging to $\mathscr{P}$ are defined on $\mathscr{A}$. The elements of $\mathscr{A}$ are called *events*.

In choosing the σ-field $\mathscr{A}$ and the set $\mathscr{P}$ of distributions it should be remembered that $\mathscr{A}$ has to contain all events which are of interest and that the properties of $\mathscr{P}$ have to adequately reflect the properties of the realization generation mechanism. Whenever possible (in particular, when $\boldsymbol{\Omega}$ is at most countable) it is convenient to choose $\mathscr{A}$ as the σ-field of all subsets of the set $\boldsymbol{\Omega}$. However, when $\boldsymbol{\Omega} = \boldsymbol{R}^n$ this choice would involve a considerable restriction of the possible probability distributions, and hence $\mathscr{A}$ is then taken as the σ-field of all Borel sets.

The basic formal description of a phenomenon is a triple

$$M = (\boldsymbol{\Omega}, \mathscr{A}, \mathscr{P})$$

called a *statistical space* or a *statistical model*. In particular, when we have full information about the probability distribution according to which a realization is performed, i.e., when $\mathscr{P} = \{P\}$, the basic formal description reduces to the probability space $(\boldsymbol{\Omega}, \mathscr{A}, \mathscr{P})$.

Distributions belonging to the set $\mathscr{P}$ are often given by the densities with respect to a fixed measure. Distribution sets (and corresponding statistical spaces) given in this form are called *dominated*. For instance, the families of continuous distributions in $\boldsymbol{R}^k$ are dominated by the Lebesgue measure in $\boldsymbol{R}^k$.

Usually the set $\mathscr{P}$ of distributions is defined as an indexed family $(P_\theta, \theta \in \boldsymbol{\Theta})$, i.e., $\mathscr{P} = \{P_\theta, \theta \in \boldsymbol{\Theta}\}$. The family of all distributions $P^{(s)}_{\lambda, p_1, p_2}$ for a fixed s (cf. formula (7) in the Introduction) in the proof-reading process serves as an example. In this example θ is a triple (λ, p_1, p_2) and the set $\boldsymbol{\Theta}$ is the Cartesian product of the set of positive reals and of two intervals $(0, 1)$.

In many practical problems the index set $\boldsymbol{\Theta}$ has a natural interpretation and the indexed family $\mathscr{P}$ can be introduced in a natural way.

Those descriptions of phenomena in which the set of distributions may be defined by an indexed family with the set $\boldsymbol{\Theta}$ contained in $\boldsymbol{R}^k$ deserve special attention. Such descriptions (statistical spaces) and such families of distributions are called *parametric*. Otherwise, we say that a description of a phenomenon is nonparametric. The family $\big(P^{(s)}_{\lambda, p_1, p_2}\colon \lambda > 0,\ p_1, p_2 \in (0, 1)\big)$ is parametric.

*

In practical situations we mostly deal with sets of objects (items), called *populations*, and we investigate different features of those objects. To each object we assign the numerical value of the feature in question. If a given feature has a countable set of values (e.g., a finite set or the set of natural numbers), then the population can be described by a discrete probability distribution. In particular, if the set of objects and the set of all possible values of a given feature are both finite, the probability of each value can be taken to be equal to the relative frequency of occurrence of objects with this value. Generally, to describe a population with respect to a certain feature one needs a probability distribution defined on the set of values of that feature. This probability distribution is only known to belong to a certain selected set of probability distributions. Thus, a formal description of a population with respect to a given feature (or a set of features) forms a statistical space.

If the population and the set of the possible values of a given feature are both finite and we draw an object according to a uniform distribution on the set of objects, then the distribution of the feature of the drawn object and the distribution describing the population are identical. In general, in an arbitrary (not necessarily finite) population the identity of the above distributions is guaranteed when the choice of an object is independent of the value of the feature in question. Hence, such a random selection of an object is formally described by means of a statistical space.

A similar situation arises when the choice of an object is repeated n-times; if the choice of an object is independent of the value of the feature, then the joint distribution of the values of the feature of the objects drawn equals the product of the distribution of the values of that feature in the population. In the case of a finite populations this equality holds when n drawings (with replacements) of an object are performed according to a uniform probability distribution on the set of objects.

A sample formed in this way is called a *simple random sample of size n* with respect to a given feature. A simple random sample of size n is formally a statistical space. In this space each element of the set of distributions is the product of n feature distributions. If a population is described by $(\Omega, \mathscr{A}, \mathscr{P})$, then a simple random sample of size n is of the form

$$(\Omega^n, \mathscr{A}^n, \{P^n, P \in \mathscr{P}\}).$$

In practice, a simple random sample of size n often describes n independent repetitions of an experiment performed in the same circumstances.

1.2. Parameters and indices of probability distributions

Consider the family of continuous and symmetric probability distributions on $\boldsymbol{R}$. In this family the location of a distribution is determined in a natural way by the centre of symmetry of its density. Thus, we have a function assigning to each distribution from the family in question a certain real number describing location of the distribution. Generally, a function $\varkappa$ defined on a set $\mathscr{P}$ of probability distributions and taking on values from a given set $\boldsymbol{\varkappa}$ is called a *parameter* on $\mathscr{P}$ with values in $\boldsymbol{\varkappa}$.

By using parameters we can briefly and concisely describe families of distributions. Sometimes we can easily choose parameters having a natural meaning. But more often we find it difficult to specify which information is of interest to us. For instance, in the family of all continuous univariate distributions it is not obvious how to describe the location of a distribution by using a single number. According to our needs we may use the expected value or the median (for symmetric distributions the expected value and the median coincide with the centre of symmetry).

Similar difficulties arise when we want to find parameters describing the dispersion of univariate distributions.

One often needs parameters whose values are not real numbers. For instance, the probability distributions on $\boldsymbol{R}^k$ are described by covariance matrices or regression functions. These parameters are defined only on certain subsets of multivariate distributions.

*

Consider now an indexed family of distributions $(P_\theta, \theta \in \boldsymbol{\Theta})$. The function assigning distributions P to indices θ from the set $\boldsymbol{\Theta}$ is called the *indexing function.* An indexing function may or may not be a one-to-one function. For instance, in the proof-reading process different distributions correspond to different indices (λ, p_1, p_2) and (λ', p_1', p_2'). It happens, however, that an indexing function which is natural for a given problem is not one-to-one. In psychometry, for example, responses to random stimuli are investigated under the following assumptions: the magnitude of stimulus has normal distribution $N(\mu, \sigma^2)$ with unknown parameters μ, σ^2, the magnitude of response depends exponentially on the magnitude of stimulus with an unknown coefficient γ. In a natural way distributions of response are indexed by triples (μ, σ^2, γ) and the corresponding densities are given by the formula

$$f_{\mu, \sigma^2, \gamma}(x) = \frac{1}{\sigma\gamma x\sqrt{2\pi}} \exp\left(-\frac{(\ln x - \mu\gamma)^2}{2\sigma^2\gamma^2}\right), \quad \mu \in \boldsymbol{R}, \quad \sigma^2, \gamma \in \boldsymbol{R}^+, \quad x \in \boldsymbol{R}^+.$$

This is the density of the lognormal distribution with the expected value $\exp(\mu\gamma + \sigma^2\gamma^2/2)$ and the variance $\exp(2\mu\gamma + \sigma^2\gamma^2)(\exp \sigma^2\gamma^2 - 1)$. Clearly we can introduce a one-to-one indexing function by using pairs (α, β), $\alpha \in \boldsymbol{R}$, $\beta \in \boldsymbol{R}^+$, with $\alpha = \mu\gamma$, $\beta = \sigma^2\gamma^2$. But this indexing function, though being one-to-one, has no interpretation related directly to the response generation mechanism.

If a mapping of the set $\boldsymbol{\Theta}$ into the set of distributions on $\boldsymbol{\Omega}$ is one-to-one, then the inverse mapping is a parameter. An indexing function is called *identifiable* if it is a one-to-one function, i.e., if for different indices θ, θ', we have different distributions P_θ, $P_{\theta'}$. In such a situation we say that "index θ is identifiable".

Generally, a function $\varkappa$: $\boldsymbol{\Theta} \to \boldsymbol{\varkappa}$ is called *identifiable* if

$$\varkappa(\theta) \neq \varkappa(\theta') \quad \Rightarrow \quad P_\theta \neq P_{\theta'}. \tag{1.2.1}$$

Each identifiable function $\varkappa$ defined on the set $\boldsymbol{\Theta}$ induces on a given set of distributions a new parameter $\tilde{\varkappa}$ according to the formula

$$\tilde{\varkappa}(P) = \varkappa(\theta).$$

1.3. Random variables and statistics

As we have mentioned before, the parameters of probability distributions allows us concise descriptions of distributions. A similar role is played by functions defined on the set $\boldsymbol{\Omega}$ of realizations. By means of these functions some information about realizations is expressed. For instance, in the proof-reading process to each realization it is assigned the number m of errors detected both by the first and the second proof-reader. Another example of such a function in the proof-reading process is provided by the quotient $n_1 n_2/m$ which estimates the total number of errors in the book.

For arbitrary λ, p_1, p_2 and a fixed number of pages s the distribution of the number of errors m (cf. formula (7) in Introduction) is given by

$$Q^{(s)}_{\lambda, p_1, p_2}(i) = \sum_{n=i}^{\infty} \binom{n}{i} (p_1 p_2)^i (1-p_1 p_2)^{n-i} \frac{(\lambda s)^n}{n!} \exp(-\lambda s), \quad i = 0, 1, \dots$$

Hence, the number of errors m induces on the set $\{0, 1, \dots\}$ the distribution $Q^{(s)}_{\lambda, p_1, p_2}$ which corresponds to the distribution $P^{(s)}_{\lambda, p_1, p_2}$.

If we consider the whole set $\mathscr{P}$,

$$\mathscr{P} = \{P_{\lambda, p_1, p_2}, \lambda > 0, p_1, p_2 \in (0, 1)\},$$

then the function m induces on the set $\{0, 1, \dots\}$ the family $\mathscr{Q}$ of distributions of the form

$$\mathscr{Q} = \{Q_{\lambda, p_1, p_2}, \lambda > 0, p_1, p_2 \in (0, 1)\}. \tag{1.3.1}$$

As we see, each function defined on the set $\boldsymbol{\Omega}$ of realizations can be considered either with respect to a particular probability distribution P defined on $\boldsymbol{\Omega}$ or with respect to the set $\mathscr{P}$ of probability distributions. To represent formally both situations, we introduce the notions of a random variable and of a statistic.

Let X be a function defined on $\boldsymbol{\Omega}$ and taking on values in a certain set $\boldsymbol{X}$. If P is a distribution on $(\boldsymbol{\Omega}, \mathscr{A})$ according to which realizations are generated, then we are interested in determining the probability that the function X takes on values in a given subset $\boldsymbol{B}$ of $\boldsymbol{X}$. To this aim we introduce a σ-field $\mathscr{B}$ of subsets of $\boldsymbol{X}$. For a fixed $\boldsymbol{B} \in \mathscr{B}$, the probability that the function X takes on values in $\boldsymbol{B}$ can be determined whenever the set $\{\omega \in \boldsymbol{\Omega}: X(\omega) \in \boldsymbol{B}\}$ is an event, i.e.,

$$X^{-1}(\boldsymbol{B}) \in \mathscr{A}. \tag{1.3.2}$$

A function $X: \boldsymbol{\Omega} \to \boldsymbol{X}$ which satisfies (1.3.2) for each $\boldsymbol{B} \in \mathscr{B}$ is called a *measurable function which maps a measurable space* $(\boldsymbol{\Omega}, \mathscr{A})$ *into a measurable space* $(\boldsymbol{X}, \mathscr{B})$. If it is clear which measurable spaces are considered, we say briefly that X is a *measurable function.*

For a given distribution P, a measurable function X induces a probability distribution Q defined on $(\boldsymbol{X}, \mathscr{B})$, such that, for each $\boldsymbol{B} \in \mathscr{B}$, we have

$$Q(\boldsymbol{B}) = P(X^{-1}(\boldsymbol{B})),$$

which can be rewritten as

$$Q = PX^{-1}. \tag{1.3.3}$$

For a given pair of probability spaces $(\boldsymbol{\Omega}, \mathscr{A}, P)$ and $(\boldsymbol{X}, \mathscr{B}, Q)$, a measurable function X which maps $(\boldsymbol{\Omega}, \mathscr{A})$ into $(\boldsymbol{X}, \mathscr{B})$ and satisfies condition (1.3.3) is called a *random variable mapping* $(\boldsymbol{\Omega}, \mathscr{A}, P)$ *into* $(\boldsymbol{X}, \mathscr{B}, Q)$, or shortly, a *random variable.*

The probability distribution Q is called the *probability distribution of the random variable* X. Since Q is determined by P, to define a random variable it is enough to give a measurable function X and a distribution P.

A random variable taking on values in $\boldsymbol{R}^k$, $k > 1$, is called a *k-dimensional vector* or a *k-dimensional random variable.*

In the proof-reading process with s fixed, the number m of errors can be regarded as a random variable if we fix λ, p_1, p_2. The distribution of this random variable can be written as

$$Q^{(s)}_{\lambda, p_1, p_2} = P^{(s)}_{\lambda, p_1, p_2} m^{-1}.$$

For different indices (λ, p_1, p_2), (λ', p_1', p_2'), $p_1 p_2 \neq p_1' p_2'$ the number m of errors yields different random variables since distributions $Q^{(s)}_{\lambda, p_1, p_2}$ and $Q^{(s)}_{\lambda', p_1', p_2'}$ are different.

Now we introduce the notion of a statistic. As follows from our previous considerations, for a given statistical space $(\boldsymbol{\Omega}, \mathscr{A}, \mathscr{P})$, a measurable function X induces the set $\mathscr{Q}$ of distributions defined on $\boldsymbol{X}$,

$$\mathscr{Q} = \{PX^{-1}\colon P \in \mathscr{P}\}. \tag{1.3.4}$$

Just as formerly, we say that a measurable function X: $\boldsymbol{\Omega} \to \boldsymbol{X}$ is a statistic which maps a statistical space $(\boldsymbol{\Omega}, \mathscr{A}, \mathscr{P})$ into a statistical space $(\boldsymbol{X}, \mathscr{B}, \mathscr{Q})$, if for a given sets $\mathscr{P}$ and $\mathscr{Q}$ of probability distributions defined on $(\boldsymbol{\Omega}, \mathscr{A})$ and $(\boldsymbol{X}, \mathscr{B})$, respectively, the function X satisfies (1.3.4). If it is clear which statistical spaces are considered, we say briefly that X is a *statistic*. The set $\mathscr{Q}$ is called the *set of distributions of the statistic* X. Since $\mathscr{Q}$ is defined by $\mathscr{P}$, a statistic is determined by a measurable function X and a family of distributions $\mathscr{P}$.

In the proof-reading process with a fixed s, the number m of errors can be regarded as a statistic with the set of distributions (1.3.1).

Traditionally the same symbols are used to denote measurable functions, random variables, and statistics when related to the same function X on $\boldsymbol{\Omega}$. This may lead and sometimes leads to confusion. But a consistent use of distinct symbols would require the introduction of additional subscripts (to denote the respective measurable spaces, probability spaces, statistical spaces). This, in turn, would make the whole text so cumbersome as to be practically unreadable.

*

Consider a simple random sample with the set of distributions $\{P^n, P \in \mathscr{P}\}$ where $\mathscr{P}$ is a set of distributions on $\boldsymbol{R}^k$. Hence, the realizations of any phenomenon are of the form

$$\omega = (\omega_1, \ldots, \omega_n), \quad \omega_i \in \boldsymbol{R}^k, \quad i = 1, \ldots, n.$$

Each realization $\omega = (\omega_1, \ldots, \omega_n)$ defines a discrete probability distribution on $\boldsymbol{R}^k$, denoted by $P^{(n)}_\omega$ and called an *empirical distribution* (or a *sample distribution*). This distribution is defined by the formula

$$P^{(n)}_\omega(A) = \frac{1}{n} \#\{i\colon \omega_i \in A\}, \quad A \in \mathscr{B}(\boldsymbol{R}^k). \tag{1.3.5}$$

The distribution function of the distribution $P_\omega^{(n)}$ is called an *empirical distribution function* (or a *sample distribution function*). The parameters of the distribution $P_\omega^{(n)}$ are called *empirical parameters* (or *sample parameters*). They include, e.g., the sample mean and the sample variance (for $k = 1$) and the sample correlation coefficient (for $k = 2$).

If the sample size tends to infinity, then, for each $P \in \mathscr{P}$ and almost all([1]) $\omega = (\omega_1, \dots, \omega_n)$ we have

$$\lim_{n\to\infty} \sup_{x \in R^k} \left| P^{(n)}_{\omega_1, \dots, \omega_n}((-\infty, x]) - P((-\infty), x] \right| = 0.$$

The parameters of empirical distributions are statistics defined on $\boldsymbol{\Omega}^n$. These statistics are of particular importance in statistical inference. For many parameters $\varkappa$ we can show that, for each $P \in \mathscr{P}$, the sequence of empirical parameters converges to $\varkappa(P)$ for almost all ω. In particular, this is true for moments (if they exist), and consequently also for continuous functions of moments.

As mentioned before, the values of functions defined on $\boldsymbol{\Omega}$ provide information about the realizations of a given phenomenon and the values of functions defined on $\mathscr{P}$ provide information about the probability distributions which generate the realizations. Hence, the values of functions defined on the set $\boldsymbol{\Omega} \times \mathscr{P}$ give joint information about the realizations and about the probability distributions.

If the set $\mathscr{P}$ of distributions is indexed by a one-to-one indexing function defined on a given set $\boldsymbol{\Theta}$, then functions defined on $\boldsymbol{\Omega} \times \mathscr{P}$ can be replaced by functions defined on $\boldsymbol{\Omega} \times \boldsymbol{\Theta}$. The likelihood function can serve as an example here. Namely, for an arbitrary family of probability distributions dominated by a measure ν, we have function L, called the *likelihood function*, which is defined by the formula

$$L(\omega, \theta) = p_\theta(\omega), \tag{1.3.6}$$

where p_θ denotes the density of the probability distribution P_θ with respect to ν.

The function L is defined by a family of density functions and itself defines a family of functions $(f_\omega\colon \boldsymbol{\Theta} \to \boldsymbol{R}^+,\ \omega \in \boldsymbol{\Omega})$, where

$$f_\omega(\theta) = L(\omega, \theta) = p_\theta(\omega). \tag{1.3.7}$$

If, for each $\omega \in \boldsymbol{\Omega}$, the function f_ω attains its maximum at exactly one point of $\boldsymbol{\Theta}$, then the function

$$\hat{\theta}\colon \boldsymbol{\Omega} \to \boldsymbol{\Theta}$$

which assigns to each ω the element of $\boldsymbol{\Theta}$ realizing the maximum of f_ω over $\boldsymbol{\Theta}$ is well defined. If $\nu(\omega) > 0$, then $\hat{\theta}(\omega)$ is the index of a probability distribution from $\mathscr{P}$ for which the chance of drawing the realization ω is greatest. If $\hat{\theta}$ is a measurable function with respect to a certain σ-field of subsets of the set $\boldsymbol{\Theta}$, then $\hat{\theta}$, regarded as a statistic, is called the *maximum likelihood statistic* (shortly, *ML statistic*).

([1]) More precisely, this is the uniform almost everywhere convergence with respect to the measure induced by the distribution P on a countable product of the measurable space $(\boldsymbol{\Omega}, \mathscr{A})$.

For a simple random sample of size n with the probability distribution from the set $\{P_\theta, \theta \in \boldsymbol{\Theta}\}$ the likelihood function is the product of densities

$$L(\omega_1, \dots, \omega_n, \theta) = \prod_{i=1}^{n} p_\theta(\omega_i), \tag{1.3.8}$$

and the ML statistic can be determined either by the maximum of functions

$$f_\omega(\theta) = \prod_{i=1}^{n} f_{\omega_i}(\theta) \tag{1.3.9}$$

for every $\omega = (\omega_1, \dots, \omega_n) \in \boldsymbol{\Omega}$, or by the maximum of functions which are monotone transformations of f_ω, e.g.,

$$\ln f_\omega(\theta) = \sum_{i=1}^{n} \ln f_{\omega_i}(\theta).$$

*

The parameters of the probability distribution of a random variable are often called *parameters of that random variable*. A conventional notation is $\varkappa(X)$ where $\varkappa$ stands for a parameter and $\varkappa(X)$ is the value of this parameter for the distribution of the random variable X. For instance, we are writing $E(X)$, $\mathrm{var}(X)$, etc. If we want to stress the fact, that we refer to the distribution of X corresponding to the distribution P on $\boldsymbol{\Omega}$, we write $\varkappa_P(X)$, e.g., $E_P(X)$ or $\mathrm{var}_P(X)$. If we deal with a set of distributions on $\boldsymbol{\Omega}$ indexed by $\boldsymbol{\Theta}$, we are writing $\varkappa_\theta(X)$ instead of $\varkappa_{P_\theta}(X)$.

This notation is convenient for representing the parameters of functions of random variables (e.g., $E(X+Y)$, $\mathrm{var}\,(\sum_{i=1}^{n} X_i)$). For consistency, the notation $\varkappa(P)$ for the distribution P on $\boldsymbol{\Omega}$ is often replaced by $\varkappa(X)$, understood as a parameter of a random variable X which identically transforms $\boldsymbol{\Omega}$ onto $\boldsymbol{\Omega}$ and has the distribution P.

The expression "the random variable X has a probability distribution P" is formally written as $X \sim P$. We also write $X \in \mathscr{P}$, which means that the random variable X has a probability distribution belonging to the family $\mathscr{P}$.

We use this notation in statistical descriptions of phenomena. If the set $\boldsymbol{\Omega}$ and the σ-field $\mathscr{A}$ of subsets of $\boldsymbol{\Omega}$ are already defined, then the description of a phenomenon by means of the statistical space $(\boldsymbol{\Omega}, \mathscr{A}, \mathscr{P})$ can be replaced by the following formulation: "let X be a random variable with a distribution $P \in \mathscr{P}$." Speaking about X here, we mean the identity mapping of the set $\boldsymbol{\Omega}$ onto itself and a probability distribution $P \in \mathscr{P}$. Statisticians usually describe empirical phenomena in this way and tacitly identify this form of description with the corresponding statistical space. For instance, the equivalent of the expression: "let X be a normal random variable with unknown parameters μ and σ^2" is the statistical space

$$(\boldsymbol{R}, \mathscr{B}, (N(\mu, \sigma^2), \mu \in \boldsymbol{R}, \sigma^2 > 0)),$$

and the equivalent of the expression: "let $X_1, \ldots, X_n$ be a sequence of n independent random variables with the same distribution $P \in \mathscr{P}$" is a simple random sample of size n

$$(\Omega^n, \mathscr{A}^n, \{P^n, P \in \mathscr{P}\}).$$

However, when describing phenomena by random variables one can easily overlook some relevant information. For instance, the following description of the proof-reading process could be given:

> "Two proof-readers perform independently of each other the proof-reading of a book. Let n be the total number of errors in the book. Let $X_1, \ldots, X_n$ and $Y_1, \ldots, Y_n$ be sequences of independent random variables taking on values 0 and 1. The random variable X_i is equal to 1 if the first proof-reader detects the i-th error and 0 otherwise, and Y_i is defined in the same way for the second proof-reader. Let p_1, p_2 denote the unknown probabilities of success for the X's and the Y's, respectively".

Such a traditional description is concise and gives a straightforward insight into the problem itself and its formalization. Moreover, this description does note require a complex notation; but it leaves some aspects of the problem not quite clear. One question is how we should treat the number n. If n is a random variable what can be said about its probability distributions? It is not clear, too, that the probabilities p_1 and p_2 are constant irrespective of the error actually detected and that they take on values from the whole interval $(0, 1)$.

In this book we often describe phenomena by random variables since a consistent use of statistical spaces would make the text hard to be read and lacking associations with traditional notation. However, we attempt to give all the information necessary for the reconstruction of the relevant statistical spaces.

Finally, let us note that in colloquial use statistics are often called random variables and are treated as random variables. For instance, the term "probability distribution of a statistic X" is used instead of the "set of probability distributions of a statistic X". By the "probability distribution of a statistic X" we usually mean a certain particular element of the family of its probability distributions, namely, that one which is an unknown ("true") probability distribution from $\mathscr{P}$ generating the realizations of a phenomenon. Here a statistic is understood as a random variable with that particular probability distribution. The parameters of this probability distribution are usually called *parameters of a statistic*. This terminology is commonly used.

1.4. Observable events

A statistical space describing a given phenomenon provides information about the realizations of a phenomenon and about the probability distribution which generates these realizations. However, in investigating a particular phenomenon, it often happens that we cannot state definitely which realization we are dealing with because realizations are not fully observable. For instance, after a proof-reading process is completed what we know are only numbers n_1, n_2 and m of errors and the number s of pages, while we do not know n, and the values of indicators x_i and y_i for consecutive errors. The numbers n_1, n_2 and m constitute the observable data.

In textbooks on statistics frequent use is made of the terms "observable data", "observable events", and "observable statistic" and the intuitive sense of these terms is clear. It is also clear that in statistical inference we can only use observable data and thus the observability of data must be taken into account in the formalization process.

Formalization of observability consists simply in distinguishing a certain subset of the set of events $\mathscr{A}$. The elements of this subset are called *observable events*. In practice, an event is considered observable if one knows whether it has occurred or not at the very moment of decision making. In other words, for every observable event we have to know whether the current realization of the phenomenon belongs to it or not.

In the proof-reading process one knows, for instance, whether the number of errors detected by the first proof-reader was greater than, say, 10; thus this event is observable. On the other hand, it is not observable whether the number of errors undetected by the first proof-reader was greater than 10.

Another illustration refers to experiments in which one observes aggregated results. The set $\boldsymbol{\Omega}$ is then partitioned into disjoint observable subsets corresponding to this aggregation and the only thing we know about the current realization is to which subset of $\boldsymbol{\Omega}$ it belongs.

As we have mentioned before, when formalizing observability we should distinguish those events which are interpreted as observable. According to this interpretation, the complement of an observable event is observable, the union of two observable events is also observable and the sure event is observable as well. Hence, the set of all observable events should form a field. For technical reasons, as the set of all observable events we assume the least σ-field containing that field.

A measurable function defined on the set of realizations is called *observable* if the inverse images of its values are observable events. In practice, this means that at the very moment of decision making the values of a given function (e.g., the function $n_1 n_2/m$ in the proof-reading process) are known.

Given probability distribution or a family of probability distributions defined on $\boldsymbol{\Omega}$, an observable measurable function can be regarded as an observable random variable or an observable statistic.

We are particularly interested in a function which assigns to realizations all observable data. In the proof-reading process this role was played by the function with values (n_1, n_2, m). A statistic which, for a given phenomenon, can be interpreted in this way has the property that each observable statistic is a function of that statistic. Such a statistic will be called *maximal observable statistic*.

Clearly, maximal observable statistic is not unique. However, in practical problems, e.g., in the proof-reading process, there is but one *natural representation* of observable data.

If a problem is fully observable, then the maximal observable statistic is the mapping of the set $\boldsymbol{\Omega}$ onto itself.

The specification of a maximal observable statistic is equivalent to the specification of the set of observable events. This set is equal to the least σ-field containing the inverse images of the values of any maximal observable statistic. Thus, a full statistical description of an empirical phenomenon is a pair of which the first element is a statistical space and the second element is a set of observable events or a maximal observable statistic. Clearly, when a maximal observable statistic T is indicated, one has to specify the set T from which it takes values and the corresponding σ-field of subsets of T, which must contain all the one-element sets.

As it is in the proof-reading process for the statistic (n_1, n_2, m), the families of probability distributions for maximal observable statistics are sometimes difficult to determine.

We now present a few examples of phenomena with limited observability. In these examples we give only the function T: $\boldsymbol{\Omega} \to \boldsymbol{T}$ and we do not specify the families of probability distributions $\mathscr{P}$ defined on $\boldsymbol{\Omega}$. In fact, these families can be chosen in different ways.

(i) We observe the results x of the measurement of a random variable with a random error e whose magnitude does not depend on the results of the measurement. Hence,

$$\omega = (x, e), \quad T(\omega) = x+e.$$

In the description of the phenomenon we take into account the information about the probability distributions of x and e.

In examples (ii), (iii), (iv) we consider objects randomly chosen from a certain population and described by a pair of real-valued features. The realizations are of the form

$$\omega = (x, y).$$

(ii) We observe the values of the first feature:

$$T(\omega) = x.$$

(iii) We observe the minimum of the values of both features:

$$T(\omega) = \min(x, y).$$

(iv) We observe the minimum of the value of both features, knowing moreover which of the features takes the smaller value

$$T(\omega) = \big(\min(x, y), \varepsilon\big), \tag{1.4.1}$$

where

$$\varepsilon = \begin{cases} 1 & \text{if} \quad x > y, \\ 0 & \text{if} \quad x \leqslant y. \end{cases}$$

This kind of observability is typical for problems in which x denotes the life-span of an object, and for some objects x is fully observable, while for others only the part truncated by means of the feature y is observable (cf. Chapter 9).

(v) Suppose we deal with a realization of a random quantity x of positive values which can be observed only in certain situations. These situations are described by a binary indicator ε. Hence,

$$\omega = (x, \varepsilon), \quad T(\omega) = \begin{cases} 0 & \text{if} \quad \varepsilon = 0, \\ x & \text{if} \quad \varepsilon = 1. \end{cases}$$

This type of observability appears in "missing value" problems.

Problems (ii) – (v) can also be formulated for sets of realizations which are Cartesian products. In fact, in problems often encountered in practice in which the sequences $\omega_1, \ldots, \omega_n$ are realized, we observe sequences of values $T(\omega_1), \ldots, T(\omega_n)$.

In next examples the sets of realizations are Cartesian products of n components, and the limitations of observability are related to the ordering of the components of the realizations $(\omega_1, \ldots, \omega_n)$.

(vi) Let $\boldsymbol{\Omega} = \boldsymbol{R}$. We observe the components of a sequence $(\omega_1, \ldots, \omega_n)$ ordered nondecreasingly

$$T(\omega_1, \ldots, \omega_n) = (\omega_{\pi(1)}, \ldots, \omega_{\pi(n)}), \tag{1.4.2}$$

where π is a permutation of the set $\{1, \ldots, n\}$ satisfying the condition

$$\omega_{\pi(1)} \leqslant \ldots \leqslant \omega_{\pi(n)}. \tag{1.4.3}$$

Statistic T is the *vector of order statistics.*

(vii) Let $\boldsymbol{\Omega} = \boldsymbol{R}$. For each sequence $(\omega_1, \ldots, \omega_n)$, let π be a permutation of the set $\{1, \ldots, n\}$ satisfying (1.4.3). If π is determined uniquely, then, for each $i \in \{1, \ldots \ldots, n\}$, we define the *rank $r(i)$ of the element* ω_i as the position number of ω_i in the sequence $(\omega_1, \ldots, \omega_n)$ ordered increasingly. Then we have $\pi(r(i)) = i$. If π is not determined uniquely, then the rank $r(i)$ of ω_i is defined as the arithmetic mean of all the possible position numbers of ω_i in the sequence $(\omega_1, \ldots, \omega_n)$ ordered nondecreasingly. We observe the vector of ranks:

$$T(\omega_1, \ldots, \omega_n) = (r(1), \ldots, r(n)). \tag{1.4.4}$$

(viii) Let $\boldsymbol{\Omega} = \boldsymbol{R}^2$, i.e., a realization is given as $(\omega_1, \ldots, \omega_n) = (x_1, y_1, \ldots, x_n, y_n)$. If $(x_{(\pi)1}, \ldots, x_{\pi(n)})$ is a sequence of order statistics for $(x_1, \ldots, x_n)$, then $(y_{\pi(1)}, \ldots, y_{\pi(n)})$ is called the *sequence of concomitants* of the order statistics. We observe a sequence of values of the concomitants of order statistics. If π is uniquely determined, then

$$T(\omega_1, \ldots, \omega_n) = (y_{\pi(1)}, \ldots, y_{\pi(n)}).$$

Otherwise, the definition of the statistic T is not complete. However, if we assume that the family of distributions on $\boldsymbol{\Omega}^n$ consists of continuous distribution functions, then the sequence of concomitants is defined uniquely for almost all realizations.

1.5. Final remarks

Classical statistics deals mostly with inference about the parameters of distributions of observable features in populations of objects on the basis of simple random samples. This affects not only the methodology of research but also the formalization of the description of phenomena, notation, and terminology.

For simple random samples the statistical space is of the form $(\boldsymbol{\Omega}^n, \mathscr{A}^n, (P^n, P \in \mathscr{P}))$ and is defined by $(\boldsymbol{\Omega}, \mathscr{A}, \mathscr{P})$ and n. In classical statistics, $\boldsymbol{\Omega}$ is called the *set of elementary events*, $(\omega_1, \ldots, \omega_n)$ is the *sample*, $\boldsymbol{\Omega}^n$ is the *sample space*, and n is the *sample size*. Research situations which cannot be accomodated within this scheme are considered separately and have their own nomenclature. In the design of experiments we speak of the *results of an experiment*, in the inference on stochastic processes of the *realizations of the process*; use is also made of the term "realization of a phenomenon" and this term is adopted in this book.

Features of objects are called *random variables* and the term *statistic* is usually applied to a function defined on the sample space. The term statistic is also used to denote a data reducing function in data analysis in which no statistical description of phenomena is given.

A consistent treatment of random variables as mappings of probability spaces, and of statistics — as mappings of statistical spaces was proposed by Barra (1983).

In textbooks of classical statistics the set $\boldsymbol{\Theta}$ is sometimes called the *set of "states of nature"* and practical problems are analyzed in which, for a given set $\boldsymbol{\Theta}$, a certain random experiment is devised. The observable result of such an experiment is a random variable X with the distribution depending upon the current state of nature $\theta \in \boldsymbol{\Theta}$. Hence, a formal description of the phenomenon consists of sets $\boldsymbol{\Theta}$, $\boldsymbol{\Omega}$, and a mapping of $\boldsymbol{\Theta}$ into the set of distributions on $\boldsymbol{\Omega}$. Sometimes such a description appears together with a certain distribution or a family of distributions on $\boldsymbol{\Theta}$, called *prior distributions*. The term "prior" refers to the fact that first a certain state of nature θ, $\theta \in \boldsymbol{\Theta}$, is chosen and then, a random experiment is performed according to a distribution P_θ. Such a description of a phenomenon is called *Bayesian*. The Bayesian description is equivalent to a description involving a statistical space in which $\boldsymbol{\Theta} \times \boldsymbol{\Omega}$ is the set of realizations and the distribution on $\boldsymbol{\Theta} \times \boldsymbol{\Omega}$ is given by the marginal distribution on $\boldsymbol{\Theta}$ and the family of conditional probability distributions on $\boldsymbol{\Omega}$ for $\theta \in \boldsymbol{\Theta}$. States of nature are considered to be unobservable. What is observable is the result of the random experiment.

In its early days statistics dealt exclusively with parametric models. The term "parametric model" is motivated by the fact that, in early models, families of probability distributions were indexed in a natural way only by a vector with real components, called *parameter*. Later on, when more complex models were formulated the term "nonparametric model" was introduced. Thus, terms "parametric model" and "nonparametric model" do not reflect the intuitions which lead to such discrimination of phenomena description.

According to the definition given in Section 1.1 a parametric model is a statistical space with a family of probability distributions whose cardinality is at most continuum. Clearly, such a family can always be indexed by a one-to-one indexing function with indices belonging to a subset of $\boldsymbol{R}^k$ (in particular, to $(0, 1)$). Such an index, however, may have no natural interpretation.

In the proof-reading process, the natural index of distributions generating realizations is $\theta = (\pi, p_1, p_2)$. If π is an arbitrary Poisson distribution, then the index could be $(\lambda, p_1, p_2) \in \boldsymbol{R}^+ \times (0, 1)^2$. If π is an arbitrary probability distribution defined on the set of natural numbers, then the set $\boldsymbol{R}^+ \times (0, 1)^2$ may still be used as the set of indices, since the set of all probability distributions defined on the set of natural numbers can be mapped by a one-to-one mapping onto $\boldsymbol{R}^+$. In each case, the statistical space describing the proof-reading process is parametric. In both cases the sets of indices are identical, only the way of indexing is different. In the first case the index is a natural parameter of the probability distribution; in the second case the index is also a parameter of the probability distribution (since the corresponding indexing function is one-to-one) but it has no natural interpretation in the proof-reading process.

In statistics the term "parameter" is ambiguous. First of all, a parameter is usually identified with an index (according to the current practice adopted not only in statistics). Hence, in a family $\mathscr{P} = (P_\theta, \theta \in \boldsymbol{\Theta})$, θ is called a *parameter* (regardless of the manner in which family is indexed) and $\boldsymbol{\Theta}$ is called the *parameter set*. Moreover, also functions defined on $\mathscr{P}$ are called *parameters*; for instance, it is said that the variance is a *dispersion parameter* in the family of probability distributions defined on $\boldsymbol{R}$. What is more, the values of such functions are sometimes called the parameters of the probability distributions. In early statistical texts this terminology was motivated by the fact that investigations concern almost exclusively families of normal or Poisson distributions indexed by the expected value and the variance.

This ambiguity of the term "parameter" interferes with the clarity of statistical texts. Moreover, also the term "parametric function" used for functions defined on $\boldsymbol{\Theta}$ is misleading.

In classical statistics indices θ are unobservable and samples ω are observable. The notions of observable events, observable random variables, and observable statistics appear in statistical texts devoted to nonclassical problems where they are treated as natural and appealing only to intuition.

Maximal observable statistic has been defined by Dąbrowska and Pleszczyńska (1980). In the same paper they define events and statistics observed conditionally with respect to a given observable event A with positive probability. These are events and statistics which become observable after truncating $\boldsymbol{\Omega}$ to A. For instance, consider the action of tossing a die whose faces with an odd number of pips are effaced. Then an event B in which at most three pips are thrown in not observable but it is conditionally observable with respect to an event A in which an even number of pips is thrown.

Conditional observability occurs in many practical problems (e.g., in problems of father recognition discussed in Chapter 7).

*

In this chapter the relationships between measurement theory and statistical descriptions of empirical phenomena have been omitted. Brief remarks on the subject will be given now.

Measurement theory is of particular importance in statistics: it deals with measurements of feature values in populations. The method of measurement should be taken into account in the description of a given phenomenon.

In measurement theory a relational structure $\mathscr{R}_0$ on a population $\boldsymbol{\Omega}_0$ is considered together with a relational structure $\mathscr{R}$ on a certain subset $\boldsymbol{\Omega} \subset \boldsymbol{R}^k$ and the two structures are related. The main objects investigated are measurement scales, i.e., homomorphisms of $\mathscr{R}_0$ into $\mathscr{R}$, and mappings $f\colon \boldsymbol{\Omega} \to \boldsymbol{\Omega}$, called *admissible functions*, which transform one scale into another.

The set of admissible mappings $\boldsymbol{\Phi}$ defines the type of measurement scale. In particular, when $\boldsymbol{\Omega} = \boldsymbol{R}$, the most common types of scales are nominal, ordinal, interval, and ratio scales, for which $\boldsymbol{\Phi}$ is set of injections, increasing mappings, linear increasing mappings and linear increasing homogeneous mappings, respectively. The absolute scale is a scale for which the only admissible function is identity.

It seems to be a natural requirement imposed on formalizations of statistical problems that the properties of the solution of a problem should remain unchanged no matter which particular scale of a given type is applied. Thus, in the description of phenomena, information about measurements (usually limited to the set of admissible mappings) should be preserved. Sometimes it is also necessary to take into account the method of measurement of observable data.

In practical statistical problems the postulate of solution invariance with respect to measurement scales of a given type is often satisfied without formal reference to the measurement theory, and measurement scales are usually discussed on the basis of intuition only. One of the consequences of this fact is inconsistent terminology. For instance, "qualitative" and "quantitative" features are distinguished. The former term corresponds to measurements to a nominal scale and the latter refers to measurements on a scale which is at least ordinal or even at least interval. In this book we avoid using terms whose meaning is not fully established but occasionally we refer to measurement scales, specifying each time the particular type of the scale in question.

In the literature there are no monographs on measurement theory in statistics. Basic information about measurement theory can be found in the monographs of Pfanzagl (1968), Roberts (1979) and the paper of Bromek *et al.* (1984).

Chapter 2

A scheme of a statistical problems

Key words: *decision rule, deferred decision, standard class of statistical problems, test, estimator, predictor, sufficient statistic, prediction sufficient statistic, reduction of a statistical problem*

2.1. Formalization of the goal of research

In statistical investigations we aim at obtaining some information about the phenomenon which interests us by exploiting that information about it which is available. In Chapter 1 we have presented a formalism permitting a description of the available information; now we shall concentrate on a formal description of the required information.

Let us recall the model of the proof-reading process given in the Introduction. If we aim at finding the total number of errors in a book, the description of the goal is clear. In estimating the proof-reader's performance, the information we seek is naturally described by the probability of error detection. In comparing the quality of work of two proof-readers there are more than one natural formalization of the required information, e.g., $|p_1 - p_2|$ or $\max(p_1, p_2)/\min(p_1, p_2)$.

Suppose, however, that we consider the model too simplified and therefore we abandon the assumption that all errors are equally difficult to detect and the assumption that the detection of a given error is independent of the detection of any other errors. Then neither the estimating the proof-reader's performance nor the comparison of the quality of work of two proof-readers has a natural unique description.

In any case formalizations of goals are more or less subjective.

The description of the goal of research is bound up with the selection of a set $\boldsymbol{I}$ whose elements describe the required information. For instance, if we aim at finding the total number of errors in a book, $\boldsymbol{I}$ is the set of all natural numbers, if we are interested in finding the probability of error detection, $\boldsymbol{I}$ is the interval $(0, 1)$ and if we want to know the absolute value of the difference between p_1 and p_2, $\boldsymbol{I}$ is the interval $[0, 1)$. In the first case what we seek is information about the realizations of the proof-reading process. In the next two cases we seek information about the parameters of the probability distribution generating the realizations. Of course, we may happen to be interested both in the number n and in the probabilities p_1, p_2, i.e., we may seek information about realizations of the proof-reading together with information about the parameters of the probability distribution generating these realizations.

Hence, in general, we look for information either about the realizations of a phenomenon or about the probability distributions generating realizations or about both. Such goals of research are usually described formally by a function C defined on $\boldsymbol{\Omega} \times \mathscr{P}$ with values in $\boldsymbol{I}$,

$$C\colon \boldsymbol{\Omega} \times \mathscr{P} \to \boldsymbol{I}. \tag{2.1.1}$$

In the sequel we restrict ourselves to goals of research formulated in this form. If

$$C(\omega, P) = I(\omega), \quad \omega \in \boldsymbol{\Omega}, \quad P \in \mathscr{P} \tag{2.1.2}$$

for a certain function $I\colon \boldsymbol{\Omega} \to \boldsymbol{I}$, then we seek information only about the realizations of the phenomenon, namely we want to know the value of the function I for the current realization ω. Such a goal is posed when the function I is not observable. For instance, in the proof-reading process we have

$$I(n, x_1, y_1, \ldots, x_n, y_n) = n, \quad n \in \boldsymbol{N}, \quad x_i, y_i \in \{0, 1\}.$$

Problems with goals formulated in the form (2.1.2) are often called *prediction* or *prognosis problems*. On the other hand, if

$$C(\omega, P) = \varkappa(P), \quad \omega \in \boldsymbol{\Omega}, \quad P \in \mathscr{P}, \tag{2.1.3}$$

for a certain function $\varkappa\colon \mathscr{P} \to \boldsymbol{I}$, then it means that we seek information only about the probability distribution P generating the current realization ω, namely, we want to know the value of the parameter $\varkappa$ for the distribution P. In this way the goals of the classical problems of estimating parameters (e.g., the expected value, the variance, the correlation coefficient, etc.) can be formulated. The same formulation extends also to problems of hypothesis testing since in such problems the goal of research is formally expressed as the characteristic function of a certain subset of $\mathscr{P}$ which contains, in accordance with the investigated hypothesis, the probability distribution generating the current realization of the phenomenon in question. This function is clearly a parameter of the family $\mathscr{P}$.

Any function $C\colon \boldsymbol{\Omega} \times \mathscr{P} \to \boldsymbol{I}$ can be replaced by a family of functions

$$(I_P\colon \boldsymbol{\Omega} \to \boldsymbol{I}, P \in \mathscr{P}),$$

where

$$I_P(\omega) = C(\omega, P), \quad \omega \in \boldsymbol{\Omega}, \quad P \in \mathscr{P}. \tag{2.1.4}$$

If we distinguish a certain σ-field $\mathscr{I}$ of subsets of the set $\boldsymbol{I}$, we can regard the function I_P as a random variable with the probability distribution induced by P. All prediction problems have the same function I_P, while in *problems of parameter inference*, I_P is a constant function for each $P \in \mathscr{P}$.

Problems of prediction and problems of inference on distribution parameters are the most common problems considered in statistics. Less frequently we encounter problems in which the goal of research is expressed in the form

$$C(\omega, P) = \big(I(\omega), \varkappa(P)\big), \quad \omega \in \boldsymbol{\Omega}, \quad P \in \mathscr{P}, \tag{2.1.5}$$

where I is a function defined on $\boldsymbol{\Omega}$, and $\varkappa$ is a function defined on $\mathscr{P}$. Occasionally, there also occur problems in which the goal of research cannot be expressed in the form (2.1.5). We present two examples of such problems.

In the first example we assume that the phenomenon is described by three random variables X_1, X_2, X_3. The random variables X_1, X_2 represent the levels of certain stimulae which produce nonobservable responses. The magnitude of the response can be estimated by the results of an observable test X_3. We assume only that for $i = 1, 2, 3$, the variance σ_i^2 of the random variable X_i exists and is positive, and that the joint distribution of X_1, X_2, X_3 is continuous. The goal of research is to obtain information about the response value I_P, which depends on P and on X_1, X_2 according to the formula

$$I_P(X_1, X_2, X_3) = \frac{X_1}{\sigma_1} + (1-\varrho_{12})\frac{X_2}{\sigma_2}, \tag{2.1.6}$$

where ϱ_{12} is the correlation coefficient of (X_1, X_2).

The second example is the estimation of the value of the regression function. We assume that P belongs to a family $\mathscr{P}$ of distributions of random variables (X, Y), where Y is observable. For any $\omega = (x, y)$ and any P, the required information is the value of the regression function of X on Y at y:

$$I_P(x, y) = E_P(X \mid Y = y), \quad (x, y) \in \boldsymbol{R}^2, P \in \mathscr{P}. \tag{2.1.7}$$

*

Whenever a parameter $\varkappa(P)$ is estimated under restricted observability, the goal of research has to satisfy a certain requirement. It concerns the distributions PT^{-1} of a maximal observable statistic T. Namely, for any P, $P' \in \mathscr{P}$, we must have

$$\varkappa(P) \neq \varkappa(P') \quad \Rightarrow \quad PT^{-1} \neq P'T^{-1}. \tag{2.1.8}$$

Indeed, if this requirement were not satisfied we could not expect that the observable data in the realizations generated by P or P' would permit making any tenable distinction between $\varkappa(P)$ and $\varkappa(P')$.

A parameter $\varkappa$ which satisfies condition (2.1.8) will be called *observably reducible.*

An example of a parameter which is not observably reducible is provided by the correlation coefficient, cor(P), in a simple random sample of size n with the family $\mathscr{P}$ of bivariate distributions with finite second moments, the maximal observable statistic $T^{(n)}$ being of the form

$$T^{(n)}(x_1, y_1, \ldots, x_n, y_n) = \big(T(x_1, y_1), \ldots, T(x_n, y_n)\big), \tag{2.1.9}$$

where

$$T(x, y) = \big(\min(x, y), \max(x, y)\big), \quad x, y \in \boldsymbol{R}.$$

The distributions induced by $T^{(n)}$ are different if and only if the bivariate distributions induced by T are different. Thus, we confine ourselves to the case $n = 1$. Let us consider two distributions: P—a uniform distribution on the square $[0, 1]^2$, and P'—a uniform distribution on the triangle $\{(x, y)\colon 0 \leqslant x \leqslant y,\ 0 \leqslant y \leqslant 1\}$. The two distributions induce the same distribution of the statistic T, and at the same time $\mathrm{cor}(P) = 0$ and $\mathrm{cor}(P') = \frac{1}{2}$.

A parameter which is the characteristic function of a certain subset $\mathscr{P}_0$ of $\mathscr{P}$ is observably reducible if for any $P_0 \in \mathscr{P}_0$ and $P_1 \in \mathscr{P} \setminus \mathscr{P}_0$ we have

$$P_0 T^{-1} \neq P_1 T^{-1}. \tag{2.1.10}$$

In this case we say that the hypothesis $H_0\colon P \in \mathscr{P}_0$ is observably reducible. In the case of a simple random sample of size n observed according to (2.1.9) the hypothesis of independence is not observably reducible; for instance, a uniform distribution on a square, i.e., having independent components, is transformed by T into a uniform distribution on a triangle, i.e., having dependent components. But this transformation does not change uniform distributions on triangles.

The goal of research is sometimes chosen in a natural way as a function defined on the set of indices $\boldsymbol{\Theta}$. As formerly, we require that for any θ, θ'

$$\varkappa(\theta) \neq \varkappa(\theta') \quad \Rightarrow \quad P_\theta T^{-1} \neq P_{\theta'} T^{-1}. \tag{2.1.11}$$

In the case of complete observability, if T is the identity mapping on $\boldsymbol{\Omega}$, the condition (2.1.11) means that the function $\varkappa$ is identifiable (cf. (1.2.1)).

For instance, the co-called *simple linear model* is described by a family of k-dimensional normal distributions $N_k(A\boldsymbol{r}, \sigma^2\boldsymbol{E})$, where A is a given matrix of dimension $k \times s$, $\boldsymbol{E}$ is the identity matrix of dimension $k \times k$, and $\boldsymbol{r}$ is a vector of dimension s, $\sigma^2 > 0$. Hence,

$$\theta = (\boldsymbol{r}, \sigma^2) \in \boldsymbol{\Theta} = \boldsymbol{R}^s \times \boldsymbol{R}^+.$$

Suppose that we aim at obtaining information about the values of the components of the vector $\boldsymbol{r}$. The function defining this goal is of the form

$$\varkappa(\boldsymbol{r}, \sigma^2) = \boldsymbol{r}. \tag{2.1.12}$$

Clearly, $\varkappa$ is identifiable if and only if the rank of the matrix A is equal to s.

On the other hand, if we choose

$$\varkappa(\boldsymbol{r}, \sigma^2) = \boldsymbol{C}\boldsymbol{r},$$

where $\boldsymbol{C}$ is a given matrix of dimension $m \times s$ for any m satisfying $1 \leqslant m \leqslant s$, then $\varkappa$ is identifiable if and only if the set of values of the transformation $\boldsymbol{C}^T$ is contained in the set of values of the transformation $\boldsymbol{A}^T$.

*

To close this section let us introduce the notions of the main parameter and the nuisance parameter. They appear in statistical problems concerning indexed families of distributions, where index θ is a pair (θ_1, θ_2) and our goal is to find the value of θ_1. We then have

$$\varkappa(\theta_1, \theta_2) = \theta_1.$$

The component θ_1 is called the *main parameter* (or *index*) and the component θ_2 is called the *nuisance parameter* (or *index*). In our example concerning linear models with the goal defined by (2.1.12), the vector $\boldsymbol{r}$ is the main parameter while σ^2 is the nuisance parameter.

2.2. Decision rules

Basing ourselves on available information about the phenomenon we make a decision concerning the required information. The way of choosing a decision of this kind is called a *decision rule*. A formal definition of a decision rule requires, first of all, a definition of the set $\boldsymbol{D}$ consisting of all the decisions in question or rather of symbols associated with those decisions.

In the simplest and most common case we assume the set $\boldsymbol{D}$ to be equal to the set $\boldsymbol{I}$ of the required information. In this case the element $d \in \boldsymbol{D}$ is interpreted as the statement that the required information equals d. Interval estimation problems are examples of problems in which the sets $\boldsymbol{D}$ and $\boldsymbol{I}$ do not coincide. In these problems we seek information concerning a certain real-valued parameter $\varkappa$, and hence $\boldsymbol{I}$ is a subset of the set of real numbers. Statements are of the form: "$\varkappa(P) \in [a, b]$" which gives the set $\boldsymbol{D}$ equal to the set of pairs $[a, b]$, $a \leqslant b$.

In many problems we include in the set $\boldsymbol{D}$ the decision "there is not enough information to formulate a statement concerning the required information". This kind of decision is commonly called a *deferred decision*. In everyday life we often encounter situations in which decisions are deferred. For instance, on the basis of the available information we should make a choice between two means of transport: a train and a car. It may happen, however, that the available information is insufficient and—at least for a while—we should defer our decision.

A typical example of the use of a deferred decision is the problem of testing a hypothesis H in which we make one of two alternative decisions: "reject the hypothesis H", or "there is no grounds to reject the hypothesis H". The latter decision is a deferred decision.

In the simplest and most common case we assume that a decision rule is an observable statistic with values in $\boldsymbol{D}$,

$$\delta\colon \boldsymbol{\Omega} \to \boldsymbol{D}.$$

Such rules appear in problems related to the proof-reading process when we estimate the total number of errors n, the probabilities p_1 and p_2, and the parameter λ. More precisely, in these problems the estimators are defined directly as functions of the observable data (n_1, n_2, m). In general, instead of defining a decision rule as an observable statistic it is more convenient to regard a decision rule as a function defined on the set $\boldsymbol{T}$ of values of a certain maximal observable statistic T. The rule

$$\delta\colon \boldsymbol{T} \to \boldsymbol{D}$$

can be regarded as a statistic for a family of distributions induced by T.

However, in the general case the requirement that to each $t \in \boldsymbol{T}$ there should correspond only one decision $d \in \boldsymbol{D}$ narrows too much the possible ways of choosing a decision. Thus, instead of a single decision we assign to each $t \in \boldsymbol{T}$ a certain probability distribution on the set $\boldsymbol{D}$, called a *randomized decision.* The assignement of randomized decisions defined on $\boldsymbol{D}$ to observable data $t \in \boldsymbol{T}$ is called a *randomized decision rule.* In this context the rule $\delta\colon \boldsymbol{T} \to \boldsymbol{D}$ is called a *nonrandomized rule.*

To each nonrandomized rule δ there corresponds exactly one randomized rule according to which, for each $t \in \boldsymbol{T}$, the randomized decision coincides with the probability distribution concentrated at $\delta(t) \in \boldsymbol{D}$.

In order to give a formal definition of a decision rule we must introduce a σ-field $\mathscr{D}$ of subsets of the set $\boldsymbol{D}$. A nonrandomized decision rule is a measurable function acting between measurable spaces $(\boldsymbol{T}, \mathscr{T})$ and $(\boldsymbol{D}, \mathscr{D})$ ($\mathscr{T}$ is a σ-field related to the statistic T). A randomized decision rule is formally a function

$$\delta\colon \boldsymbol{T} \times \mathscr{D} \to [0, 1] \tag{2.2.1}$$

such that the families of functions

$$(\delta_B\colon \boldsymbol{T} \to [0, 1], \boldsymbol{B} \in \mathscr{D}), \quad (\delta_t\colon \mathscr{D} \to [0, 1], t \in \boldsymbol{T}),$$

defined by

$$\delta_B(t) = \delta_t(\boldsymbol{B}) = \delta(t, \boldsymbol{B}) \tag{2.2.2}$$

satisfy the following conditions:

(i) for each $\boldsymbol{B} \in \mathscr{D}$ the function δ_B is measurable with respect to the σ-field of Borel sets on $[0, 1]$;

(ii) for each $t \in \boldsymbol{T}$ the function δ_t is a probability distribution on $(\boldsymbol{D}, \mathscr{D})$.

A function of the form (2.2.1) satisfying conditions (i) and (ii) is called, in probability theory, a *transition probability function* (from the measurable space $(\boldsymbol{T}, \mathscr{T})$ into the measurable space $(\boldsymbol{D}, \mathscr{D})$). Thus, a randomized decision rule is formally a transition probability function.

It is seen that in the case of a randomized decision rule a statement is assigned to observable data in two stages: we first fix a probability distribution δ_t on the set $\boldsymbol{D}$ and then draw according to that distribution a decision $d \in \boldsymbol{D}$. The use of randomized decision rules simplifies mathematical considerations and sometimes improve the quality of decision making.

In some cases one uses a set of decision rules which are functions of a given statistic $g\colon \boldsymbol{T} \to \bar{\boldsymbol{T}}$. This implies, in the case of a randomized decision rule $\delta\colon \boldsymbol{T} \times \mathscr{D} \to [0, 1]$, the existence of a function $\bar{\delta}\colon \bar{\boldsymbol{T}} \times \mathscr{D} \to [0, 1]$ such that

$$\delta(t, \boldsymbol{B}) = \bar{\delta}\big(g(t), \boldsymbol{B}\big), \quad t \in \boldsymbol{T}, \quad \boldsymbol{B} \in \mathscr{D}. \tag{2.2.3}$$

A decision rule δ satisfying condition (2.2.3) is said to be *measurable with respect to statistic g.*

Let us define, for instance, a randomized decision rule in the case where what we observe are the values of a sequence of random variables $X_1, \ldots, X_n$, and the set $\boldsymbol{D}$ is equal to $\{0, 1\}$. We define

$$\delta((x_1, \ldots, x_n), \{1\}) = \begin{cases} 0 & \text{if} \quad \min(x_1, \ldots, x_n) < 0, \\ 0.5 & \text{if} \quad \min(x_1, \ldots, x_n) = 0, \\ 1 & \text{if} \quad \min(x_1, \ldots, x_n) > 0. \end{cases}$$

Obviously, rule δ is measurable with respect to statistic $g(x_1, \ldots, x_n) = \min(x_1, \ldots, x_n)$ and the decision rule $\bar{\delta}$ is given by the formula

$$\bar{\delta}(y, \{1\}) = \begin{cases} 0 & \text{if} \quad y < 0, \\ 0.5 & \text{if} \quad y = 0, \\ 1 & \text{if} \quad y > 0. \end{cases}$$

A nonrandomized decision rule δ: $\boldsymbol{T} \to \boldsymbol{D}$ is measurable with respect to g if

$$\delta = \bar{\delta} \circ g \tag{2.2.4}$$

for a certain function $\bar{\delta}$: $\bar{\boldsymbol{T}} \to \boldsymbol{D}$. The notation (2.2.4) will be also used for δ, $\bar{\delta}$, g when δ and $\bar{\delta}$ are randomized and satisfy (2.2.3).

*

For a given probability distribution $P \in \mathscr{P}$ each decision rule δ induces a certain distribution on $\boldsymbol{I} \times \boldsymbol{D}$ which describes the simultaneous appearance of the research goal I_P and of decisions taken on it. This distribution will be denoted by $Q_{P,\delta}$. For a nonrandomized decision rule δ: $\boldsymbol{T} \to \boldsymbol{D}$ we define the distribution $Q_{P,\delta}$ as a distribution of the pair $(I_P, \delta \circ T)$. If $I_P = \varkappa(P)$, i.e., if the parameter $\varkappa$ constitutes the goal of research, then obviously only the distribution of the second component (i.e., the distribution of a decision rule) is of importance.

For a randomized decision rule δ the distribution $Q_{P,\delta}$ is constructed as follows:

(i) we determine a distribution on $\boldsymbol{I} \times \boldsymbol{T} \times \boldsymbol{D}$ such that its marginal distribution on $\boldsymbol{I} \times \boldsymbol{T}$ coincides with that of the pair (I_P, T) corresponding to P, and the conditional distribution on $\boldsymbol{D}$ for a fixed pair $(i, t) \in \boldsymbol{I} \times \boldsymbol{T}$ coincides with δ_t determined by the decision rule δ (cf. (2.2.2));

(ii) we determine the marginal distribution of the first and the third component of the distribution built in (i).

Once the goal of research is formulated, it is desirable to take into account only those properties of rules δ which depend exclusively on the corresponding distributions $Q_{P,\delta}$, for $P \in \mathscr{P}$. In the sequel we discuss classes of statistical problems whose solutions are determined solely by the conditions imposed on families of distributions $Q_{P,\sigma}$. For such problems it is natural to introduce the following equivalence of decision rules:

$$\delta \approx \delta' \quad \Leftrightarrow \quad \forall P \in \mathscr{P} \; Q_{P,\delta} = Q_{P,\delta'}, \tag{2.2.5}$$

i.e., two rules are considered to be equivalent if for every $P \in \mathscr{P}$ the consequences of decisions taken according to these rules are identical.

We illustrate this equivalence of decision rules by the following example.

Consider sequences of independent random variables $X_1, \ldots, X_n$ with the same uniform distributions on integers from the interval $[-k, k]$, where k is an unknown parameter assuming natural values. Hence, for a fixed sample size n $(n > 1)$ we deal with the family $\mathscr{P}$ of distributions P_k, $k = 1, 2, \ldots$, where P_k is the product of n uniform distributions on $[-k, k]$. We want to test the hypothesis that $k \leqslant K$ for a certain fixed number K. Consequently, I_{P_k} takes on two values, say 0 for $k \leqslant K$ and 1 for $k > K$, so that $\boldsymbol{I} = \{0, 1\}$. Further, let $\boldsymbol{D}$ be equal to $\boldsymbol{I}$. We define, for an arbitrary but fixed natural number a, two decision rules:

$$\delta_1\big((x_1, \ldots, x_n), \{1\}\big) = \begin{cases} 0 & \text{if} \quad \min(x_1, \ldots, x_n) > -a, \\ 0.5 & \text{if} \quad \min(x_1, \ldots, x_n) = -a, \\ 1 & \text{if} \quad \min(x_1, \ldots, x_n) < -a, \end{cases}$$

$$\delta_2\big((x_1, \ldots, x_n), \{1\}\big) = \begin{cases} 0 & \text{if} \quad \max(x_1, \ldots, x_n) < a, \\ 0.5 & \text{if} \quad \max(x_1, \ldots, x_n) = a, \\ 1 & \text{if} \quad \max(x_1, \ldots, x_n) > a. \end{cases}$$

For $k \leqslant K$ the distribution Q_{P_k, δ_1} is of the form

$$\begin{aligned} Q_{P_k,\delta_1}(1, 0) &= Q_{P_k,\delta_1}(1, 1) = 0, \\ Q_{P_k,\delta_1}(0, 0) &= P\big(\min(x_1, \ldots, x_n) > -a\big) + \tfrac{1}{2} P\big(\min(x_1, \ldots, x_n) = -a\big), \\ Q_{P_k,\delta_1}(0, 1) &= P\big(\min(x_1, \ldots, x_n) < -a\big) + \tfrac{1}{2} P\big(\min(x_1, \ldots, x_n) = -a\big). \end{aligned}$$

Hence,

$$Q_{P_k,\delta_1}(0, 0) = \begin{cases} 1 & \text{if} \quad k < a, \\ \dfrac{(k+a+1)^n + (k+a)^n}{2(2k+1)^n} & \text{if} \quad k \geqslant a, \end{cases}$$

$$Q_{P_k,\delta_1}(0, 1) = 1 - Q_{P_k,\delta_1}(0, 0).$$

For $k > K$, we have

$$Q_{P_k,\delta_1}(0, 0) = Q_{P_k,\delta_1}(0, 1) = 0$$

and the probabilities concentrated in $(1, 0)$ and $(1, 1)$ coincide with the probabilities concentrated in $(0, 0)$ and $(0, 1)$ for $k \leqslant K$, respectively.

According to the symmetry of the distributions P_k and of decisions rules δ_1, δ_2, the distribution Q_{P_k,δ_1} is equal to Q_{P_k,δ_2} for any k; therefore, δ_1 and δ_2 are equivalent.

2.3. Standard classes of statistical problems

In the set of all decision rules of a given statistical problem we distinguish a non-empty subset Δ of rules initially admitted for considerations (for instance Δ may contain all nonrandomized rules). In the set Δ we distinguish a subset Δ^* of rules considered to be satisfactory in the context of the particular investigation. Any element of Δ^* is called a *solution of the problem* in question and can be used to formulate a statement (i.e., to deliver a decision). Usually, Δ^* is the solution set of a certain optimization problem.

We shall now single out certain classes of problems called *standard classes.* Roughly speaking, in those classes of problems the set Δ^* is given only by the conditions imposed on the joint distribution of the research goal and the decision taken, i.e., by the conditions imposed on the family $(Q_{P,\delta},\ P \in \mathscr{P},\ \delta \in \Delta)$. In the sequel this family will be identified with an *indexing function* q which maps the set $\mathscr{P} \times \Delta$ into the set of all distributions on $\boldsymbol{I} \times \boldsymbol{D}$:

$$q(P, \delta) = Q_{P,\delta}, \quad P \in \mathscr{P},\ \delta \in \Delta. \tag{2.3.1}$$

Function q is derived in statistical problems from the descriptions of the phenomenon and of the goal of research.

We now give examples of classes of statistical problems in which the set Δ^* is defined by the conditions imposed on the family of distributions $Q_{P,\delta}$. The first example is a class of hypothesis testing problems. Namely, in a family $\mathscr{P}$ of distributions we distinguish a subset $\mathscr{P}_0$ and we aim at finding out whether P belongs to $\mathscr{P} \setminus \mathscr{P}_0$. We admit only two statements which refer to the hypothesis H_0: $P \in \mathscr{P}_0$. They are: "reject H_0" and "there is not enough information to reject H_0". These decisions correspond to the following asymmetry in our approach to the sets $\mathscr{P}_0$ and $\mathscr{P} \setminus \mathscr{P}_0$: we consider investigations in which it is vital to know whether P belongs to $\mathscr{P} \setminus \mathscr{P}_0$; if this fact cannot be established with sufficient reliability, we abstain from any attempt to form an opinion about P. A possible formalization of such a problem may be as follows. Let $I_P \equiv 0$ if $P \in \mathscr{P}_0$ and $I_P \equiv 1$ if $P \in \mathscr{P} \setminus \mathscr{P}_0$. Thus, $I = \{0, 1\}$. Let $\boldsymbol{D} = \{0, 1\}$, where 0 denotes the deferred decision ("there is not enough information to reject H_0") and 1 denotes the decision "reject H_0". A randomized decision rule in this problem is a function assigning to an observation t a distribution on $\boldsymbol{D} = \{0, 1\}$. Thus, this rule is uniquely determined by a function

$$\Phi\colon \boldsymbol{T} \to [0, 1], \tag{2.3.2}$$

where $\Phi(t)$ is the probability of taking decision 1 when t is observed. The function Φ is called a *test*.

The distribution $Q_{P,\Phi}$ (i.e., the distribution $Q_{P,\delta}$ for δ defined by Φ) is specified for any P and Φ by the probability of taking decision 1. This probability is called the *power of the test* Φ and is denoted by $\beta_\Phi(P)$. We want to find a test maximizing, for each $P \in \mathscr{P} \setminus \mathscr{P}_0$, the power $\beta_\Phi(P)$ in the set of tests satisfying the condition:

$$\beta_\Phi(P) \leqslant \alpha, \quad P \in \mathscr{P}_0,$$

for a given $\alpha \in (0, 1)$. Such tests are said to be *uniformly most powerful* in the class of tests on the significance level α.

According to (2.2.5), two tests are *equivalent* if they have the same power function $\beta_\Phi(P)$, $P \in \mathscr{P}$.

To sum up, the significance level protects the test against the false statement that P belongs to $\mathscr{P} \setminus \mathscr{P}_0$ when in fact it belongs to $\mathscr{P}_0$. Thus, the decision "reject H_0" seems reliable under any test admitted for consideration. Among those tests a test is accepted as satisfactory (i.e., belongs to Δ^*) if no other test rejects H_0 with a higher probability.

The second example is a particular case of the following problem. Let Δ be a set of nonrandomized decision rules and let $L\colon \boldsymbol{\Omega} \times \mathscr{P} \times \boldsymbol{D} \to \boldsymbol{R}$ be a function whose value $L(\omega, P, d)$ represents a loss incurred through a decision d for a realization ω drawn according to a distribution P. We assume that, for each fixed $P \in \mathscr{P}$, the function L is measurable with respect to the σ-field $\mathscr{B}(\boldsymbol{R})$. We want to find decision rules minimizing, for each $P \in \mathscr{P}$, the expected loss, called the *risk*, under an additional constraint on decision rules δ. This constraint is given by a certain function $W\colon \boldsymbol{\Omega} \times \mathscr{P} \times \boldsymbol{D} \to \boldsymbol{R}$, measurable for any fixed P with respect to the σ-field $\mathscr{B}(\boldsymbol{R})$. Hence, we minimize

$$R(\delta, P) = \int_{\Omega} L\big(\omega, P, \delta(T(\omega))\big)\, \mathrm{d}P \tag{2.3.3}$$

over decision rules δ belonging to $\varDelta$ and satisfying the condition

$$\int_{\Omega} W\big(\omega, P, \delta(T(\omega))\big)\, \mathrm{d}P = 0.$$

Since decision rules serve to formulate conclusions $\delta\big(T(\omega)\big)$ concerning the required information $I_P(\omega)$, we are particularily interested in cases where the divergence between conclusions and goals is evaluated by losses and where constraints deal with statements and goals. Formally, we describe such cases by functions $\mathscr{L}, \mathscr{W}\colon \boldsymbol{I} \times \boldsymbol{D} \to \boldsymbol{R}$, measurable with respect to $\mathscr{B}(\boldsymbol{R})$ and satisfying

$$\begin{aligned} L(\omega, P, d) &= \mathscr{L}\big(I_P(\omega), d\big), \\ W(\omega, P, d) &= \mathscr{W}\big(I_P(\omega), d\big). \end{aligned} \tag{2.3.4}$$

Then

$$\begin{aligned} R(\delta, P) &= \int_{\Omega} \mathscr{L}\big(I_P(\omega), \delta(T(\omega))\big)\, \mathrm{d}P = \int_{I \times D} \mathscr{L}(i, d)\, \mathrm{d}Q_{P,\delta}, \\ \int_{\Omega} \mathscr{W}\big(I_P(\omega), \delta(T(\omega))\big)\, \mathrm{d}P &= \int_{I \times D} \mathscr{W}(i, d)\, \mathrm{d}Q_{P,\delta}. \end{aligned} \tag{2.3.5}$$

Conditions on the basis of which any rule $\delta \in \varDelta$ is included into the solution set $\varDelta^*$ can be represented in the form of conditions imposed on the family of distributions $(Q_{P,\delta},\ P \in \mathscr{P})$.

If

$$\mathscr{L}(i, d) = (i-d)^2, \quad \mathscr{W}(i, d) = i-d, \tag{2.3.6}$$

and $I_P \equiv \varkappa(P)$ for a certain parameter $\varkappa$, then the problem considered consists in finding an unbiased estimator of $\varkappa$ with uniformly minimal variance. But if (2.3.6) holds and the family $\mathscr{P}$ consists of one element only and the goal is to predict the values of a random variable I, then we deal with the probabilistic problem of finding the *best mean-square predictor*.

In the present example we do not have to restrict ourselves to nonrandomized decision rules. For a randomized decision rule δ and given functions $\mathscr{L}$, $\mathscr{W}$, we define the risk and the constraints by the formulae

$$R(\delta, P) = \int_{\Omega} \int_{D} \mathscr{L}\big(I_P(\omega), d\big)\, \mathrm{d}\delta_{T(\omega)}\, \mathrm{d}P = \int_{I \times D} \mathscr{L}(i, d)\, \mathrm{d}Q_{P,\delta}, \tag{2.3.7}$$

$$\int_{\Omega \times D} \mathscr{W}\big(I_P(\omega), d\big)\, \mathrm{d}\delta_{T(\omega)}\, \mathrm{d}P = \int_{I \times D} \mathscr{W}(i, d)\, \mathrm{d}Q_{P,\delta}. \tag{2.3.8}$$

The set Δ^* in the first example is defined as

$$\Delta^* = \{\delta^* \in \Delta:\ S_\alpha(\delta^*)\}, \tag{2.3.9}$$

where the property $S_\alpha(\delta^*)$ is of the form

$$\forall P \in \mathscr{P}\ \big[Q_{P,\delta^*}(0) \leqslant \alpha \ \wedge\ \forall \delta \in \Delta\ \big(Q_{P,\delta}(0) \leqslant \alpha \Rightarrow Q_{P,\delta}(1) \leqslant Q_{P,\delta^*}(1)\big)\big]. \tag{2.3.10}$$

Analogously, the set Δ^* in the second example is defined by the property $S_{\mathscr{L},\mathscr{W}}(\delta^*)$ of the form

$$\forall P \in \mathscr{P}\ \big[E_{P,\delta^*}(\mathscr{W}) = 0 \ \wedge\ \forall \delta \in \Delta\big(E_{P,\delta}(\mathscr{W}) = 0 \Rightarrow E_{P,\delta}(\mathscr{L}) \geqslant E_{P,\sigma^*}(\mathscr{L})\big)\big], \tag{2.3.11}$$

where $E_{P,\delta}(\mathscr{W})$ is the expected value of the random variable $\mathscr{W}$ for the distribution $Q_{P,\delta}$, and $E_{P,\delta}(\mathscr{L})$ is the risk for this distribution.

The symbols P, δ and δ^* appear in formulae (2.3.10), (2.3.11) only in the expressions $Q_{P,\delta}$, Q_{P,δ^*}.

In both examples formulae (2.3.10) and (2.3.11) are not influenced either by the description of the phenomenon and the research goal or by the set Δ. Thus, the formula S_α defines the whole class of testing problems in which we want to find tests uniformly most powerful among the tests on a given significance level α. The formula $S_{\mathscr{L},\mathscr{W}}$ defines, in turn, the whole class of problems of unbiased estimation with minimal variance. Particular problems belonging to any of these classes can be obtained by substituting specific sets $\mathscr{P}$, Δ and function q in (2.3.10), (2.3.11) and determining in this way the property of decision rule δ^* which defines the set Δ^* in a given problem. In the first example this class of problems consists of all problems for which the image $q(\mathscr{P} \times \Delta)$ is contained in the set of all distributions on $\{0, 1\} \times \times \{0, 1\}$. In the second example this class of problems is formed by all problems with the image $q(\mathscr{P} \times \Delta)$ contained in the set of all distributions on $\boldsymbol{R}^2$ for which the expected values of random variables $\mathscr{W}$ and $\mathscr{L}$ are defined.

The third example is the class of statistical problems for which $\boldsymbol{I} = \boldsymbol{D} = \boldsymbol{R}$ and

$$\Delta^* = \{\delta^* \in \Delta:\ S_\varepsilon(\delta^*)\},$$

where $S_\varepsilon(\delta^*)$ is of the form

$$\forall P \in \mathscr{P}\ \operatorname{cor}(Q_{P,\delta^*}) \geqslant 1-\varepsilon \tag{2.3.12}$$

for a given positive number $\varepsilon \leqslant 0.5$. In this example we require the image $q(\mathscr{P} \times \Delta)$ to be contained in the set of bivariate distributions with finite second moments since only for such distributions the correlation coefficient is defined.

The fourth example is the class of problems in which, for fixed sets $\boldsymbol{I}$, $\boldsymbol{D}$, and a function $\mathscr{L}$: $\boldsymbol{I} \times \boldsymbol{D} \to \boldsymbol{R}$, we have

$$\Delta^* = \{\delta^* \in \Delta:\ S_{\mathscr{L}}(\delta^*)\},$$

where $S_{\mathscr{L}}(\delta^*)$ is of the form

$$\forall \delta \in \Delta\ \forall P_1 \in \mathscr{P}\ \forall a > 0\ \exists P_2 \in \mathscr{P}\quad R(P_1, \delta^*) < R(P_2, \delta) + a, \tag{2.3.13}$$

and $R(P, \delta) = E_{P,\delta}(\mathscr{L})$. The property $S_{\mathscr{L}}(\delta^*)$ can be equivalently rewritten as

$$\forall \delta \in \Delta \sup_{P \in \mathscr{P}} R(\delta^*, P) \leqslant \sup_{P \in \mathscr{P}} R(\delta, P). \tag{2.3.14}$$

A decision rule δ^* satisfying this requirement is said to be a *minimax rule* in Δ with respect to the loss function $\mathscr{L}$. Here, as in the second example, we require the image $q(\mathscr{P} \times \Delta)$ to be contained in the set of distributions on $\boldsymbol{I} \times \boldsymbol{D}$ for which the expected value of the random variable $\mathscr{L}$ exists.

Basing ourselves on the preceding four examples, we now introduce the notion of a *standard class of statistical problems.*

Suppose we are given sets $\boldsymbol{I}$ and $\boldsymbol{D}$ and nonnegative integer numbers k, m. Consider an arbitrary property(1)

$$S(\delta^*, \mathscr{P}, \Delta, q, P_1, \ldots, P_k, \delta_1, \ldots, \delta_m), \tag{2.3.15}$$

satisfying the following assumptions:

1° the symbols $P_1, \ldots, P_k$ are placed after quantifiers of the form $\forall P_i \in \mathscr{P}$, $\exists P_i \in \mathscr{P}$, $i = 1, \ldots, k$;

2° the symbols $\delta_1, \ldots, \delta_m$ are placed after quantifiers of the form $\forall \delta_i \in \Delta$, $\exists \delta_i \in \Delta$, $i = 1, \ldots, m$;

3° the symbols q, $P_1, \ldots, P_k, \delta_1, \ldots, \delta_m, \delta^*$ appear only in terms $q(P_i, \delta_j)$, $q(P_i, \delta^*)$, for $i = 1, \ldots, k$, $j = 1, \ldots, m$;

4° the symbols $\mathscr{P}$ and Δ appear only as domains of quantifiers preceding symbols $P_1, \ldots, P_k$ and $\delta_1, \ldots, \delta_m$.

To a pair of sets $\boldsymbol{I}$, $\boldsymbol{D}$ and to any formula S satisfying conditions 1° – 4° for certain k and m we assign a class of statistical problems as follows. A problem belongs to this class if the set Δ^* consists of rules $\delta^* \in \Delta$ such that δ^* together with a triple $(\mathscr{P}, \Delta, q)$ which is determined by that problem satisfy the formula S. Obviously it is required that the substitution of the triple $(\mathscr{P}, \Delta, q)$ in S should not lead to undefined expressions. For each such triple, the formula S defines Δ^*. Every such class of problems is called a *standard class defined by* $\boldsymbol{I}, \boldsymbol{D}$, *and* S. Each specific problem from a given standard class is determined by $(\mathscr{P}, \Delta, q)$.

In probabilistic problems the set $\mathscr{P}$ consists of one element and thus each problem is determined by P, Δ, and q, where q is identical with a function on Δ.

In formulae (2.3.10) and (2.3.11) we have one symbol P and one symbol δ, i.e., $k = m = 1$, in formula (2.3.12) $k = 1$, $m = 0$, in formula (2.3.13) $k = 2$, $m = 1$. Each of these formulae together with the sets $\boldsymbol{I}$ and $\boldsymbol{D}$ defines a certain standard class of statistical problems. Standard classes of probabilistic discriminant problems and screening problems will be dealt with in Chapters 3 and 4.

Formulae defining classes are usually indexed by certain parameters (e.g., by the significance level α in (2.3.10), by functions $\mathscr{L}$ and $\mathscr{W}$ in (2.3.11)). Thus, we deal, in fact, with *families* of standard classes.

---

(1) A precise definition of the notion of property requires the apparatus of mathematical logic (e.g., the notions of language, formula, etc). Readers who are interested in such a definition should consult textbooks of mathematical logic (e.g., Schoenfeld, 1967). However, the intuitive meaning of the notion seems to be sufficiently clarified by means of the preceding examples.

Suppose now that we deal with a family of statistical problems and we want to know whether this family is included in a certain standard class of statistical problems. Thus either we have to find for the set Δ^* a defining formula which satisfies the requirements standing in the definition of a standard class, or we have to prove that such a formula does not exist. Either of these cases may turn out to be difficult.

As an example of a family which is not standard will serve a family of testing problems concerning a hypothesis $P \in \mathscr{P}_0$, $\mathscr{P}_0 \subset \mathscr{P}$. The solutions are required to be tests on a prescribed significance level α with their power exceeding some prescribed value for any P sufficiently far from $\mathscr{P}_0$. Thus, set Δ^* is defined by the formula

$$\forall P \in \mathscr{P} \; [Q_{P,\delta*}(0) \leqslant \alpha \quad \wedge \quad (\eta(P, \mathscr{P}_0) \geqslant \eta_0 \quad \Rightarrow \quad Q_{P,\delta*}(1) \geqslant 1-\varepsilon)],$$

where η is a chosen distance in $\mathscr{P}$.

On the other hand, the family of maximum likelihood (briefly ML) estimation problems is an example of a family of problems for which we are not able either to prove or to disprove that the family is contained in a standard class of problems. This family contains problems such that 1° family $\mathscr{P}$ is indexed injectively by elements of a set $\boldsymbol{\Theta}$ and is dominated by a measure ν; 2° any realization $\omega \in \boldsymbol{\Omega}$ is fully observable; 3° the goal of research is to find the value of a parameter $\varkappa(\theta)$; 4° Δ is the set of nonrandomized rules; 5° Δ^* is the set of estimators $\hat{\varkappa}$ defined by the formula

$$\hat{\varkappa}(\omega) = \varkappa(\hat{\theta}(\omega)), \quad \omega \in \boldsymbol{\Omega}, \tag{2.3.16}$$

where $\hat{\theta}$ is an ML statistic (cf. Section 1.3).

*

In any decision problem we want to know whether it is possible to make a preliminary reduction of observable data in a way which does not influence the solutions of the problem. Thus, the question is whether, for a given problem, it is possible to define another problem with data reduced by a statistic g: $\boldsymbol{T} \to \overline{\boldsymbol{T}}$ and such that the two problems are equivalent.

Let Δ_g be the set of rules from Δ which are measurable with respect to a statistic g. To each problem from a standard class, determined by $\mathscr{P}$, Δ, and a function q indexing distributions on $\boldsymbol{I} \times \boldsymbol{D}$, there corresponds another problem from the same class, determined by $\mathscr{P}$, Δ_g, and q_g, which is a restriction of q to the set $\mathscr{P} \times \Delta_g$. A problem restricted in this way is said to be equivalent to the original problem if, for each solution δ^* of the original problem, there exists an equivalent solution of the restricted problem. Clearly, this equivalence holds, in particular, for statistics g with the property that

for each rule $\delta \in \Delta$ there exists a decision rule $\delta_g \in \Delta$, measurable with respect to g and such that $\delta_g \approx \delta$. (2.3.17)

2.4. Sufficient and prediction sufficient statistics

We begin with a discussion of admissible reductions of information contained in data in the process of inference about the parameters of distributions under complete observability. To begin with we reject the information which is redundant no matter which parameter of family $\mathscr{P}$ is the current goal of research.

Consider a set $\mathscr{P}$ of distributions indexed by a one-to-one function defined on a set $\boldsymbol{\Theta}$. Let g be an arbitrary statistic. For each $\theta \in \boldsymbol{\Theta}$, a random choice of a realization ω according to the probability P_θ can be replaced by a two-stage random choice. Namely, first we draw a value $\bar{\omega}$ of statistic g according to the distribution $P_\theta g^{-1}$, and then we draw a realization ω according to the conditional distribution obtained from P_θ under the condition that $g(\omega)$ equals $\bar{\omega}$. For a fixed $\bar{\omega}$, this conditional distribution is usually different for different $\theta_1, \theta_2 \in \boldsymbol{\Theta}$. However, if the conditional distributions happen to be the same for all $\theta \in \boldsymbol{\Theta}$, then the second drawing does not give any further information about distribution P_θ. The whole information about P_θ is contained in $\bar{\omega}$. Such a statistic g is called *sufficient for the family* $\mathscr{P}$, or shortly *sufficient*, if it is clear from the context which family is meant.[1]

Trivial sufficient statistics are injective mappings of $\boldsymbol{\Omega}$, in particular the identity mapping. Generally, each measurable composition of a sufficient statistic and an injection is also a sufficient statistic.

Clearly, a measurable function g: $\boldsymbol{\Omega} \to \bar{\boldsymbol{\Omega}}$ which is a sufficient statistic for a certain family of distributions defined on $\boldsymbol{\Omega}$ is not necessarily a sufficient statistic for the enlarged family. For example, consider n independent random variables $X_1, \ldots$ $\ldots, X_n$ with the same distribution belonging to

$$\mathscr{P} = \{U(a, b)\colon\ a, b \in \boldsymbol{R},\ a < b\}. \tag{2.4.1}$$

Let $X_{(i)}$ denote the i-th order statistic, $i = 1, \ldots, n$, in the sequence $X_1, \ldots, X_n$. The pair $(X_{(1)}, X_{(n)})$ is a sufficient statistic for $\mathscr{P}$ (more precisely, for $\{P^n, P \in \mathscr{P}\}$) since for arbitrary a, $b \in \boldsymbol{R}$, the value at the point $(x_1,\ldots,x_n)$ the density of the conditional distribution of $(X_1, \ldots, X_n)$ under the condition $X_{(1)} = y$, $X_{(n)} = z$ is equal to

$$\prod_{i=1}^{n} \chi_{[y,z]}(x_i)\,\bigl(2n(z-y)^{n-1}\bigr)^{-1}. \tag{2.4.2}$$

at the point $(x_1, \ldots, x_n)$. If a is known, then for each $a \in \boldsymbol{R}$, the statistic $X_{(n)}$ is sufficient for the family

$$\mathscr{P}_a = \{U(a, b)\colon\ b > a\},$$

since, for an arbitrary b greater than a, the value of the density at the point $(x_1,\ldots,x_n)$ of the conditional distribution of $X_1, \ldots, X_n$ under the condition $X_{(n)} = z$ is equal to

$$\chi_{[a,+\infty)}(z) \prod_{i=1}^{n} \chi_{[a,z]}(x_i)\,\bigl(n(z-a)^{n-1}\bigr)^{-1}. \tag{2.4.3}$$

---

[1] According to our previous considerations on statistics (cf. Section 1.3), one should specify a statistical space $M = (\Omega, \mathscr{A}, \mathscr{P})$; hence, any particular function defined on Ω leads to different statistics for different families of distributions. This explains the expressions: "statistic sufficient for a family $\mathscr{P}$", and "statistic sufficient for a space M".

It can be seen from (2.4.3) that $X_{(n)}$ is not a sufficient statistic for the family (2.4.1).

In general, it is not easy to find a conditional distribution under the condition that a given statistic has a given value. However, for a dominated family of distributions there exists a simple criterion for the sufficiency of a statistic, known as the *factorization criterion*:

a function $g\colon \boldsymbol{\Omega} \to \bar{\boldsymbol{\Omega}}$ *is a sufficient statistic for a dominated family* $\{P_\theta\colon \theta \in \boldsymbol{\Theta}\}$ *if and only if, for each* $\theta \in \boldsymbol{\Theta}$, *the density* p_θ *of the distribution* P_θ *with respect to a chosen dominating me asure can be represented in the form* (2.4.4)

$$p_\theta(\omega) = f(\theta, g(\omega))\,k(\omega)$$

for some functions $f\colon \boldsymbol{\Theta} \times \bar{\boldsymbol{\Omega}} \to \boldsymbol{R}^+$, $k\colon \boldsymbol{\Omega} \to \boldsymbol{R}^+$.

For instance, in the previous example of a simple random sample with the family of distributions given by (2.4.1) for any a, b, $a < b$, the density can be represented as

$$p_{a,b}(x_1, \dots, x_n) = \chi_{[a,+\infty)}(y)\,\chi_{(-\infty,b]}(z)\,((b-a)^n)^{-1}, \tag{2.4.5}$$

where y and z are values of $X_{(1)}$ and $X_{(n)}$, respectively. Thus, we obtain the sufficiency of the pair of statistics $(X_{(1)}, X_{(n)})$ for family (2.4.1) without calculating conditional density (2.4.2). Moreover, we see that if a is known $X_{(n)}$ is a sufficient statistic for $\mathscr{P}_a$ since, for an arbitrary but fixed a, $a < b$, the density $p_{a,b}$ is the product of two terms:

$$f_a(b, z) = \chi_{(-\infty,b]}(z)\,((b-a)^n)^{-1}$$

and

$$k_a(x_1, \dots, x_n) = \chi_{[a,+\infty)}(y).$$

Now we apply the factorization criterion to any two-element family $\{P_1, P_2\}$ of distributions with positive densities p_1, p_2 with respect to a given measure. The ratio h of densities, called the *likelihood ratio*,

$$h(\omega) = \frac{p_2(\omega)}{p_1(\omega)}, \tag{2.4.6}$$

is a sufficient statistic since

$$p_i(\omega) = f(i, h(\omega))k(\omega),$$

where

$$f(i, x) = \begin{cases} 1 & \text{if} \quad i = 1, \\ x & \text{if} \quad i = 2, \end{cases} \qquad k(\omega) = p_1(\omega).$$

In the same way we can verify that the product of likelihood ratios

$$g(\omega_1, \dots, \omega_n) = h(\omega_1) \cdot \ldots \cdot h(\omega_n) \tag{2.4.7}$$

is a sufficient statistic for the family $(P^n\colon P \in \{P_1, P_2\})$, since

$$\prod_{j=1}^{n} p_i(\omega_j) = f\Big(i, \prod_{j=1}^{n} h(\omega_j)\Big) \cdot \prod_{j=1}^{n} k(\omega_j). \tag{2.4.8}$$

Now we formulate the *fundamental property of sufficient statistics*:

> *In any problem of inference concerning a parameter of a family $\mathscr{P}$, if $g\colon \boldsymbol{T} \to \bar{\boldsymbol{T}}$ is a sufficient statistic for the family of distributions of a maximal observable statistic $T\colon \boldsymbol{\Omega} \to \boldsymbol{T}$, then for each decision rule δ there exists an equivalent rule δ_g which is measurable with respect to g.* (2.4.9)

Indeed, if we denote by $P_{\bar{t}}$ the conditional distribution of T under the condition $g(\boldsymbol{T}) = \bar{t}$, then by the sufficiency of statistic g, this distribution is the same for all distributions PT^{-1}. Hence, for each decision rule $\delta\colon \boldsymbol{T}\times\mathscr{D} \to [0, 1]$ we can define $\bar{\delta}\colon \bar{\boldsymbol{T}}\times\boldsymbol{D} \to [0, 1]$ by the formula

$$\bar{\delta}(\boldsymbol{B}, \bar{t}) = \int_{\boldsymbol{T}} \delta(\boldsymbol{B}, t)\, \mathrm{d}P_{\bar{t}}, \quad \boldsymbol{B} \in \mathscr{D}, \quad \bar{t} \in \bar{\boldsymbol{T}}. \tag{2.4.10}$$

Rule $\delta_g\colon \boldsymbol{T}\times\mathscr{D} \to [0, 1]$ can be defined as

$$\delta_g(\boldsymbol{B}, t) = \bar{\delta}\big(\boldsymbol{B}, g(t)\big), \quad \boldsymbol{B} \in \mathscr{D}, \quad t \in \boldsymbol{T}. \tag{2.4.11}$$

Clearly, the δ_g just defined is measurable with respect to g, and we can prove the equivalence $\delta \approx \delta_g$.

Therefore, if in an inference concerning a distribution parameter the set $\boldsymbol{\Delta}$ is closed under the relation $\approx$, i.e.,

$$\forall \delta, \delta' \quad \delta \in \boldsymbol{\Delta} \wedge \delta \approx \delta' \Rightarrow \delta' \in \boldsymbol{\Delta}, \tag{2.4.12}$$

then any sufficient statistic g has the property (2.3.17). This implies that solving any parameter inference problem belonging to a standard class and satisfying (2.4.12) can be reduced to finding solutions in the set $\boldsymbol{\Delta}_g$. This makes it possible to reduce the set of feasible solutions to rules measurable with respect to g.

In the first example of Section 2.3, which concerns finding uniformly most powerful tests, all possible tests are taken into consideration. Thus, condition (2.4.12) is satisfied and we can restrict ourselves to tests measurable with respect to an arbitrary sufficient statistic. In particular, in seeking the most powerful test of the hypothesis that the density is equal to p_1 against the alternative hypothesis that it is equal to p_2 ($p_2 > 0$), considerations can be restricted to tests which are functions of the likelihood ratio of p_1 and p_2 (formula (2.4.6)). Recall that this ratio is a sufficient statistic for the two-element family $\{p_1, p_2\}$. Observe that by the necessary condition of the Neyman–Pearson lemma (cf. Lehmann, 1959, Section 3.2), there is no other solution of this problem.

In the second example of Section 2.3, which concerns seeking unbiased estimators with uniformly minimal variance, the set of nonrandomized decision rules does not satisfy condition (2.4.12). However, we can start by admitting all rules, which allows us to restrict Δ to the set of rules measurable with respect to an arbitrary sufficient statistic g. In this set, by the convexity of the loss function, for each rule there exists a nonrandomized rule of no greater risk. Thus, we can choose Δ as the set of nonrandomized rules measurable with respect to an arbitrary sufficient statistic. By the Rao–Blackwell theorem (cf. Bickel and Doksum, 1978, Th. 4.2.1) there are no other solutions to this problem.

*

Reduction of irrelevant information can also be made in problems of predicting the value of a nonobservable statistic I: $\boldsymbol{\Omega} \to \boldsymbol{I}$. It seems reasonable to consider only those prediction problems in which all relevant information is contained in the family of distributions of the pair (I, T), where T is a chosen maximal observable statistic. In such problems the reduction of observability by a function g: $\boldsymbol{T} \to \bar{\boldsymbol{T}}$ causes no loss of information if the distribution of (I, T) can be reconstructed from the distribution of $(I, g(T))$ in an identical way for each $P \in \mathscr{P}$. This holds true if the following two conditions are satisfied:

1° g is sufficient for the family of distributions of T;

2° for any $t \in \boldsymbol{T}$ and any $P \in \mathscr{P}$ the conditional distribution of I under the condition $T = t$ is equal to the conditional distribution of I under the condition $g(T) = \bar{t}$ for $\bar{t} = g(t)$.

Indeed, let $\bar{t}$, t', and i' be drawn according to the distribution of $g(T)$, the conditional distribution of T under the condition $g(T) = \bar{t}$, and the conditional distribution of I under $g(T) = \bar{t}$, respectively. The last drawing is clearly equivalent to the drawing according to the conditional distribution of I under the condition $g(T) = g(t')$. Let (I', T') denote a pair of random variables with realization (i', t') drawn in the way just described. By 2°, (I', T') is distributed as (I, T). As we see, in order to draw (i', t') one only needs to know the distribution of $(I, g(T))$ and the conditional distribution of T under the condition $g(T) = \bar{t}$. By 1°, this conditional distribution is the same for all $P \in \mathscr{P}$. It follows that the observable data can be transformed by g with no loss of information.

A statistic g satisfying 1° and 2° is called *prediction sufficient for the family of distributions of the pair* (I, T).

In probabilistic problems (i.e., when P is known) condition 1° is always satisfied, and hence a statistic (a random variable) is prediction sufficient if it satisfies 2°. We now give an example of a prediction sufficient statistic in the class of probabilistic problems of predicting the value of a binary random variable I. Assume that a phenomenon is described by a pair (I, Z), where I takes on values 1 and 2, and Z takes on values in a certain set $\boldsymbol{Z}$. Denote by π the probability that $I = 1$, and by f_i, $i = 1, 2$, the densities (with respect to a certain measure) of the conditional distributions of Z under the conditions $I = i$, $i = 1, 2$. Suppose that f_1 and f_2 are positive on $\boldsymbol{Z}$. Moreover, the probability π and the densities f_1, f_2 are assumed to be known (i.e., the distribution of (I, Z) is known).

In the realization $\omega = (i, z)$ we are interested in the nonobservable component

i and we observe the component z. The maximal observable statistic T is of the form $T(i, z) = z$, and I is the predicted variable. Denote by h the likelihood ratio

$$h(z) = \frac{f_2(z)}{f_1(z)}, \quad z \in Z. \tag{2.4.13}$$

The conditional distribution of I under the condition that we observe z is given by

$$P(I = 1 \mid T = z) = \frac{\pi f_1(z)}{\pi f_1(z)+(1-\pi)f_2(z)} = \frac{\pi}{\pi+(1-\pi)h(z)}. \tag{2.4.14}$$

Hence, condition 2° is satisfied, i.e., for given π, f_1, f_2, the likelihood ratio is a prediction sufficient statistic.

Suppose now that the probability π is not known, i.e., that for fixed f_1, f_2 we have the family of distributions of (I, Z) for $\pi \in (0, 1)$. Again, the ratio h is a prediction sufficient statistic for this family of distributions of (I, Z) since 2° is satisfied in view of (2.4.14) and 1° is satisfied by the factorization of density p_π of the distribution of T of the form

$$p_\pi(z) = \pi f_1(z)+(1-\pi)f_2(z) = (\pi+(1-\pi)h(z))f_1(z). \tag{2.4.15}$$

*

Now we formulate the *fundamental property of prediction sufficient statistics*:

If $g\colon T \to \overline{T}$ is a prediction sufficient statistic for the family of distributions of (I, T) in any problem of predicting the value of I when one observes the value of T, then for each rule δ there exists an equivalent rule δ_g, measurable with respect to g, which is given by the formulae (2.4.10) *and* (2.4.11). (2.4.16)

If in a given prediction problem the set Δ satisfies condition (2.4.12), then any prediction sufficient statistic g has property (2.3.17). Hence, solving any prediction problem which belongs to a standard class and satisfies (2.4.12) we can restrict the set Δ to Δ_g. This makes it possible to reduce the set of feasible solutions to rules measurable with respect to g.

Standard classes of probabilistic prediction problems related to binary variable I are discussed in Chapter 3. Solutions of problems from these classes are measurable with respect to the likelihood ratio (2.4.13).

2.5. Reduction of a statistical problem

In previous sections we have discussed the reduction of observable data which does not affect the solution of the problem. Now we consider the reduction of the entire statistical problem and we formulate the conditions ensuring a kind of equivalence of the original and of the reduced problems.

A reduction of a problem is a transformation of the realizations of a phenomenon and the observable data. This transformation should preserve the goal of research and the properties of the solutions. In the sequel, for each $P \in \mathscr{P}$, phenomenon realizations are reduced by a suitably measurable functions f_P: $\boldsymbol{\Omega} \to \bar{\boldsymbol{\Omega}}$, and observable data are reduced by a suitably measurable function g: $\boldsymbol{T} \to \bar{\boldsymbol{T}}$. Thus, the reduction of a problem is formally given by the family of functions

$$((f_P, P \in \mathscr{P}), g). \tag{2.5.1}$$

Typically, problems are reduced in such a way that either the realizations are not subjected to any transformation (i.e., $f_P = \mathrm{id}$) or the transformations f_P are identical for each $P \in \mathscr{P}$. For instance, in any inference concerning distributions parameters under partial observability, realizations are usually reduced by a maximal observable statistic T (i.e., $f_P = T$, $g = \mathrm{id}$). In this case the reduced realizations are fully observable. Further, in prediction problems, when the value of a random variable I is sought for and T is the maximal observable statistic, realizations are usually reduced by the pair (I, T) (i.e., $f_P = (I, T)$, $g = \mathrm{id}$). Prediction problems are usually formulated directly in the reduced form in which ω is a pair (i, t).

In the sequel we describe reductions of the phenomenon descriptions, of the research goal, and of the sets Δ and Δ^*, performed by means of the family of functions (2.5.1).

(i) Reduction of the description of a phenomenon

The family $(f_P, P \in \mathscr{P})$ reduces the statistical space $M = (\boldsymbol{\Omega}, \mathscr{A}, \mathscr{P})$ to the statistical space

$$\bar{M} = (\bar{\boldsymbol{\Omega}}, \bar{\mathscr{A}}, \bar{\mathscr{P}}), \tag{2.5.2}$$

where

$$\bar{\mathscr{P}} = \{Pf_P^{-1}\colon P \in \mathscr{P}\}. \tag{2.5.3}$$

The probability distribution $\bar{P} = Pf_P^{-1}$ is a distribution of the random variable f_P.

Observable data are assumed to be reduced by a function g: $\boldsymbol{T} \to \bar{\boldsymbol{T}}$. Maximal observable statistic $\bar{T}$: $\bar{\boldsymbol{\Omega}} \to \bar{\boldsymbol{T}}$ assigns reduced data $\bar{t} \in \bar{\boldsymbol{T}}$ to reduced realizations $\bar{\omega} \in \bar{\boldsymbol{\Omega}}$. However, the function $\bar{T}$ can be defined if and only if the following *condition of reduction of phenomenon description* is satisfied:

$$f_P(\omega) = f_{P'}(\omega') \quad \Rightarrow \quad gT(\omega) = gT(\omega') \tag{2.5.4}$$

for any $\omega, \omega' \in \boldsymbol{\Omega}$, $P, P' \in \mathscr{P}$. Then we define $\bar{T}$ as

$$\bar{T}(\bar{\omega}) = gT(\omega) \quad \text{for} \quad \bar{\omega} = f_P(\omega). \tag{2.5.5}$$

The mutual relations between f_P, g, T, and $\bar{T}$ are represented in the following diagram

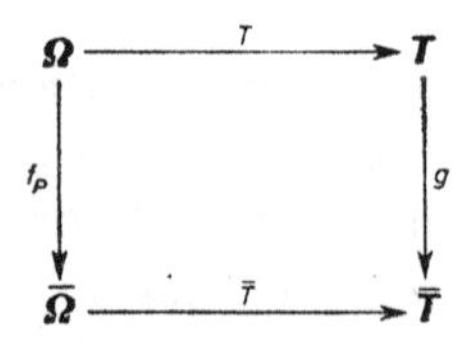

The reduced description of a phenomenon is given by the pair $(\bar{M}, \bar{T})$.

We will illustrate the construction of $(\overline{M}, \overline{T})$ for the problem of predicting the value of a binary variable I (cf. Section 2.4). The phenomenon is primarily described by the statistical space M for bivariate statistic (I, Z) and by the maximal observable statistic T defined by $T(i, z) = z$. The distribution of (I, Z) is given by known densities f_1, f_2 and an unknown probability $\pi \in (0, 1)$. Suppose for instance that f_i, $i = 1, 2$, is the density of $N_k(\mu_i, \Sigma)$, where μ_1, μ_2, Σ are known. For each $\pi \in (0, 1)$, we reduce (M, T) by using (f, g), where f and g are defined as

$$g(z) = \ln \frac{f_2(z)}{f_1(z)} = \left(z - \tfrac{1}{2}(\mu_1 + \mu_2)\right)^T \Sigma^{-1}(\mu_2 - \mu_1). \tag{2.5.6}$$
$$f(i, z) = \left(i, g(z)\right).$$

Pair (f, g) satisfies condition (2.5.4). After reduction we obtain a space $\overline{M}$ for the pair

$$(\overline{I}, \overline{Z}) = f(I, Z) = \left(I, g(Z)\right).$$

Set $\boldsymbol{\Omega}$ is equal to $\{0, 1\} \times \boldsymbol{R}$. The distributions of $(\overline{Z} \mid \overline{I} = 1)$ and $(\overline{Z} \mid \overline{I} = 2)$ are equal to $N(-\gamma^2/2, \gamma^2)$, and $N(\gamma^2/2, \gamma^2)$, respectively, where

$$\gamma = \left((\mu_1 - \mu_2)^T \Sigma^{-1} (\mu_1 - \mu_2)\right)^{1/2}. \tag{2.5.7}$$

Set $\overline{\mathscr{P}}$ consists of distributions of pairs $(\overline{I}, \overline{Z})$ given by known densities $\overline{f}_1, \overline{f}_2$ of $N(-\gamma^2/2, \gamma^2)$ and $N(\gamma^2/2, \gamma^2)$, and by an unknown $\pi \in (0, 1)$ (where π denotes the probability of the event $\{\overline{I} = 1\}$).

Maximal observable statistic $\overline{T}$ is of the form $\overline{T}(i, \overline{z}) = \overline{z}$.

The parameter γ is called the *Mahalanobis distance* (between two distributions on $\boldsymbol{R}^k$ with means μ_1, μ_2, and the same covariance matrix Σ). It is usually denoted by $\varrho_{\text{Mah}}(\mu_1, \mu_2, \Sigma)$.

(ii) Reduction of the goal of research.

Functions $(f_P, P \in \mathscr{P})$ reducing space M to space $\overline{M}$ should be such as to make it possible to define the reduced goal of research $(I_{\overline{P}}, \overline{P} \in \overline{\mathscr{P}})$ for space $\overline{M}$, equivalent to the original goal $(I_P, P \in \mathscr{P})$. Therefore, we require the implication

$$\left(f_P(\omega) = f_{P'}(\omega')\right) \wedge \left(Pf_P^{-1} = P'f_{P'}^{-1}\right) \Rightarrow I_P(\omega) = I_{P'}(\omega') \tag{2.5.8}$$

to be satisfied for all $\omega, \omega' \in \boldsymbol{\Omega}$ and $P, P' \in \mathscr{P}$.

Now we define

$$I_{\overline{P}}(\overline{\omega}) = I_P(\omega) \quad \text{for} \quad \overline{P} = Pf_P^{-1}, \quad \overline{\omega} = f_P(\omega). \tag{2.5.9}$$

If the goal of research depends only on distribution P, i.e., $I_P(\omega) \equiv \varkappa(P)$, then the condition (2.5.8) reduces to

$$Pf_P^{-1} = P'f_{P'}^{-1} \Rightarrow \varkappa(P) = \varkappa(P'), \quad P, P' \in \mathscr{P} \tag{2.5.10}$$

and (2.5.9) can be written as

$$I_{\overline{P}} = \varkappa(P) \quad \text{for} \quad \overline{P} = Pf_P^{-1},$$

where $I_{\overline{P}}$ is a parameter of the reduced family $\overline{\mathscr{P}}$. Moreover, if ω is reduced by a maximal observable statistic T (i.e., $f_P = T$ for $P \in \mathscr{P}$), then (2.5.10) is the condition of observable reducibility of $\varkappa$ (cf. (2.1.8)).

On the other hand, if the goal of research depends solely on ω, then (2.5.8) reduces to

$$f_P(\omega) = f_{P'}(\omega') \;\Rightarrow\; I(\omega) = I(\omega') \quad \text{for} \quad \omega, \omega' \in \boldsymbol{\Omega}, \quad P, P' \in \mathscr{P}. \tag{2.5.11}$$

Then (2.5.9) reduces to $\bar{I}(\bar{\omega}) = I(\omega)$ for some $\omega \in \boldsymbol{\Omega}$ and $P \in \mathscr{P}$ such that

$$f_P(\omega) = \bar{\omega}.$$

Condition (2.5.11) ensures that $\bar{I}$ is well defined.

Conditions (2.5.8), (2.5.10) and (2.5.11) will be called *goal conservation conditions.*

The reduction of problems of predicting a random variable I by a maximal observable statistic T, given by

$$f_P = (I, T), \quad g = \mathrm{id},$$

satisfies conditions (2.5.4) and (2.5.11) concerning reduction of phenomenon description and goal conservation. In particular, this holds true for predicting a binary variable I as previously discussed; then we have $\bar{I}(i, \bar{z}) = i$. Generally, a reduction may be given by

$$f_P = (I_P, T), \quad g = \mathrm{id}. \tag{2.5.12}$$

Then, the goal conservation condition given by (2.5.8) and the condition (2.5.4) related to phenomenon description are satisfied. As the result of reduction we get

$$\bar{\omega} = (y_1, y_2), \quad \text{where} \quad y_1 = \bar{I}(\bar{\omega}), y_2 = \bar{T}(\bar{\omega}),$$

so that the reduced problem concerns predicting the first component of $\bar{\omega}$ on the ground of the second one.

Reduction of the form (2.5.12) can be applied, for instance to the problem of responses to stimuli, where the goal is defined by (2.1.6), as well as to estimating the value of the regression function with the goal defined by (2.1.7).

(iii) Reduction of set Δ.

First, let us consider the subset of the set of rules of the original problems containing all rules measurable with respect to a chosen statistic g. To each rule δ from this subset there corresponds a rule $\bar{\delta}$ in the reduced problem such that $\delta = \bar{\delta} \circ g$, i.e.,

$$\bar{\delta}(\bar{t}, B) = \delta(t, B) \quad \text{for} \quad B \in \mathscr{D}, \; g(t) = \bar{t}$$

(cf. (2.2.3)). The set of rules $\bar{\delta}$ is the set of all rules in the reduced problem in which the phenomenon is described by $(\bar{M}, \bar{T})$ and the goal of research is defined by $(\bar{I}_{\bar{P}}, \bar{P} \in \bar{\mathscr{P}})$. Clearly,

$$\delta = \bar{\delta} \circ g \;\Rightarrow\; (Q_{P,\delta} = Q_{\bar{P},\bar{\delta}} \text{ for } \bar{P} = Pf_P^{-1}), \tag{2.5.13}$$

and hence, for any rules $\bar{\delta}$, $\bar{\delta}'$, we have

$$\bar{\delta} \approx \bar{\delta}' \;\Leftrightarrow\; \delta \approx \delta' \quad \text{for} \quad \delta = \bar{\delta} \circ g, \; \delta' = \bar{\delta}' \circ g. \tag{2.5.14}$$

Let Δ_g denote the subset of the set Δ of rules of the original problem containing all rules measurable with respect to g. We require that

$$\Delta_g \neq \emptyset. \tag{2.5.15}$$

Then we put

$$\bar{\Delta} = \{\bar{\delta}: \delta = \bar{\delta} \circ g, \delta \in \Delta_g\}. \tag{2.5.16}$$

Observe that if set Δ is closed under relation $\approx$ (cf. condition (2.4.12)), then, by (2.5.14), set $\bar{\Delta}$ is also closed under $\approx$.

(iv) Reduction of set Δ^*.

Under conditions (2.5.4) and (2.5.8) necessary for reducing both the phenomenon description and the goal of research and under the condition $\Delta_g \neq \emptyset$ we obtain the elements of the reduced problem which have been discussed before, namely $\bar{M}$, $\bar{T}$, $(\bar{I}_{\bar{P}}, \bar{P} \in \bar{\mathscr{P}})$, $\bar{\Delta}$. Clearly, sets $\boldsymbol{I}$, $\boldsymbol{D}$ of the original problem remain unchanged. In defining the solution set of the reduced problem we shall restrict ourselves to the case where the original problem belongs to a certain standard class given by a formula S. In view of (2.5.13) we define the indexing function

$$\bar{q}: \bar{q}(\bar{P}, \bar{\delta}) = Q_{P,\delta} \quad \text{for} \quad \bar{P} = Pf_{\bar{P}}^{-1}, \ \delta = \bar{\delta} \circ g.$$

Now, the set $\bar{\Delta}^*$ is defined as the set of those rules $\bar{\delta}^*$ which together with $\bar{\mathscr{P}}$, $\bar{\Delta}$, $\bar{q}$ satisfy formula S.

Thus, the primary problem has been reduced to the problem $(\bar{M}, \bar{T}, \boldsymbol{I}, (\bar{I}_{\bar{P}}, \bar{P} \in \bar{\mathscr{P}}), \boldsymbol{D}, \bar{\Delta}, \bar{\Delta}^*)$ by means of (2.5.1).

To illustrate this, let us continue the reduction of the problem of predicting a binary variable I with f and g given by (2.5.6). We then have $\boldsymbol{I} = \{1, 2\}$. We put $\boldsymbol{D} = \{0, 1, 2\}$, where 0 denotes the deferred decision, and j, for $j = 1, 2$, denotes the decision "I is equal to j". Furthermore, we assume that Δ is the set of all rules and we define Δ^* as the set of all minimax rules in Δ, given by (2.3.13) for a certain loss function $\mathscr{L}$. Thus, Δ_g is not empty, $\bar{\Delta}$ is the set of all rules in the reduced problem, and $\bar{\Delta}^*$ is the set of minimax rules from $\bar{\Delta}$ (with respect to $\mathscr{L}$).

*

We say that a reduced problem is equivalent to the original one if, for each solution $\delta^* \in \Delta^*$, there exists a rule $\bar{\delta}^* \in \bar{\Delta}^*$ such that, for each $P \in \mathscr{P}$,

$$Q_{P,\delta^*} = Q_{\bar{P},\bar{\delta}^*}, \quad \text{where} \quad \bar{P} = Pf_{\bar{P}}^{-1}.$$

If a family $((f_P, P \in \mathscr{P}), g)$ satisfies conditions (2.5.4), (2.5.8) and (2.3.17), then the reduced problem and the original problem are equivalent. In case of equivalence the composition of solution $\bar{\delta}^*$ with statistic g solves the original problem, and for each solution δ^* of the original problem there exists a solution $\bar{\delta}^*$ such that $\delta^* \approx \bar{\delta}^* \circ g$. Hence, for solving a problem belonging to a standard class use can be made of a solution of any equivalent reduced problem.

Recall that in any inference concerning a distribution parameter any sufficient statistic satisfies condition (2.3.17) whenever the set Δ is closed under $\approx$ (cf. (2.4.12)). This is a consequence of the fundamental property of sufficient statistics (2.4.9). Hence, in any inference problem concerning a distribution parameter, if Δ is closed under $\approx$, then the reduction of observability by sufficient statistics leads to reduced problems which are equivalent to original ones.

In particular, in a standard class any inference problem concerning an observably reducible parameter and admitting all decision rules can be reduced to an equivalent problem with complete observability by a pair (f, g), where $f = g \circ T$, and g is an arbitrary sufficient statistic for a given family of distributions of T. Indeed, the goal conservation condition (2.5.10) becomes the conditions of observable reducibility of the parameter in question, whereas the remaining conditions (2.5.4), (2.5.15) are obviously satisfied.

In parameter estimation problems the set Δ usually consists of nonrandomized decision rules. Thus, Δ is not closed under the relations $\approx$, i.e., condition (2.4.12) is not satisfied. However, we can use sufficient statistics to reduce observability in problems extended in such a way that all rules are taken into consideration. Hence, after reduction we get an equivalent problem in which, too, all rules are taken into consideration. In optimal estimation problems with a convex loss function (for instance, in an unbiased estimation problem with uniformly minimal variance) we can also exploit the fact that for each rule there exists a nonrandomized rule which is not worse. Therefore, in such problems we can restrict our considerations to nonrandomized solutions measurable with respect to an arbitrary sufficient statistic.

Recall, in turn, that in problems of predicting values of a statistic I any sufficient statistic satisfies condition (2.3.17) whenever the set Δ is closed under $\approx$ (cf. (2.4.12)). This is a consequence of the fundamental property of prediction sufficient statistics. In such problems the reduction of observability by means of the prediction sufficient statistic g yields an equivalent problem. Moreover, if such a problem is reduced by means of a pair (f, g) such that

$$f = (I, g \circ T),$$

then, for each statistic g, the conditions concerning reduction of phenomenon description and goal conservation are satisfied. In addition, if g is a prediction sufficient statistic, then $\Delta_g \neq \emptyset$, and the reduced problem is equivalent to the original one.

The above arguments can be applied to the prediction of a binary variable I if this problem is reduced by f, g given by (2.5.6). As g is prediction sufficient (since it is an injective function of the likelihood ratio), the reduced problem is equivalent to the original one. Hence, we can restrict ourselves to solutions measurable with respect to the likelihood ratio or with respect to g. In Chapter 3 we will show that there is no other solution.

*

We often have to deal with the reduction of problems from standard classes leading to problems which are not equivalent to the original ones. We are then interested in evaluating the consequences of nonequivalent reductions and in choosing such reductions which can be regarded as optimal. For binary prediction problems these topics will be considered in Section 3.4.

Let us also mention the reduction of problems from ML class. Here, observability can be reduced by means of an arbitrary sufficient statistic, since, in view of the factorization criterion (2.4.4), the ML estimators are measurable with respect to this statistic.

2.6. Final remarks

In Section 2.1 we have considered goals of research represented by means of a function $C: \boldsymbol{\Omega} \times \mathscr{P} \to \boldsymbol{I}$. It may happen, however, that such a representation does not exist. Suppose, for instance, that we intend to determine tolerance intervals. We consider a model of a simple random sample corresponding to a family $\mathscr{P}$ of univariate distributions. Thus, for a chosen number $p \in (0, 1)$, we want to determine an interval (a, b) in which the mass p of the unknown distribution P from $\mathscr{P}$ is concentrated. The set $\boldsymbol{I}$ is the set of pairs (a, b), where $a < b$, and the tolerance intervals are related to distributions by means of a certain relation on the set $\mathscr{P} \times \boldsymbol{I}$. Thus, when inferring about the parameter of the distribution P we formalize the goal of research by a relation on $\mathscr{P} \times \boldsymbol{I}$. In general, goals of research are relations on $\boldsymbol{\Omega} \times \mathscr{P} \times \boldsymbol{I}$.

*

When the description of a phenomenon and the goal of research are determined statisticians may proceed in two ways. They can either establish a priori preferable properties of the solutions of the problem, i.e., define sets Δ and Δ^*, or adopt a certain decision rule as a potential solution of the problem and investigate its properties in detail. The latter approach was applied in the investigation of the proof-reading process (cf. Introduction). Basing ourselves on intuition only we first choose an estimator of the number of errors in a book. Next, the properties of the estimator were carefully examined and consequently the initial definition was modified in order to refine the properties of the estimator.

The ML estimation is a typical example of the latter approach. Statisticians make a considerable effort in determining and investigating ML estimators in various problems of particular interest. It is obvious that under restricted observability ML estimators are determined for observable data only, i.e., initial reduction is made by a suitable maximal observable statistic. Sometimes initial reduction is also made in the case of full observability, e.g., when it is computationally impossible to determine the ML estimator for the nonreduced problem.

In a problem reduced by a statistic g the ML estimation is equivalent to that in the original (nonreduced) problem only if g is a sufficient statistic (cf. (2.4.7)).

Suppose a certain index θ consists of (θ_1, θ_2), where θ_1 is the main parameter, i.e., it is the goal of research, and θ_2 is the nuisance parameter. In such a case a preliminary reduction by a statistic $g: \boldsymbol{T} \to \bar{\boldsymbol{T}}$ is sometimes performed provided that, for each $\bar{t} \in \bar{\boldsymbol{T}}$ and each θ_1, the conditional distribution of maximal observable statistic T under the condition $g(t) = \bar{t}$ is the same for all θ_2. Such a statistic is said to be *sufficient for the nuisance parameter* θ_2. After the reduction, for each observed, $\bar{t}$, the ML estimate of θ_1 is determined. This procedure, called the *conditional maximum likelihood method*, is another example of a rule intuitively accepted as worth considering.

Applications of the conditional maximum likelihood method will be discussed in Chapters 9 and 10.

In the case of ML estimators it is mainly the limit properties of sequences of ML estimators based on simple random samples of size n that are investigated. In general, suppose that the phenomenon realizations are infinite sequences $\omega = (\omega_1, \omega_2, \ldots)$. Then the involved statistical problems usually concern some asymptotic properties of various sequences of rules δ_n, $n = 1, 2, \ldots$, where δ_n depends solely upon the first n components of the realization. In a colloquial way one says that rule δ_n possesses a certain asymptotic property meaning in fact that this property applies to a given sequence of rules.

In the present chapter we have not considered problems with conditions imposed on sequences of rules.

*

The definition of a standard class of problems given in Section 2.3 is connected with what is known in the literature as typical methods of statistical inference. Barra (1982, Section 4.2) says: "In mathematical statistics all we can learn about the practical value of a strategy is through its image". In the terminology used by Barra, a *strategy* is a randomized decision rule and its *image* is the distribution induced on $\boldsymbol{D}$ by this strategy under a given distribution $P \in \mathscr{P}$, the goal of research being a distribution parameter. The image of a strategy δ and a parameter of P jointly determine the distribution $Q_{P,\delta}$, where $P \in \mathscr{P}$ and $\delta \in \Delta$. A similar approach can be found in Blackwell (1953). In Takeuchi and Akahira (1975) distributions $Q_{P,\delta}$ are considered in relation to prediction problems, and the family $(Q_{P,\delta},\ P \in \mathscr{P})$ is called the *performance characteristic of the decision rule* δ.

The notion of a standard class of problems allows us to consider in a general way the preliminary reduction of sets of rules and the reduction of the entire statistical problem. When a parameter of distribution is the goal of inference, the equivalence of an arbitrary rule δ and the rule δ' which is derived from δ by formulae (2.4.10) and (2.4.11) and is measurable by any sufficient statistic g is a consequence of Theorem 5.1 in Bahadur (1954). An analogous equivalence in the case of any prediction sufficient statistic g was proved by Takeuchi and Akahira (1975, Theorem 1).

The first references to sufficiency appeared in the literature in the thirties and were concerned with parameter distribution inference. Then, investigations followed two trends. The first, initiated by Halmos and Savage (1949), dealt with purely theoretical probabilistic questions: the connections between sufficient and distribution-free statistics[1], and minimal sufficient statistics (Basu, 1958). The second

[1] A statistic in the space $(\Omega, \mathscr{A}, \mathscr{P})$ is *distribution-free* if its distributions are the same for all $P \in \mathscr{P}$.

trend dealt with the reduction of observable data considered from the point of view of decision theory (complete classes of decision rules). The relationships between these two trends were discussed by Blackwell (1951, 1953).

Conditions 1° and 2° appearing in the definition of a prediction sufficient statistic were based on the conditions formulated by Bahadur (1954), who investigated the reducibility of sequential decision problems. The definition of the prediction sufficient statistics and its relationship to the reduction of prediction problems were presented by Skibinsky (1967), Takeuchi and Akahira (1975), and Torgersen (1977).

In contrast with sufficient statistics, prediction sufficient statistics appear rather rarely in the literature. For example, the relationship between discriminant functions and prediction sufficient statistics (cf. Chapters 3 and 4) seems to have been overlooked up to the present.

In standard classes the equivalence of reduced and original problems can be established in a natural way. It is the first step in investigating equivalence of statistical problems. Roughly speaking, we want to treat as equivalent problems which supply the same information about the common goal of research and have solutions with identical properties but may differ in phenomenon description. Some special problems of phenomenon descriptions were discussed by Morse and Sacksteder (1966), Sacksteder (1967), Moszyńska and Pleszczyńska (1983), Bromek and Moszyńska (1983). Definitions of a standard class and equivalent reduction in such a class are given in Bromek (1986).

*

At the end of Section 2.5 we mentioned problems involved in choosing a proper reduction of observable data from within the family of non-sufficient reductions. The main point is then the choice of a scheme of observation of phenomenon realizations. As a further step in this direction we can also make a choice between different problems which have the same goal but differ in description of phenomenon. In a colloquial way we say that we can deal with various designs of experiment. For instance, in the proof-reading process one may choose among different methods of proof-reading. A very simple example of experiment planning is the choice of the sample size in investigations based on simple random samples.

Solving problems of this type relies on the choice of a plan of experiment and of a decision rule (suitable for that plan) such that all the requirements imposed on them in the problem in question are satisfied. In particular, the costs and the precision of inference can both be taken into account.

Formalization of such problems is not considered in the present book; some relevant practical examples can be found in Part Three.

*

In the formal schemes presented in this chapter the measurement scales corresponding to realizations of a phenomenon and observable data were not mentioned. Information about the type of the scale should be taken into account both in formulating the goal of research and in choosing the set of admitted rules. In the literature we sometimes find requirements on the set of rules to be invariant under groups of transformations; in particular, such a group may consist of transformations admissible for a given type of measurement scale (cf. the end of Chapter 1). Such invariance requirements are sometimes imposed also on the goals of research. In general, however, measurement theory is only loosely connected with statistics. More comments on this topic will be found in Chapters 5 and 10.

Part Two

SELECTED THEORETICAL TOPICS

Chapter 3

Discriminant analysis

Key words: *prediction, classification, diagnosis, identification, learning sample, class of objects, prior probability, posterior probability, decision deferment, adapted rule, threshold rule, discriminant function, risk, Mahalanobis distance, consistent sequence of rules, sample splitting, leaving-one-out, cross validation, class separability measure, prediction power, likelihood ratio.*

3.1. INTRODUCTION

Can we recognize the sex of a child before its birth? What medical tests should be made and what information should be obtained from the parents in order to make the right prediction? How can we exploit information about the accuracy of the guesses already made? Is the chance of the right prediction the same for boys and for girls?

Questions of this kind arise when one investigates an unobservable feature of an object, e.g. the sex of a child not yet born. The part of statistics dealing with such problems is called *discriminant analysis.* The name comes from the word *discriminaire*—discriminate. Other terms having the same meaning are: *prediction, diagnosis, identification, classification.*

Practical discriminant problems concern an object or a set of objects selected from a population which is divided into a number, say m, $m = 2.3, \dots,$ of disjoint classes. The class membership of those selected objects is not observable and has to be established. To this aim one often makes use of a learning sample, i.e., a sample consisting of objects with known class membership. Then all those objects are investigated with respect to their observable features.

If no learning sample is available and we are interested in classifying exactly one object, our problem can be formally represented as a pair (I, Z), where I is a random variable, with values $i = 1, \dots, m$, defining the class membership of the object in question and Z is a random vector whose components represent observable features of the object. We assume that the class-conditional distributions of what is observed, i.e., the distributions of the vectors

$$Z_i = (Z \mid I = i), \quad i = 1, \dots, m,$$

are given by their densities $f_1, \ldots, f_m$ with respect to a measure ν. Hence, the joint distribution of the pair (I, Z), say P, is given by the measure ν, the densities $f_1, \ldots, f_m$, and the numbers $\pi_1, \ldots, \pi_m$ standing for the prior probabilities of the respective class memberships of the object, i.e.,

$$P(I = i) = \pi_i.$$

In solving this problem it is necessary to determine what information about P is available, i.e., to define the family $\mathscr{P}$ to which P belongs. This family depends on the nature of the problem itself but also, as should be stressed, on the more or less arbitrary choice made by the statistician. In the simplest case where P is known we obtain a probabilistic problem.

Vector Z represents the observable data, and the values of I constitute the goal of research.

If a learning sample is to be used in the considerations and thus introduced to the description of a phenomenon, we should determine whether it is a simple random sample drawn from the whole population or a set of simple random samples drawn from each class separately. This depends on experiment planning but sometimes is imposed by the particular circumstances of the study in question.

Thus the descriptions of the phenomenon in various discriminant problems differ from one another with regard to the occurrence of learning samples and to the assumptions about the description of those samples. Further differentiation of discriminant problems results from the diversity of the requirements imposed on the credibility and the costs of the estimate obtained.

Of special interest are problems in which the statistician may refrain from making a definite statement concerning the membership of the object. This makes it possible to avoid losses due to wrong classification, at the cost of deferring the statement in doubtful cases. If the deferrence is not allowed (*forced problems*), it may happen that the expected loss or some other index of quality may prove to be intolerably high from the point of view of practical applications.

In this chapter we discuss discriminant problems for two classes of objects ($m = 2$). In Section 3.2 we solve probabilistic problems from some standard classes. They are mainly classes of problems admitting decision deferrence. We refer to binary prediction problems formulated in the preceding chapter and exploit prediction sufficient statistics as defined in Section 2.4. In Section 3.3 we discuss methods of solving statistical discriminant problems, which—apart from a number of special cases—amount to adapting solutions of probabilistic problems. Here stress is laid upon ranking procedures. In Section 3.4 we deal with the reduction of observable data according to different class separability measures. It should be noted here that some statisticians regard data reduction as a problem in itself and aim at determining those features which discriminate "optimally" between classes and try to find a practical interpretation for these features. We also mention prediction sufficient reduction and point out the relations between class separability measures and prediction power indices for optimal decision rules.

3.2. Probabilistic problems

Suppose that $m = 2$ and P is a known distribution of the pair (I, Z). This distribution is given by the probability π that the random variable I will assume the value 1 and by the densities f_1, f_2 (with respect to a measure ν) of the conditional distributions of the random vector Z in classes 1 and 2. Hence, f_1 and f_2 are the densities of the random vectors Z_1, Z_2,

$$Z_1 = (Z \mid I = 1), \quad Z_2 = (Z \mid I = 2),$$

respectively.

Denote by $\boldsymbol{Z}$ the set of those values of the random vector Z for which $f_1(z)+ +f_2(z) > 0$(1). Let

$$A = \{z \in \boldsymbol{Z}\colon f_1(z) \cdot f_2(z) = 0\}. \tag{3.2.1}$$

The random variable I is conditionally observable with respect to A with probability 1 since for $i = 1, 2$ we have

$$P\big(I = i \mid Z \in A,\ f_i(Z) > 0\big) = 1.$$

Thus, we can restrict our considerations to the set $\boldsymbol{Z} \setminus A$ by replacing f_1 and f_2 with the densities of conditional distributions restricted to $\boldsymbol{Z} \setminus A$ and by taking

$$P(I = 1 \mid Z \in \boldsymbol{Z} \setminus A) = \frac{\pi\big(1 - \int_A f_1(z)\, d\nu(z)\big)}{1 - \pi \int_A f_1(z)\, d\nu(z) - (1-\pi) \int_A f_2(z)\, d\nu(z)} \tag{3.2.2}$$

instead of π. In this way an arbitrary discriminant problem is reduced to a discriminant problem with positive densities f_1, f_2. Hence, without loss of generality, we assume f_1, f_2 to be positive on $\boldsymbol{Z}$.

Consider the set Δ of randomized decision rules of the form

$$\delta = (\delta_0, \delta_1, \delta_2),$$

where $\delta_0(z)$ denotes the probability of decision deferrence for the observed value $z \in \boldsymbol{Z}$, and $\delta_i(z)$ $(i = 1, 2)$ is the probability of classifying the objects as a member of the i-th class. Clearly, $\delta_0(z)+\delta_1(z)+\delta_2(z) = 1$ for each $z \in \boldsymbol{Z}$. In this way we obtain a function $\delta\colon \boldsymbol{Z} \to [0, 1]^3$ which determines uniquely the distribution on the set of decisions $\{0, 1, 2\}$. Hence, the function δ can also be regarded as a randomized decision rule(2) defined by formula (2.2.1).

To each statistic $g\colon \boldsymbol{Z} \to \boldsymbol{R}$ there correspond some rules called *threshold rules with respect to g*. By using a threshold rule the object is classified: as a member of

(1) Since density is defined only up to the sets of ν-measure 0, the set $\boldsymbol{Z}$ (as well as other sets considered below) depend on the choice of densities f_1 and f_2. In the sequel, we shall always tacitly assume that we selected some fixed version of densities of random variables under study. Typically there is a "natural" choice, e.g. standard normal density defined as $f(x) = e^{-x^2/2}/\sqrt{2\pi}$. In other cases there may be more than one natural choice, e.g. $f(x) = \lambda e^{-\lambda x}$ for $x>0$ or $x \geqslant 0$ for exponential density. For unifom distribution $f(x) = 1/\theta$ for $0<x<\theta$ or $0 \geqslant x \geqslant \theta$. The choice here is not quite irrelevant. In one case maximum likelihood estimator of θ exists, in the other it does not.

(2) In a similar way we deal with tests in hypothesis testing problems.

class 1 whenever the value $g(z)$ is less than a certain lower threshold $\varkappa_1$, as a member of class 2 if $g(z)$ is greater than a certain upper threshold $\varkappa_2$; for intermediate values of $g(z)$ the decision is deferred. The randomization of the decision may appear only if $g(z)$ is equal to one of the thresholds.

Strictly speaking, $\delta \in \Delta$ is a *threshold rule with respect to* g: $\boldsymbol{Z} \to \boldsymbol{R}$ if there exist thresholds $\varkappa_1, \varkappa_2 \in \boldsymbol{R}$ and randomization constants $p_1^1, p_2^1, p_1^2, p_2^2 \in [0, 1]$ such that $\varkappa_1 \leqslant \varkappa_2$,

$$p_i^1 = p_i^2 \quad \text{if} \quad \varkappa_1 = \varkappa_2 \; (i = 1,2),$$
$$p_2^1 = p_1^2 = 0 \quad \text{if} \quad \varkappa_1 < \varkappa_2,$$

and

$$\delta(z) = \begin{cases} (0,1,0) & \text{if} \quad g(z) < \varkappa_1, \\ (1-p_1^1-p_2^1, p_1^1, p_2^1) & \text{if} \quad g(z) = \varkappa_1, \\ (1,0,0) & \text{if} \quad \varkappa_1 < g(z) < \varkappa_2, \\ (1-p_1^2-p_2^2, p_1^2, p_2^2) & \text{if} \quad g(z) = \varkappa_2, \\ (0,0,1) & \text{if} \quad g(z) > \varkappa_2, \end{cases} \tag{3.2.3}$$

for each $z \in \boldsymbol{Z}$.

If $\varkappa_1 = \varkappa_2$ there is only one threshold denoted by $\varkappa$, and only two randomization constants denoted by p_1 and p_2 which are the probabilities of decisions 1 and 2 when $g(z) = \varkappa$. In problems in which decision deferment is a priori excluded there is also only one threshold and $p_1+p_2 = 1$.

If $P(g(Z) = \varkappa_i) = 0$ for $i = 1, 2$, no randomization at thresholds is needed and therefore a threshold rule can be replaced by a suitable nonrandomized rule.

The above construction of a threshold rule meets the postulate that decision deferment has to take place only if there is some doubt as to how the object should be classified. Hence, a deferred decision is something in between decisions 1 and 2. A suitable randomization appears at most at the thresholds between decisions 1 and 0 or decisions 0 and 2.

If δ is a threshold rule with respect to g: $\boldsymbol{Z} \to \boldsymbol{R}$, then it is also a threshold rule with respect to $\phi \circ g$, where ϕ: $\boldsymbol{R} \to \boldsymbol{R}$ is an increasing function. The thresholds $\varkappa_1, \varkappa_2$ are then replaced by $\phi(\varkappa_1)$, $\phi(\varkappa_2)$, and the randomization constants remain unchanged.

In discriminant problems a real-valued statistic g used to define a threshold rule is called a *discriminant function.* A particular role is played here by those statistics which are prediction sufficient for distribution P of (I, Z). It was shown in Section 2.4 that the likelihood ratio h_P defined by

$$h_P(z) = \frac{f_2(z)}{f_1(z)}, \quad z \in \boldsymbol{Z} \tag{3.2.4}$$

is a prediction sufficient statistic. The posterior probability of the event that I equals 2 when we observe a value z of Z is given by

$$P(I = 2 \mid Z = z) = \frac{(1-\pi)f_2(z)}{\pi f_1(z)+(1-\pi)f_2(z)} \tag{3.2.5}$$

and is an increasing function of the ratio h_P:

$$P(I = 2 \mid Z = z) = 1 - \left(1 + \frac{1-\pi}{\pi} h_P(z)\right)^{-1}. \tag{3.2.6}$$

Hence, this probability is also a prediction sufficient statistic for P.

If $\boldsymbol{Z} \subset \boldsymbol{R}^k$, we define a statistic which is called a *quadratic discriminant function* and is denoted by QDF. This is a statistic depending only on the means $\boldsymbol{\mu}_1$ and $\boldsymbol{\mu}_2$ and the covariance matrices Σ_1 and Σ_2 of vectors Z_1 and Z_2. For nonsingular matrices Σ_1, Σ_2, the QDF statistic is given by the formula

$$\mathrm{QDF}(z; \boldsymbol{\mu}_1, \boldsymbol{\mu}_2, \Sigma_1, \Sigma_2)$$
$$= \frac{1}{2}\left[(z-\boldsymbol{\mu}_1)^T \Sigma_1^{-1}(z-\boldsymbol{\mu}_1) - (z-\boldsymbol{\mu}_2)^T \Sigma_2^{-1}(z-\boldsymbol{\mu}_2) + \ln\frac{|\Sigma_1|}{|\Sigma_2|}\right]. \tag{3.2.7}$$

For some classes of distributions of Z_1 and Z_2 the function QDF is an increasing function of the ratio h_P (i. e., it is a prediction sufficient statistic). For instance, if for nonsingular matrices Σ_1 and Σ_2 we have

$$Z_i \sim N_k(\boldsymbol{\mu}_i, \Sigma_i), \quad i = 1, 2, \tag{3.2.8}$$

then

$$\mathrm{QDF}(z: \boldsymbol{\mu}_1, \boldsymbol{\mu}_2, \Sigma_1, \Sigma_2) = \ln h_P(z), \quad z \in \boldsymbol{Z}.$$

Denote by Σ the covariance matrix of vector Z. Then

$$\Sigma = \pi\Sigma_1 + (1-\pi)\Sigma_2. \tag{3.2.9}$$

If $\boldsymbol{Z} \subset \boldsymbol{R}^k$ and Σ is nonsingular, we introduce a statistic called the *linear discriminant function* (LDF) and defined by the formula

$$\mathrm{LDF}(z; \boldsymbol{\mu}_1, \boldsymbol{\mu}_2, \Sigma)$$
$$= (z - \tfrac{1}{2}(\boldsymbol{\mu}_1 + \boldsymbol{\mu}_2))^T \Sigma^{-1}(\boldsymbol{\mu}_2 - \boldsymbol{\mu}_1), \quad z \in \boldsymbol{Z}. \tag{3.2.10}$$

This statistic depends only on $\boldsymbol{\mu}_1$, $\boldsymbol{\mu}_2$ and Σ.

If $\Sigma_1 = \Sigma_2$, then QDF = LDF. Hence, for $Z_i \sim N_k(\boldsymbol{\mu}_i, \Sigma)$, we have

$$\mathrm{LDF}(z; \boldsymbol{\mu}_1, \boldsymbol{\mu}_2, \Sigma) = \ln h_P(z), \quad z \in \boldsymbol{Z},$$

and consequently LDF is a prediction sufficient statistic.

*

To each rule δ there corresponds a distribution P_δ defined on $\{1, 2\} \times \{0, 1, 2\}$t It is a joint distribution of the predicted variable I and of the decision taken about I when the rule δ is applied. This is a special case of the joint "goal-decision" distributions which were introduced in Section 2.2 and denoted by $Q_{P,\delta}$. Here we shall use a simpler symbol P_δ, which is not misleading in the case of probabilistic problems. For clarity we also introduce the symbol $(I, \hat{I}_\delta)$ to denote a pair of random variables with distribution P_δ which identically transform the set $\{1, 2\} \times \{0, 1, 2\}$ onto itself[1]. The random variable $\hat{I}_\delta$ represents the decisions on I taken according to rule δ. Thus, we have

[1] Thus, the same symbol I appears in the pair (I, Z) defined on $\{1, 2\} \times \boldsymbol{Z}$ and the pair $(I, \hat{I}_\delta)$ defined on $\{1, 2\} \times \{0, 1, 2\}$. The first components in the pairs (I, Z) and $(I, \hat{I}_\delta)$ have the same set of values and the same distribution. In this sense they are "the same random variable I".

$$P_\delta(I = 1) = \pi,$$
$$P_\delta(I = i, \hat{I}_\delta = j) = a_{ij}(\delta)P_\delta(I = i), \quad i = 1, 2, \quad j = 0, 1, 2,$$

where $a_{ij}(\delta)$ is the conditional probability of taking decision j under the condition that the object belongs to class i:

$$a_{ij}(\delta) = P_\delta(\hat{I}_\delta = j \mid I = i) = \int_Z \delta_j(z) f_i(z)\, d\nu(z), \quad i = 1, 2, \quad j = 0, 1, 2. \tag{3.2.11}$$

The misclassification of an object from class 1 as a member of class 2 is briefly called the *first error* and, analogously, by the *second error* we mean the misclassification of an object from class 2 as a member of class 1. Now $a_{12}(\delta)$ and $a_{21}(\delta)$ are, respectively, the probabilities of the first error and the second error (for rule δ).

The properties of δ are usually expressed in terms of π and $a_{ij}(\delta)$. For instance, the probability of taking an incorrect decision (the so-called *misclassification rate*) and the probability of decision deferment are equal respectively to

$$\begin{aligned} P_\delta(\hat{I}_\delta \neq \hat{I} \wedge \hat{I}_\delta \neq 0) &= \pi a_{12}(\delta) + (1-\pi) a_{21}(\delta), \\ P_\delta(\hat{I}_\delta = 0) &= \pi a_{10}(\delta) + (1-\pi) a_{20}(\delta). \end{aligned} \tag{3.2.12}$$

Probabilities $a_{ij}(\delta)$ allow us to order the set Δ of rules. Namely, we say that a rule $\delta \in \Delta$ is *not worse* than $\delta' \in \Delta$ if

$$a_{ij}(\delta) \leqslant a_{ij}(\delta') \quad \text{for} \quad i \neq j, \quad i = 1, 2, \quad j = 0, 1, 2. \tag{3.2.13}$$

If there is at least one strict inequality in (3.2.13) then δ is said to be *better* than δ'. The inequalities (3.2.13) entail

$$a_{ii}(\delta) \geqslant a_{ii}(\delta'), \quad i = 1, 2. \tag{3.2.14}$$

This ordering seems reasonable irrespective of which discriminant problem has to be solved. Therefore it would be advisable to search for solutions of such problems among those rules which are admissible with respect to this ordering. A rule $\delta \in \Delta$ is said to be *admissible* with respect to the ordering (3.2.13) if there is no rule in Δ which is better than δ.

Admissible rules can be characterized in the following way. *A rule δ is admissible with respect to the ordering* (3.2.13) *if and only if δ is a threshold rule with respect to the likelihood ratio h_P* (Bromek and Niewiadomska, 1987). This result stresses the importance of the likelihood ratio and other prediction sufficient discriminant functions, e.g., posterior probability.

A similar result holds if, instead of Δ, we consider the set Δ_F,

$$\Delta_F = \{\delta \in \Delta:\ \delta_0(z) \equiv 0\}\text{(1)}. \tag{3.2.15}$$

Instead of (3.2.13) we then have

$$\begin{aligned} a_{12}(\delta) &\leqslant a_{12}(\delta'), \\ a_{21}(\delta) &\leqslant a_{21}(\delta'). \end{aligned} \tag{3.2.16}$$

(1) The index F used in the symbol Δ_F comes from the expression "forced discrimination", i.e., discrimination with a forced decision to assign the object to class 1 or class 2.

Inequalities (3.2.16) entail inequalities (3.2.14) in Δ_F. A rule δ is admissible with respect to the ordering in Δ_F defined by (3.2.16) if and only if δ is a threshold rule (with only one threshold) with respect to h_P.

The prediction sufficiency of the ratio h_P implies that in any probabilistic discriminant problem from a standard class we can restrict ourselves to rules measurable with respect to h_P (cf. Sec. 2.4). Now, let us consider some standard classes of problems in which the solution is a threshold rule with respect to h_P. These problems concern risk minimization with suitably formulated restrictions.

Let L_{ij} ($i = 1, 2, j = 0, 1, 2$) be losses resulting from taking decision j when the object belongs to class i. Assume $L_{ij} > 0$ for $i \neq j$, and $L_{11} = L_{22} = 0$. According to (2.3.7) the *risk* (i.e., the expected loss) is of the form

$$\begin{aligned} R_P(\delta) &= \pi L_{12} a_{12}(\delta) + (1-\pi) L_{21} a_{21}(\delta) \\ &\quad + \pi L_{10} a_{10}(\delta) + (1-\pi) L_{20} a_{20}(\delta). \end{aligned} \tag{3.2.17}$$

Let Δ_P denote a set of rules satisfying one or more constraints of the form

$$c_1 a_{12}(\delta) + c_2 a_{21}(\delta) + c_3 a_{10}(\delta) + c_4 a_{20}(\delta) \leqslant 1 \tag{3.2.18}$$

with nonnegative coefficients c_i ($i = 1, \ldots, 4$). If the set Δ_P is nonempty, then a rule that minimizes the risk over Δ_P is a threshold rule with respect to h_P. This follows from the admissibility of threshold rules (with respect to the ordering (3.2.13)) and from the fact that the problem under consideration is that of minimizing a linear function with nonnegative coefficients over a convex set

$$\{(a_{12}(\delta), a_{21}(\delta), a_{10}(\delta), a_{20}(\delta)) \colon \ \delta \in \Delta_P\}. \tag{3.2.19}$$

A similar problem arises when we exclude the possibility of decision deferrence from our considerations. The risk and the constraint functions defining Δ_P are then both linear functions of probabilities $a_{12}(\delta)$, $a_{21}(\delta)$. If Δ_P is non empty, then a threshold rule (with one threshold) is the solution of the problem of risk minimization over Δ_P.

Now we give examples of this kind of classes of problems. In each case thresholds and randomization constants for solutions (regarded as threshold rules with respect to h_P) are derived. Thresholds and randomization constants depend in general on P, and sometimes on some parameters of the class of problems as well. However, we do not indicate this in the notation: sometimes we even omit the index P in h_P. □

EXAMPLE 3.2.1A. *Minimization of the risk over the set Δ of all rules.*

For arbitrarily chosen (L_{ij}) we minimize the risk over $\Delta_P = \Delta$. If

$$\frac{L_{10}}{L_{12}} + \frac{L_{20}}{L_{21}} < 1, \tag{3.2.20}$$

then

$$\varkappa_1 = \frac{\pi}{1-\pi} \frac{L_{10}}{L_{21}-L_{20}}, \qquad \varkappa_2 = \frac{\pi}{1-\pi} \frac{L_{12}-L_{10}}{L_{20}}. \tag{3.2.21}$$

Otherwise $\varkappa_1 = \varkappa_2 = \varkappa$, where

$$\varkappa = \frac{\pi}{1-\pi} \frac{L_{12}}{L_{21}}. \tag{3.2.22}$$

Condition (3.2.20) is satisfied whenever the losess caused by decision deferment are sufficiently small in comparison with the losses caused by misclassification. The probability of decision deferment is then positive; decisions 0 and 1 (excluding decision 2) can be randomized quite arbitrarily for values z satisfying $h(z) = \varkappa_1$; decisions 0 and 2 (excluding decision 1) can be randomized arbitrarily for any value z satisfying $h(z) = \varkappa_2$. If $L_{10}/L_{12}+L_{20}/L_{21} = 1$, then, for z satisfying $h(z) = \varkappa$, decisions 0, 1, 2 can be randomized arbitrarily. If $L_{10}/L_{12}+L_{20}/L_{22} > 1$, then the losses caused by decision deferment are sufficiently high (in comparison with the losses caused by misclassification) to eliminate the possibility of decision deferment. For z satisfying $h(z) = \varkappa$, we then randomize arbitrarily decisions 1 and 2.

If condition (3.2.20) is satisfied, then

$$R_P(\delta) = \int_Z \min\big(\pi L_{12} f_1(z),\, (1-\pi) L_{21} f_2(z),\\ \pi L_{10} f_1(z) + (1-\pi) L_{20} f_2(z)\big)\, d\nu(z). \tag{3.2.23}$$

If (3.2.20) is not satisfied, then

$$R_P(\delta) = \int_Z \min\big(\pi L_{12} f_1(z),\, (1-\pi) L_{21} f_2(z)\big)\, d\nu(z). \tag{3.2.24}$$

□

EXAMPLE 3.2.1B. *Minimization of the risk over the set Δ_F.*

Formula (3.2.22) defines the threshold for the solution. The randomization of decisions 1, 2 at the threshold is arbitrary. Hence, if (3.2.20) does not hold, then risk minimization over Δ leads to the same solution as in the case of risk minimization over Δ_F.

When $\pi L_{12} = (1-\pi) L_{21}$, the threshold $\varkappa$ is equal to 1. When $L_{12} = L_{21} = 1$, the solution can be represented as a threshold rule with respect to the posterior probability $P(I = 2 \mid Z = z)$. By (3.2.6) and (3.2.22), the threshold is then equal to 1/2, which means that the object is classified as a member of the class with higher posterior probability. □

In the literature risk minimization over the set Δ or Δ_F is called ***Bayesian discrimination*** and the optimal rule is called the ***Bayesian solution.***

A rule which is optimal in the sense of risk minimization over Δ will be called a ***Bayesian rule*** and will be denoted by δ^B, and the thresholds corresponding to δ^B will be denoted by $\varkappa_1^B$, $\varkappa_2^B$ (cf. formula (3.2.21)). By δ^{FB} we shall denote a rule which is optimal with respect to risk minimization over Δ_F. It will be called a *forced decision Bayesian rule* and the corresponding threshold (3.2.22) will be denoted by $\varkappa^{FB}$. Moreover, the probabilities of the first error and the second error will be denoted by a_{12}^B, a_{21}^B for rule δ^B, and by a_{12}^{FB}, a_{21}^{FB} for rule δ^{FB}.

As was to be expected, we have

$$\varkappa_1^{\mathrm{B}} \leqslant \varkappa^{\mathrm{FB}} \leqslant \varkappa_2^{\mathrm{B}}, \tag{3.2.25}$$

whence

$$a_{ij}^{\mathrm{B}} \leqslant a_{ij}^{\mathrm{FB}}, \quad i, j = 1, 2,\ i \neq j. \tag{3.2.26}$$

It follows from (3.2.25) that the classification of an object cannot be changed by admitting decision deferment into consideration. In other words $\delta^{\mathrm{B}}(z) = \delta^{\mathrm{FB}}(z)$ for $z < \varkappa_1$ or $z > \varkappa_2$, $\delta_0^{\mathrm{B}}(z) = 1$ for $\varkappa_1 < z < \varkappa_2$, $\delta_0^{\mathrm{B}}(\varkappa_i)+\delta_i^{\mathrm{B}}(\varkappa_i) = 1$, $i = 1, 2$. □

Bayesian descrimination may cause the expected conditional losses $L_{12}a_{12}^{\mathrm{B}}$, $L_{21}a_{21}^{\mathrm{B}}$ to be too large. This drawback may be avoided by putting upper bounds on $L_{12}a_{12}^{\mathrm{B}}$ and $L_{21}a_{21}^{\mathrm{B}}$, but such an approach usually enlarges the domain of decision deferment. A formal description of this approach will be given in the next example.

EXAMPLE 3.2.2A. *Risk minimization over Δ with restrictions imposed upon the probabilities of the first and the second error.*

For arbitrarily chosen (L_{ij}) and $\alpha_1, \alpha_2 \in [0, 1]$ we minimize the risk function over the set

$$\Delta_P = \{\delta \in \Delta:\ a_{12}(\delta) \leqslant \alpha_1, a_{21}(\delta) \leqslant \alpha_2\}. \tag{3.2.27}$$

If $\alpha_1 \geqslant a_{12}^{\mathrm{B}}$ and $\alpha_2 \geqslant a_{21}^{\mathrm{B}}$, the solution δ is clearly equal to δ^{B}. If $\alpha_1 < a_{12}^{\mathrm{B}}$ and $\alpha_2 < a_{21}^{\mathrm{B}}$, the thresholds and randomization constants solve the following equations

$$P\big(h(Z) > \varkappa_2 \mid I = 1\big)+p_2^2 P\big(h(Z) = \varkappa_2 \mid I = 1\big) = \alpha_1, \tag{3.2.28}$$

$$P\big(h(Z) < \varkappa_1 \mid I = 2\big)+p_1^1 P\big(h(Z) = \varkappa_1 \mid I = 2\big) = \alpha_2. \tag{3.2.29}$$

Since $P\big(h(Z) > x \mid I = 1\big)$ is a monotone function of x, the threshold $\varkappa_2$ and the constant p_2^2 are determined uniquely by (3.2.28); the same holds for $\varkappa_1, p_1^1$ and formula (3.2.29).

If $\alpha_1 < a_{12}^{\mathrm{B}}$ and $\alpha_2 \geqslant a_{21}^{\mathrm{B}}$, then $\varkappa_1 = \varkappa_1^{\mathrm{B}}$, p_1^1 is an arbitrary number, and $\varkappa_2, p_2^2$ satisfy (3.2.28). If $\alpha_1 \geqslant a_{12}^{\mathrm{B}}$ and $\alpha_2 < a_{21}^{\mathrm{B}}$, then $\varkappa_2 = \varkappa_2^{\mathrm{B}}$, p_2^2 is an arbitrary number, and $\varkappa_1, p_1^1$ satisfy (3.2.29).

The thresholds $\varkappa_1^{\mathrm{B}}$, $\varkappa_2^{\mathrm{B}}$ depend only on π and (L_{ij}). If $\alpha_2 < a_{21}^{\mathrm{B}}$, then the threshold $\varkappa_1$ depends exclusively on the distribution of the random variable $h_P(Z)$ and α_2; the same holds for $\varkappa_2$ whenever $\alpha_1 < a_{12}^{\mathrm{B}}$. □

By replacing the solution δ of 3.2.2A corresponding to some α_1, α_2 with the solution δ' corresponding to α_1', α_2' such that $\alpha_i' \leqslant \alpha_i$, $i = 1, 2$, we do not alter the classification of the object in question, since, for any z, $\delta_i(z) \geqslant \delta_i'(z)$, $i = 1, 2$.

EXAMPLE 3.2.2B. *Risk minimization over the set Δ_{F} with restriction imposed upon the probability of the first error.*

For any L_{12}, L_{21}, and $\alpha \in [0, 1]$ we minimize the risk function over the set

$$\Delta_P = \{\delta \in \Delta_{\mathrm{F}}:\ a_{12}(\delta) \leqslant \alpha\}.$$

If $\alpha \geqslant a_{12}^{FB}$, then δ^{FB} solves the problem. Otherwise, the threshold $\varkappa$ and the constant p_2 are determined by the equation

$$P(h(Z) > \varkappa \mid I = 1) + p_2 P(h(Z) = \varkappa \mid I = 1) = \alpha.$$

Hence, if $\alpha \leqslant a_{12}^{FB}$, then the solution δ of problem 3.2.2B coincides with the most powerful (Neyman–Pearson) test on the significance level α used to test the hypothesis that the density is equal to f_1 against the hypothesis that the density is equal to f_2. The power of this test is $a_{22}(\delta)$ (cf. Section 2.3). □

The constraints appearing in (3.2.27) impose upper bounds on the probabilities of the first and the second error. It is natural, however, to require that the probabilities of those errors should be sufficiently small as compared with the probabilities of correct decisions. This requirement can be formalized by imposing bounds on the ratio of the probability of an incorrect decision to that of a correct decision for each class separately. We denote this ratio for the i-th class by $b_i(\delta)$, $i = 1, 2$,

$$b_1(\delta) = \frac{a_{12}(\delta)}{a_{11}(\delta)}, \quad b_2(\delta) = \frac{a_{21}(\delta)}{a_{22}(\delta)}. \tag{3.2.30}$$

Clearly, the ratios b_1, b_2 can be large even if a_{12}, a_{21} are small numbers. Consider first the extreme case where density f_1 coincides with f_2 on $\boldsymbol{Z}$. Then both classes are not discernible, and $h_P(z) \equiv 1$. Suppose that the losses do not satisfy (3.2.20). Then the solution of the risk minimization problem with bounds on the probabilities of the first and the second error (Example 3.2.2A) is a *constant rule* δ^0 given by the formula

$$\delta^0(z) \equiv (1-p_1-p_2, p_1, p_2), \tag{3.2.31}$$

where $p_1 \leqslant \alpha_2$, $p_2 \leqslant \alpha_1$, $0 < p_1+p_2 < 1$. Hence, we have

$$b_1(\delta^0) = \frac{p_2}{p_1}, \quad b_2(\delta^0) = \frac{p_1}{p_2},$$

i.e. at least one of these quotients cannot be less than 1.

This undesirable situation will undergo a slight change if the densities f_1, f_2 differ only a little. In practice large value of one of the ratios appears very often. This fact suggests a modification of Example 3.2.2 by imposing bounds on $b_1(\delta)$, $b_2(\delta)$. Such a modification is presented in the next example.

EXAMPLE 3.2.3. *Risk minimization with restrictions imposed upon quotients of probabilities of wrong and correct decisions in class* 1 *and class* 2.

For arbitrarily chosen numbers β_1, β_2 from the interval (0, 1) satisfying the inequality

$$\beta_1 \cdot \beta_2 > \inf_Z h(z) \cdot \inf_Z \frac{1}{h(z)} \tag{3.2.32}$$

we minimize the risk over the set

$$\Delta_P = \{\delta \in \Delta\colon\ b_i(\delta) \leqslant \beta_i,\ i = 1, 2\}. \tag{3.2.33}$$

If inequality (3.2.32) is not satisfied, then there is no rule belonging to Δ_P (hence, for some distributions the numbers β_1, β_2 cannot both be chosen arbitrarily small). The solution depends on the relations between β_1, β_2 and the ratios b_1, b_2 for the Bayesian rule δ^B. Let

$$b_i^B = b_i(\delta^B), \quad b_i^{FB} = b_i(\delta^{FB}).$$

If $\beta_i \geqslant b_i^B$ for $i = 1, 2$, the solution is equal to δ^B. On the other hand, if $\beta_i < b_i^B$ for $i = 1, 2$, then the thresholds and randomization constants are uniquely determined by the following system of equations

$$\begin{aligned} \frac{P(h(Z) > \varkappa_2 \mid I = 1) + p_2^2 P(h(Z) = \varkappa_2 \mid I = 1)}{P(h(Z) < \varkappa_1 \mid I = 1) + p_1^1 P(h(Z) = \varkappa_1 \mid I = 1)} &= \beta_1, \\ \frac{P(h(Z) < \varkappa_1 \mid I = 2) + p_1^1 P(h(Z) = \varkappa_1 \mid I = 2)}{P(h(Z) > \varkappa_2 \mid I = 2) + p_2^2 P(h(z) = \varkappa_2 \mid I = 2)} &= \beta_2. \end{aligned} \tag{3.2.34}$$

We shall not discuss the remaining cases. □

If we replace solution δ of problem 3.2.3 for β_1, β_2 by solution δ' of the problem for β_1', β_2' such that $\beta_i' \leqslant \beta_i$, $i = 1, 2$, the classification of the object does not change; the only effect of taking δ' instead of δ is that the decision taken either becomes a deferred decision or remains unchanged.

Observe that any solution of problem 3.2.2A for (α_1, α_2) satisfying the conditions $\alpha_1 \leqslant a_{12}^B$, $\alpha_2 \leqslant a_{21}^B$ coincides with a solution of problem 3.2.3 for some (β_1, β_2) such that $\beta \ > \alpha\ ,\ i = 1, 2$.

So far we have considered classes of problems in which the inequality constraints are linear with respect to probabilities $a_{ij}(\delta)$ and have positive coefficients. Hence, according to the theorem given at the beginning of this section, the solutions of such problems can be represented as threshold rules with respect to the likelihood ratio h_P. Now we consider another type of constraints leading to solutions of the same form.

EXAMPLE 3.2.4A. *Risk minimization over the set Δ when the expected conditional losses in classes 1 and 2 are equal.*

For arbitrary (L_{ij}) we minimize the risk on the set

$$\Delta_P = \{\delta \in \Delta\colon L_{12}a_{12}(\delta) + L_{10}a_{10}(\delta) = L_{21}a_{21}(\delta) + L_{20}a_{20}(\delta)\}. \tag{3.2.35}$$

In Δ_P each rule which is admissible with respect to the ordering (3.2.13) is a threshold rule. Hence, the solution is also a threshold rule.

If condition (3.2.20) holds and $h(Z)$ has a continuous distribution, then

$$\varkappa_2 = \varrho\varkappa_1 \quad (\text{hence} \quad \varrho > 1) \tag{3.2.36}$$

and

$$\varrho = \frac{(L_{12} - L_{10})(L_{21} - L_{20})}{L_{10}L_{20}}.$$

The threshold rule which solves the problem is determined by (3.2.36) and by the equality of the expected conditional losses. Observe that if $h(Z)$ has a continuous distribution, then the Bayesian thresholds $\varkappa_1^B$, $\varkappa_2^B$ satisfy (3.2.36).

When (3.2.20) is not satisfied, a rule which is a solution of the problem belongs to Δ_F and is a solution of the problem discussed in the next example (3.2.4B). □

EXAMPLE 3.2.4B. *Risk minimization over the set Δ_P provided that expected conditional losses in classes 1 and 2 are equal.*

The set Δ_P is of the form

$$\Delta_P = \{\delta \in \Delta_F\colon L_{12}a_{12}(\delta) = L_{21}a_{21}(\delta)\}. \tag{3.2.37}$$

In this case there is only one threshold rule in Δ_P. For this rule the threshold $\varkappa$ and the number p_1 can be determined from the formula

$$\begin{aligned} &L_{12}\big(P(h(Z) > \varkappa \mid I = 1) + (1-p_1)P(h(Z) = \varkappa \mid I = 1)\big) \\ &= L_{21}\big(P(h(Z) < \varkappa \mid I = 2) + p_1 P(h(Z) = \varkappa \mid I = 2)\big). \end{aligned} \tag{3.2.38}$$

The threshold $\varkappa$ Belongs to the interval $[\varkappa_1, \varkappa_2]$, where $\varkappa_1, \varkappa_2$ are thresholds in the solution of the problem from Example 3.2.4A for any losses L_{10}, L_{20} satisfying (3.2.20). The Bayesian thresholds (inequality (3.2.25)) are situated in a similar way.

When $L_{12} = L_{21}$, the set Δ_P contains rules for which the probabilities of the first and second error are identical. In this case the optimal rule will be denoted by δ^{EER} (equal error rates). □

Now we discuss in detail the solutions of the problems from Examples 3.2.1 – 3.2.4 and their properties for the following families of distributions, normal and exponential

(i) $Z_i \sim N_k(\mu_i, \Sigma)$, $i = 1, 2$, $k \geqslant 1$;

(ii) $Z_i \sim P(\lambda_i)$, $\lambda_1 > \lambda_2$, $i = 1, 2$.

In both cases it is convenient to determine the solutions as threshold rules with respect to the statistic $\ln h_P$, since it is a linear function. In case (i) we have

$$\begin{aligned} &\big(\ln h_P(Z) \mid I = 1\big) \sim N(-\tfrac{1}{2}\gamma^2, \gamma^2), \\ &\big(\ln h_P(Z) \mid I = 2\big) \sim N(\tfrac{1}{2}\gamma^2, \gamma^2), \end{aligned} \tag{3.2.39}$$

where γ is the Mahalanobis distance given by (2.5.7).

In case (ii) we have

$$\begin{aligned} &\big(\ln (h_P(Z) \mid I = 1) + \ln \lambda\big) \sim P\left(\frac{\lambda}{\lambda-1}\right), \\ &\big(\ln (h_P(Z) \mid I = 2) + \ln \lambda\big) \sim P\left(\frac{1}{\lambda-1}\right), \end{aligned} \tag{3.2.40}$$

where $\lambda = \lambda_1/\lambda_2$ (i.e. $\lambda > 1$). Hence, any distribution of the pair $\big(I, \ln h_P(Z)\big)$ is determined in case (i) by π and γ and in case (ii) by π and α. Since the random variable $\ln h_P(Z)$ has a continuous distribution, we can neglect randomization at the thresholds. Thus, the solutions are determined exclusively by the thresholds, which in turn are determined by L_{ij}, π, γ or λ, and by the parameters of problems such as (α_1, α_2) or (β_1, β_2).

In Examples 3.2.1A and 3.2.1B the threshold depends only on (L_{ij}) and π. In Example 3.2.2A, in accordance with the relation between α_1, α_2 and the probabilities a_{12}^B, a_{21}^B, the thresholds either are Bayesian thresholds (i.e., depend on (L_{ij}) and π) or are appropriate quantiles of the normal distributions (3.2.39) or the exponential distributions (3.2.40) (i.e., depend on γ or λ). We have a similar situation in Example 3.2.2B. In Example 3.2.3, for $\beta_i < b_i^B$ $(i = 1, 2)$, both thresholds obtained from (3.2.34) depend on β_1, β_2, and on γ (for the normal distribution), or on λ (for the exponential distribution). For $\beta_i > b_i^B$ $(i = 1, 2)$ we obtain Bayesian thresholds, i.e., thresholds which depend on (L_{ij}) and π. In Example 3.2.4A the thresholds depend on (L_{ij}) and on γ or λ, and are independent of π. In Example 3.2.4B the thresholds depend on the quotient L_{12}/L_{21} and γ or λ.

Various properties of the solutions from Examples 3.2.1 – 3.2.4 are presented in Table 3.2.1 (case (i)) and in Table 3.2.2 (case (ii)).

TABLE 3.2.1.

Probabilities of the first and second error (a_{12}, a_{21}), quotients of probabilities of wrong and correct decision (b_1, b_2), and the probability of deferred decision (a_0) in Examples 3.2.1–3.2.4 obtained under the assumption that $L_{12} = L_{21} \leqslant L_{10}+L_{20}$, $\pi = 1/2$, $Z_i \sim N_k(\mu_i, \Sigma)$ $(i = 1, 2)$ (for $\gamma = \varrho_{\mathrm{Mah}}(\mu_1, \mu_2, \Sigma) = 0.5, 1, 2, 3$).

γ	Examples 3.2.1 and 3.2.4		Example 3.2.2 for chosen bounds α_1, α_2 $(\alpha_1 = \alpha_2 = \alpha)$			Example 3.2.3 for chosen bounds β_1, β_2 $(\beta_1 = \beta_2 = \beta)$		
	$a_{12} = a_{21}$	$b_1 = b_2$	α	$b_1 = b_2$	a_0	β	$a_{12} = a_{21}$	a_0
(1)	(2)	(3)	(4)	(5)	(6)	(7)	(8)	(9)
0.5	0.401	0.670	0.01	0.297	0.956	0.20	0.00007	0.956
			0.025	0.346	0.903			
			0.05	0.427	0.833	0.25	0.00348	0.983
			0.1	0.453	0.682			
			0.2	0.545	0.433			
1	0.309	0.446	0.01	0.109	0.898	0.1	0.0078	0.914
			0.025	0.148	0.806	0.15	0.0258	0.802
			0.05	0.204	0.705	0.20	0.0549	0.671
			0.1	0.257	0.510	0.25	0.0939	0.531
			0.2	0.355	0.237			
2	0.159	0.189	0.01	0.027	0.619	0.05	0.026	0.451
			0.025	0.049	0.459	0.1	0.070	0.230
			0.05	0.080	0.328			
			0.1	0.131	0.136	0.15	0.119	0.087
3	0.067	0.072	0.01	0.013	0.242	0.01	0.0071	0.285
			0.025	0.028	0.125	0.04	0.0353	0.081
			0.05	0.059	0.045	0.06	0.0552	0.025

TABLE 3.2.2.

Probabilities of the first and second error (a_{12}, a_{21}), quotients of probabilities of wrong and correct decision (b_1, b_2), and the probability of deferred decision (a_0) in Examples 3.2.1 – 3.2.4 obtained under the assumption that $L_{12} = L_{21} \leqslant L_{10}+L_{20}$, $\pi = 1/2$, $Z_i \sim P(\lambda_i)$ $(i = 1, 2)$ (for $\lambda_1/\lambda_2 =$ $= 1.5, 2, 3, 4, 5$).

$\frac{\lambda_1}{\lambda_2}$	Example 3.2.1				Example 3.2.2 for chosen bounds α_1, α_2 $(\alpha_1 = \alpha_2 = \alpha)$				Example 3.2.3 for chosen bounds β_1, β_2 $(\beta_1 = \beta_2 = \beta)$				Example 3.2.4	
	a_{12}	a_{21}	b_1	b_2	α	b_1	b_2	a_0	β	a_{12}	a_{21}	a_0	a_{12}, a_{21}	b_1, b_2
(1)	(2)	(3)	(4)	(5)	(6)	(7)	(8)	(9)	(10)	(11)	(12	(13)	(14)	(15)
1.5	0.297	0.557	0.423	1.256	0.12	0.703	0.5448	0.487	0.2	0.0002	0.0007	0.997	0.430	0.755
									0.25	0.0008	0.0022	0.993		
2	0.250	0.500	0.333	1	0.1	0.526	0.316	0.646	0.1	0.0004	0.0020	0.987	0.382	0.618
					0.15	0.541	0.387	0.517	0.15	0.0020	0.0067	0.967		
					0.2	0.556	0.447	0.395	0.2	0.0063	0.0159	0.933		
									0.25	0.0151	0.0308	0.885		
3	0.191	0.421	0.236	0.727	0.025	0.342	0.085	0.790	0.1	0.0051	0.0172	0.877	0.317	0.464
					0.05	0.351	0.136	0.694	0.15	0.0166	0.0382	0.790		
					0.1	0.370	0.215	0.533	0.2	0.0376	0.0670	0.686		
					0.15	0.389	0.282	0.392						
4	0.158	0.370	0.187	0.588	0.025	0.260	0.063	0.728	0.1	0.0128	0.0336	0.745	0.276	0.380
					0.05	0.270	0.106	0.621	0.15	0.0354	0.0651	0.615		
					0.1	0.291	0.178	0.448						
					0.15	0.314	0.300							
5	0.134	0.331	0.154	0.495	0.01	0.204	0.025	0.769	0.1	0.0211	0.0462	0.630	0.243	0.320
					0.025	0.210	0.052	0.679	0.15	0.0530	0.0833	0.478		
					0.05	0.221	0.091	0.662						
					0.1	0.244	0.159	0.380						

In Example 3.2.2 the bounds (α_1, α_2) have been selected so as to satisfy the condition $\alpha_1 = \alpha_2 < \min(a_{12}^{\mathrm{B}}, a_{21}^{\mathrm{B}})$. Hence, the solution δ satisfies the condition $a_{12}(\delta) = a_{21}(\delta) = \alpha$. In Example 3.2.3 the bounds (β_1, β_2) satisfy the condition $\beta_1 = \beta_2 < \min(b_1^{\mathrm{B}}, b_2^{\mathrm{B}})$, and hence the solution δ satisfies the condition $b_1(\delta) = b_2(\delta) = \beta$. Moreover, in all the examples it has been assumed that $\pi = \frac{1}{2}$ and $L_{12} = L_{21} \leqslant L_{10} = L_{20}$. Hence, the rule δ^{FB} with the threshold 1 was taken as the solution of the problem from Example 3.2.1, and the rule δ^{EER} as the solution in Example 3.2.4. These rules are identical in case (i) and differ in case (ii). However, in case (ii) the difference between the misclassification probabilities are not large; e.g. for $\lambda = 1.5$ we have $0.4301 - \frac{1}{2}(0.2971 + 0.5565) = 0.0033$, and for $\lambda = 5$ we have 0.01. The probabilities of the first and the second error for the rule δ^{FB} are clearly different, e.g. the difference equals 0.26 if $\lambda = 1.5$ and equals 0.20 if $\lambda = 5$. But it should be noted that in both cases the misclassification probability is high even for significantly different distributions (e.g. it is equal to 0.24 if $\lambda = 5$). This fact speaks for admitting deferrence of decision.

A decrease of the misclassification probability leads automatically to an increase (sometimes considerable) of the probability of deferring the decision. If in case (i) and for $\gamma = 2$ we admit on the average only one incorrect decision for each ten correct decisions (i.e., $\beta_1 = \beta_2 = 0.1$), then the expected percentage of the deferred decisions is equal to 23 (column (9) in Table 3.2.1). The same requirement in case (i) and $\gamma = 1$ leads to the expected frequency of deferred decisions for 91% of the objects classified (column (9) in Table 3.2.1) and in case (ii) and $\lambda = 3$ for 88% of the objects.

As follows from the data contained in Table 3.2.1 and 3.2.2, and corresponding to Example 3.2.2, if the classes do not differ much (small values of γ and λ), then the ratios b_1 and b_2 largely exceed α_1 and α_2. For example for $\alpha_1 = \alpha_2 = 0.05$ we have $b_1 = b_2 = 0.20$ if $\gamma = 1$ (column (5) in Table 3.2.1) and $b_1 = 0.35$, $b_2 = 0.14$ if $\lambda = 3$ (columns (7) and (8) in Tables 3.2.2). With a decrease of λ and γ or an increase of α_1 and α_2 the situation becomes even more drastic.

The results presented in Tables 3.2.1 and 3.2.2 may serve as a warning against an unjustified optimism in applying discriminant rules and neglecting the way in which the problem is formalized.

3.3. Statistical problems

In the previous section we considered discriminant problems in which the distribution P was given. In practice, however, such detailed information is not available; what we know is only that P belongs to a certain family $\mathscr{P}$ of distributions. If the only information about P is the support $\mathbf{Z}$ of the observation vector Z and the measure ν, then the family $\mathscr{P}$ consists of all distributions defined by arbitrary densities f_1, f_2 on $\mathbf{Z}$ and an arbitrary number $\pi \in [0, 1]$. In what follows we consider only families $\mathscr{P}$ in which, for any distribution $P \in \mathscr{P}$, the densities f_1, f_2 have the same support and hence, as previously, we restrict ourselves to the case of f_1, f_2 being positive on $\mathbf{Z}$.

Consider first the family

$$\mathscr{P}_0 = \{P_{\pi, f_1, f_2} \colon \pi \in [0, 1]\}$$

for fixed ν, f_1, f_2. In Example 3.2.4, the solution of the probabilistic problem is independent of π and thus coincides with the solution of the statistical problem of uniform risk minimization under the condition that the expected conditional losses are all equal in the family $\mathscr{P}_0$. In Example 3.2.1, the solution δ^{B} depends on π, and hence, there exists no rule solving the problem for each $P \in \mathscr{P}_0$. We may, however, formulate a certain minimax problem, namely the problem of finding a rule $\delta \in \Delta$ such that, for each $\delta' \in \Delta$, we have

$$\sup_{P \in \mathscr{P}_0} R(\delta, P) \leqslant \sup_{P \in \mathscr{P}_0} R(\delta', P). \tag{3.3.1}$$

Hence the minimax rule minimizes the expression

$$\max\left(L_{12} a_{12}(\delta) + L_{10} a_{10}(\delta), L_{21} a_{21}(\delta) + L_{20} a_{20}(\delta)\right). \tag{3.3.2}$$

The minimum of (3.3.2) is attained by a rule for which the expected conditional losses are equal. Therefore, the problem consists in minimizing either one of the expected conditional losses under the condition that they are equal. This, in turn, is equivalent to risk minimization (for any $\pi \in [0, 1]$) under the condition that the expected conditional losses are equal. Consequently, rule δ is a minimax rule if and only if δ solves the problem from Example 3.2.4.

In Example 3.2.2 the solution is independent of π for any $\alpha_1 < a_{12}^{\mathrm{B}}$, $\alpha_2 < a_{21}^{\mathrm{B}}$. However, to check these conditions, one must know the value of π, and hence there is no rule which would solve the problem for each $P \in \mathscr{P}_0$. The same holds for Example 3.2.2.

It usually happens in practice that not only the probability π but also the densities f_1 and f_2 are unknown. However, in general, without knowing f_1 and f_2 it is impossible to formulate problems which are solvable and satisfactory from the practical point of view.

This lack of information may be compensated for by making use of learning samples. These are collections of objects belonging to the same population as the object to be classified. For the objects from a learning sample, we have at our disposal full information about the values of Z and at least partial information about the values of I. This allows us to estimate selected parameters of the distribution of (I, Z).

In the simplest case we have two learning samples of sizes n_1 and n_2 chosen from the first and the second class. Each sample is regarded as a simple random sample with respect to Z. In this case we can estimate the parameters of the distributions of Z_1 and Z_2 but not the probability π. In a more general case we have one simple random sample of size n taken as a learning sample from the whole population such that, for each element of the sample, we know to which class it belongs. The number n_i of objects from the i-th class ($i = 1, 2$) is then a random variable. In this case we can estimate not only the parameters of distributions in both classes but also the probability π.

In the above mentioned learning samples the feature I is observable. In practice, however, one may encounter learning samples consisting of objects for which the feature I is not observable. Then one may sometimes use a certain substitute feature I', strongly stochastically dependent on I, which is not observable for the object in question but is observable for objects from the learning sample. For instance, it is so in archeological investigations, in which on the basis of the equipment found in a given grave (feature Z) we want to determine the sex of the person buried in that grave (feature I). Here, a learning sample may consist of graves for which the sex of the person buried is not known but an anthropological experts' report (feature I') is available.

Sometimes one also uses learning samples in which some values of Z are chosen in advance, and for each of those values, a sample of a given size consisting of objects with that value of Z is considered. Then the assumption is usually made that those samples are simple random samples selected from the set of objects with a given fixed value of Z.

In what follows we restrict our considerations to the case where we have a simple random sample, with full observability, chosen from the whole population. Let

$$E^{(n)} = (I^{(1)}, Z^{(1)}, I^{(2)}, Z^{(2)}, \dots, I^{(n)}, Z^{(n)}) \tag{3.3.3}$$

be such a sample and let $\boldsymbol{E}^{(n)}$ be the set of values of the sequence $E^{(n)}$. Consequently, we have a triple $(I, Z, E^{(n)})$ of random elements, where (I, Z) refers to the object in question and $E^{(n)}$ refers to the objects from the learning sample. The pair (I, Z) is independent of $E^{(n)}$, and hence the joint distribution of the triple $(I, Z, E^{(n)})$ is equal to P^{n+1}. We observe the pair $(Z, E^{(n)})$ and, on the basis of this observation we want to predict the value of I; the decision rules δ are defined on $\boldsymbol{Z} \times \boldsymbol{E}^{(n)}$.

When the distribution P is known, learning sample does not provide any information about I and can be disregarded. Formally, this follows from the fact that the reduction $g(z, e^{(n)}) = z$ is prediction sufficient for a one-element family $\mathscr{P}$.

Now, suppose that $\mathscr{P}$ is not one-element and we consider a family $(\delta_P, P \in \mathscr{P})$ of solutions of probabilistic problems belonging to one of classes discussed in Section 3.2. Each rule δ defined on $\boldsymbol{Z} \times \boldsymbol{E}^{(n)}$ determines, for a fixed $P \in \mathscr{P}$, a certain distribution on $\{1, 2\} \times \{0, 1, 2\}$, i.e., distribution of the pair $(I, \hat{I}_\delta)$. If the learning sample is large, the rule δ is required to ensure, for each $P \in \mathscr{P}$, that the distribution of $(I, \hat{I}_\delta)$ is close enough to the distribution of $(I, \hat{I}_{\delta_P})$.

In formalizing this requirement the sample $E^{(n)}$ should be replaced by a sample E with infinitely many elements; the set of the values of the sequence E will be denoted by $\boldsymbol{E}$. Hence, we impose requirements on the sequence $(\delta^{(n)})$ of rules on $\boldsymbol{Z} \times \boldsymbol{E}$ whose element $\delta^{(n)}$ depends only on $E^{(n)}$. In particular, we say that the sequence $(\delta^{(n)})$ is consistent with the family $(\delta_P, P \in \mathscr{P})$ of rules if for each $P \in \mathscr{P}$ the sequence of distributions of $(I, \hat{I}_{\delta^{(n)}})$ weakly converges to the distribution of $(I, \hat{I}_{\delta_P})$.

It follows from the consistency of $(\delta^{(n)})$ that, for each $P \in \mathscr{P}$ and an arbitrary continuous parameter $\varkappa$ of the set of distributions on $\{1, 2\} \times \{0, 1, 2\}$, the sequence $\varkappa(\hat{I}, I_{\delta^{(n)}})$ converges to $\varkappa(\hat{I}, I_{\delta_P})$.

A consistent sequence of rules is usually constructed by a suitable adaptation of the family $(\delta_P;\ P \in \mathscr{P})$. Let τ be an arbitrary parameter of the family $\mathscr{P}$, such that the solutions δ_P depend on $\mathscr{P}$ through $\tau(P)$, i.e.,

$$\delta_P(z) = f(z, \tau(P)) \tag{3.3.4}$$

for a certain function f. Let $\hat{\tau}^{(n)}$ denote an estimator of τ defined on $\boldsymbol{E}^{(n)}$. The rule $\delta^{(n)}$ defined by the formula

$$\delta^{(n)}(z, e^{(n)}) = f(z, \hat{\tau}^{(n)}(e^{(n)})), \quad e^{(n)} \in \boldsymbol{E}^{(n)}, \tag{3.3.5}$$

is called an *adapted rule with respect to the estimator* $\hat{\tau}^{(n)}$ *and the family of rules* $(\delta_P, P \in \mathscr{P})$. Clearly, the rule $\delta^{(n)}$ is defined only for those $(z, e^{(n)})$ for which $(z, \tau^{(n)}(e^{(n)}))$ belongs to the domain of the function f.

If, for each $P \in \mathscr{P}$, δ_P is a threshold rule corresponding to a discriminant function g_P with thresholds $\varkappa_1(P)$, $\varkappa_2(P)$, then we have to adapt separately the family (g_P) and the families of thresholds. Let $g^{(n)}(z)$ be the value of an adapted discriminant function at z, and let $\varkappa_1^{(n)}$, $\varkappa_2^{(n)}$ denote the adapted thresholds. Suppose that, for each $P \in \mathscr{P}$, $g_P(Z)$ has a continuous distribution and, for each $z \in \boldsymbol{Z}$, the sequences $(g^{(n)}(z))$, $(\varkappa_1^{(n)})$, $(\varkappa_2^{(n)})$ converge with probability 1 to $g_P(z)$, $\varkappa_1(P)$ and $\varkappa_2(P)$, respectively. Then the sequence $(\delta^{(n)})$ of rules is *consistent* with $(\delta_P, P \in \mathscr{P})$.

For a given family $\mathscr{P}$ the adaptation of $(\delta_P, P \in \mathscr{P})$ is not unique. First, for a fixed parameter τ different estimators can be chosen. Second, one may choose among different parameters τ. Third, in the case of threshold rules different discriminant functions g_P can be applied. To a large extent, the performance of an adapted rule depends upon the choice of parameters and their estimators.

A typical example of how a consistent sequence of rules is constructed is provided by the adaptation of the family of functions $\mathrm{LDF}(z, \boldsymbol{\mu}_1, \boldsymbol{\mu}_2, \Sigma)$ (cf. the formula (3.2.10)) and of the families of thresholds corresponding to Examples 3.2.1 – 2.3.4 where $Z_i \sim N_k(\boldsymbol{\mu}_i, \Sigma)$. Then we use the estimators $\hat{\boldsymbol{\mu}}_1, \hat{\boldsymbol{\mu}}_2, \hat{\Sigma}$ which converge with probability 1 to $\boldsymbol{\mu}_1, \boldsymbol{\mu}_2, \Sigma$. Hence, for each distribution P and each $z \in \boldsymbol{R}^k$, the sequence of random variables

$$\widehat{\mathrm{LDF}}(z) = \mathrm{LDF}(z; \hat{\boldsymbol{\mu}}_1, \hat{\boldsymbol{\mu}}_2, \hat{\Sigma}) \tag{3.3.6}$$

defined on $\boldsymbol{E}$ converges with probability 1 to $\mathrm{LDF}(z; \boldsymbol{\mu}_1, \boldsymbol{\mu}_2, \Sigma)$. In Examples 3.2.2, 3.2.3, 3.2.4 the thresholds are continuous functions of parameters $\boldsymbol{\mu}_1, \boldsymbol{\mu}_2, \Sigma$ and hence the respective adapted rules are consistent. In Example 3.2.1, the adaptation consists in replacing π by its estimator $\hat{\pi} = n_1/(n_1+n_2)$, which also converges.

Estimator $\hat{\Sigma}$ makes use of the whole of the learning sample. In the course of adapting the quadratic discriminant function QDF, covariance matrices Σ_1 and Σ_2 are estimated separately, each from the respective part of the learning sample. Then

$$\widehat{\mathrm{QDF}}(z) = \mathrm{QDF}(z; \hat{\mu}_1, \hat{\mu}_2, \hat{\Sigma}_1, \hat{\Sigma}_2). \tag{3.3.7}$$

The use of $\widehat{\mathrm{QDF}}$ is justified in particular when $Z_i \sim N_k(\mu_i, \Sigma_i)$, $i = 1, 2$.

*

Let $\mathscr{P}^{\uparrow}$ be the family of distributions of pairs (I, Z) such that Z is a univariate continuous random variable and the likelihood ratio h_P is increasing. We shall present the adaptation of family $(\delta_P, P \in \mathscr{P}^{\uparrow})$ of solutions of some selected problems.

Let F_i $(i = 1, 2)$ denote the distribution function of Z_i. By the assumption that f_1 and f_2 are positive on $\boldsymbol{Z}$ it follows that F_1 and F_2 are both increasing functions on $\boldsymbol{Z}$, and hence the sum F_1+F_2 is an increasing function of the ratio h_P, for $P \in \mathscr{P}^{\uparrow}$. Therefore, if for $P \in \mathscr{P}^{\uparrow}$ a solution of a probabilistic problem is a threshold rule with respect to h_P, it can also be represented as a threshold rule with respect to F_1+F_2. The thresholds of the new rule are parameters of the distribution of $\big(I, F_1(Z)+F_2(Z)\big)$. We determine those thresholds for the solutions of some problems from Section 3.2.

We start with determining the threshold K corresponding to F_1+F_2 in the rule δ^{EER}, which solves the problem from Example 3.2.4B whenever $L_{12} = L_{21}$. By (3.2.28), the threshold corresponding to h_P, say $\varkappa$, is determined from the equation

$$P\big(h_P(Z) > \varkappa \mid I = 1\big) = P\big(h_P(Z) < \varkappa \mid I = 2\big). \tag{3.3.8}$$

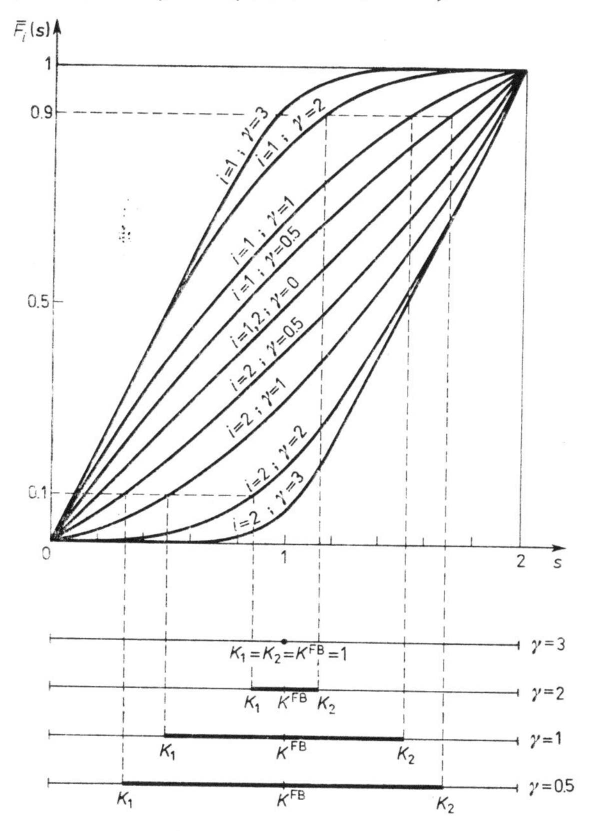

FIG. 3.3.1. Distribution functions $\bar{F}_i$ of random variables $S_i = F_1(Z_i)+F_2(Z_i)$, $i = 1, 2$, thresholds K_1, K_2, where $a_{12} \leqslant 0.1$, $a_{21} \leqslant 0.1$, and threshold K^{FB} for rule δ^{FB} under the conditions $L_{12} = L_{21} < L_{10}+L_{20}$, $\pi = 1/2$, $Z_i \sim N(\mu_i, \sigma^2)$, $|\mu_1-\mu_2|/\sigma = \gamma$ (for $\gamma = 0.5, 1, 2, 3$).

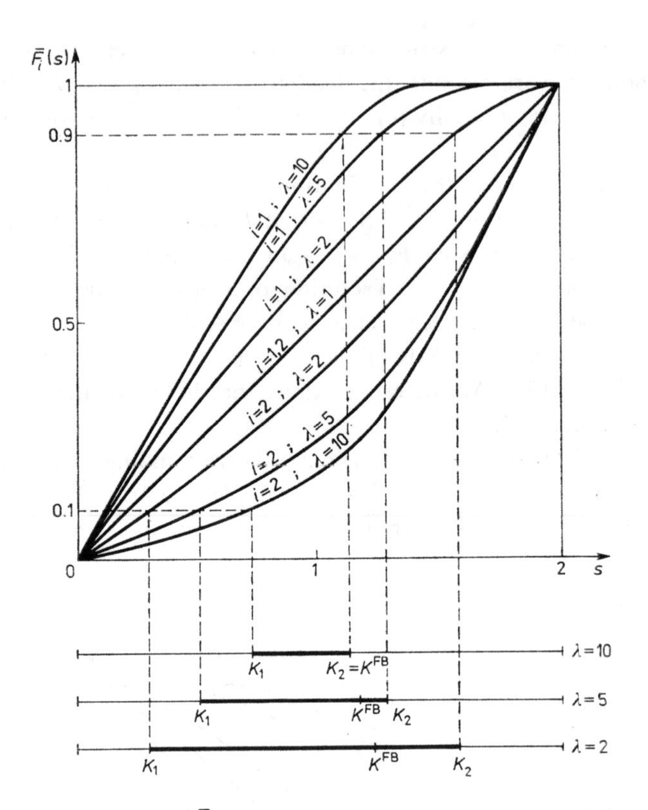

FIG. 3.3.2. Distribution functions $\bar{F}_i$ of random variables $S_i = F_1(Z_i)+F_2(Z_i)$, $i = 1, 2$, thresholds K_1, K_2, for $a_{12} \leq 0.1$, $a_{21} \leq 0.1$, and threshold K^{FB} for rule δ^{FB} under the assumption hat $L_{12} = L_{21} < L_{10}+L_{20}$, $\pi = 1/2$, $Z_i \sim P(\lambda_i)$, $\lambda_1 > \lambda_2$, $\lambda_1/\lambda_2 = \lambda$ (for $\lambda = 2, 5, 10$).

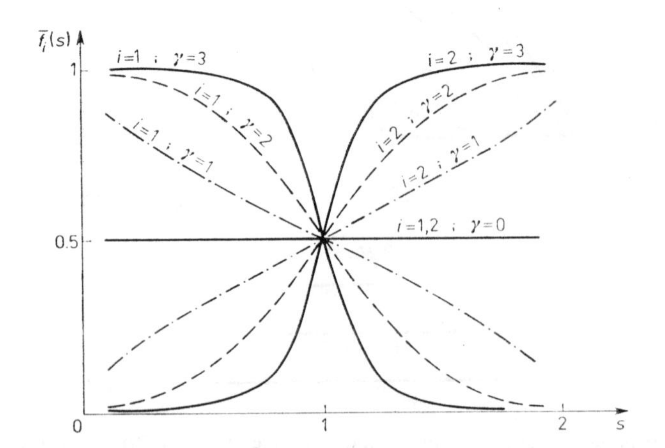

FIG. 3.3.3. Densities $\bar{f}_i$ of random variables $S_i = F_1(Z_i)+F_2(Z_i)$, $i = 1, 2$, where $Z_i \sim N(\mu_i, \sigma^2)$, $|\mu_1-\mu_2|/\sigma = \gamma$ (for $\gamma = 0, 1, 2, 3$).

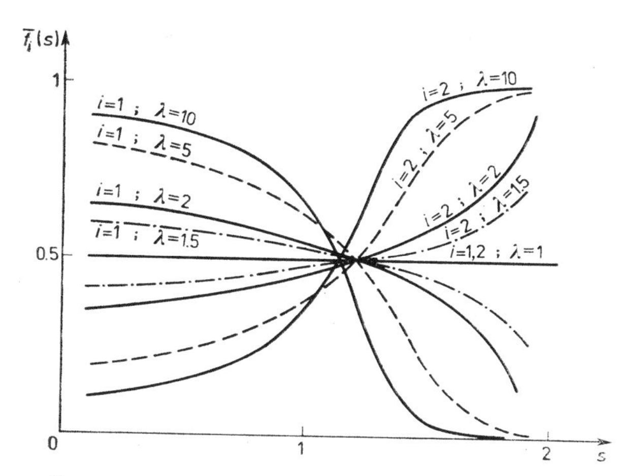

FIG. 3.3.4. Densities $\bar{f}_i$ of random variables $S_i = F_1(Z_i)+F_2(Z_i)$, $i = 1, 2$, where $Z_i \sim P(\lambda_i)$, $\lambda_1 > \lambda_2$, $\lambda_1/\lambda_2 = \lambda$ (for $\lambda = 1,\ 1.5,\ 2, 5, 10$).

Denoting by $\bar{F}_i$ the distribution function of the reduced variable

$$S_i = F_1(Z_i)+F_2(Z_i), \quad i = 1, 2,$$

we obtain from (3.3.8) the following conditions for the threshold K:

$$1-\bar{F}_1(K) = \bar{F}_2(K).$$

i.e.,

$$\bar{F}_1(K)+\bar{F}_2(K) = 1.$$

But

$$\bar{F}_i = F_i\big((F_1+F_2)^{-1}\big),$$

and hence $\bar{F}_1(K)+\bar{F}_2(K) = K$, i.e., $K = 1$. Thus, the rule δ^{EER} is a threshold rule with respect to F_1+F_2 with the threshold 1.

Figures 3.3.1 – 3.3.4 represent the distribution functions $\bar{F}_i$ and the densities $\bar{f}_i$ of the variables S_i $(i = 1, 2)$ for

(i) $Z_i \sim N(\mu_i, \sigma^2)$, $\quad \mu_1 < \mu_2$,

and

(ii) $Z_i \sim P(\lambda_i)$, $\quad \lambda_1 > \lambda_2$.

In case (i) the functions depend upon the parameter $\gamma = |\mu_2-\mu_1|/\sigma$, while in case (ii), they depend upon $\lambda = \lambda_1/\lambda_2$. Moreover, in case (i), by the symmetry of distributions of Z_1 and Z_2, we have the equality $1-\bar{F}_1(t) = \bar{F}_2(2-t)$ for $t \in (0, 2)$, while in case (ii), this equality is true only for $t = 1$ (i.e., for the threshold of the rule δ^{EER}). The values $a_{12}(\delta^{\mathrm{EER}}) = 1-\bar{F}_1(1)$, $a_{21}(\delta^{\mathrm{EER}}) = \bar{F}_2(1)$ (given in column (2) of Table 3.2.1, and column (14) of Tables 3.2.2) can be determined from Figs. 3.3.1, 3.3.2.

Consider now the family of problems from Example 3.2.1 corresponding to P from $\mathscr{P}^{\uparrow}$. Put $\bar{h} = \bar{f}_2/\bar{f}_1$, where $\bar{f}_i$ is the density of S_i. By (3.2.21), and under the assumption (3.2.20), the Bayesian thresholds K_1^B and K_2^B of the threshold rule with respect to F_1+F_2 satisfy the equalities

$$\bar{h}(K_1^B) = \frac{\pi}{1-\pi}\,\frac{L_{10}}{L_{21}-L_{20}},$$

$$\bar{h}(K_2^B) = \frac{\pi}{1-\pi}\,\frac{L_{12}-L_{10}}{L_{20}},$$

and thus they depend upon P not only through π. If the assumption (3.2.20) is not satisfied the common threshold K^{FB} satisfies

$$\bar{h}(K^{FB}) = \frac{\pi}{1-\pi}\,\frac{L_{12}}{L_{21}}.$$

The threshold K^{FB} in Example 3.2.1B is determined in the same way. Suppose that $L_{12} = L_{21}$, $\pi = 1/2$. If $\bar{f}_1$ and $\bar{f}_2$ are symmetric (for instance, in case (i)), then $K^{FB} = 1$ (i.e., $\delta^{FB} = \delta^{EER}$). In case (ii), the threshold K^{FB} depends on λ, but the sequence of thresholds K^{FB} converges to 1 whenever $\lambda \to \infty$. The probabilities a_{12}^{FB} and a_{21}^{FB} (given in columns (2) and (3) of Tables 3.2.2 for case (ii)) are clearly equal to $1-\bar{F}_1(K^{FB})$ and $\bar{F}_2(K^{KB})$, respectively. Both values can also be determined from Fig. 3.3.2.

If

$$\alpha_1 \leqslant a_{12}^B, \quad \alpha_2 \leqslant a_{21}^B, \tag{3.3.9}$$

then the threshold rule with respect to F_1+F_2 which solves the problem from Example 3.2.2 has the tresholds

$$K_1 = \bar{F}_2^{-1}(\alpha_2), \quad K_2 = \bar{F}_1^{-1}(1-\alpha_1). \tag{3.3.10}$$

If any inequality of (3.3.9) is not satisfied, then the threshold corresponding to that inequality is equal to the respective Bayesian threshold. The thresholds K_1 and K_2 are marked in Figs. 3.3.1 – 3.3.2 for $\alpha_1 = \alpha_2 = 0.1$ and for

$$L_{12} = L_{21} < L_{10}+L_{20}, \quad \pi = \tfrac{1}{2}. \tag{3.3.11}$$

The assumption (3.3.11) implies $K_1^B = K_2^B = K^{FB}$. In the case (i), for $\gamma = 0.5, 1, 2$ both inequalities (3.3.9) are satisfied, while for $\gamma = 3$ none of them is satisfied, i.e., $K_1 = K_2 = K^{FB}$. In the case (ii), for $\lambda = 2$ and 5, both inequalities (3.3.9) are satisfied; for $\lambda = 10$ we have $a_{12}^B = 0.077$, $a_{21}^B = 0.226$, i.e., $\alpha_1 > a_{12}^B$, $\alpha_2 < a_{21}^B$, whence $K_1 = \bar{F}_2^{-1}(0.1) = 0.75$, $K_2 = K^{FB} = 1.15$.

Consider now the adaptation of the families of solutions for $P \in \mathscr{P}^{\uparrow}$. As an estimator of the discriminant function F_1+F_2 we take the statistic $g^{(n)}$ defined on $Z \times E^{(n)}$ by the formula

$$g^{(n)}(z, e^{(n)}) = \frac{n_1(z)+1}{n_1+1} + \frac{n_2(z)+1}{n_2+1}, \tag{3.3.12}$$

where, $n_i(z)$, $i = 1, 2$, denotes the number of those objects from the learning sample which belong to the i-th class and for which the values of the feature Z are less than the value z observed for the object classified, while n_i denotes the number of objects from the learning sample which belong to the i-th group. The continuity of the distribution of Z guarantees (with probability 1) that none of the objects from the learning sample takes on the same value of the feature Z as the object in question. Therefore, $n_i(z)+1$ is the rank of the object in the set consisting of that object and objects from the learning sample which belong to the i-th class. Then the statistic $g^{(n)}$ is the sum of normed ranks of the object in the respective sets of objects.

For each $P \in \mathscr{P}\uparrow$ and $z \in \boldsymbol{R}$ the sequence $(g^{(n)}(z))$ converges with probability 1 to $F_1(z)+F_2(z)$.

To indicate the dependence of the rule δ^{EER} upon P we shall use the symbol $\delta_{\mathrm{P}}^{\mathrm{EER}}$ instead of δ^{EER}. The family of rules $(\delta_{\mathrm{P}}^{\mathrm{EER}}, P \in \mathscr{P}\uparrow)$ can be adapted by threshold rules with respect to $g^{(n)}(\cdot, e^{(n)})$ with threshold 1; the sequence of such rules is consistent with $(\delta_{\mathrm{P}}^{\mathrm{EER}}, P \in \mathscr{P}\uparrow)$.

For problems from the remaining examples the adaptation of families $(\delta_P, P \in \mathscr{P}\uparrow)$ of solutions is more complicated because of the forms of the thresholds. Even in Example 3.2.2. we cannot make use of the easily adaptable thresholds (3.3.10) because of condition (3.3.9) which must be satisfied for each $P \in \mathscr{P}$.

Adaptation of families of threshold rules with respect to F_1+F_2 is particularly needed when Z is not fully observable and we can observe only the ordering of the objects according to the values of Z. The statistic $g^{(n)}$ is then observable, and the distribution functions F_1, F_2 and density functions f_1, f_2 can be estimated by means of observable statistics which makes it possible to adapt thresholds in corresponding rules.

*

Learning samples are usually useful not only for the adaptation of discriminant rules, but also for the estimation of a chosen rule performance index, e.g. the probability of false decision. However, using the same learning sample for both purposes is not advisable since it leads to placing the adapted rule in a favourable position in comparison with a situation in which two independent samples would be used, one to construct the rule and the other to estimate the index. It is true that difference between the two approaches diminishes with the increase of the size of the learning sample. In practice, however, it is difficult to evaluate when the size of the learning sample has become sufficiently large to justify neglecting that difference, and hence it is customary to apply different methods of *sample splitting* to get the estimates of the performance index.

One of the most widely used estimation methods based on sample splitting is the *leaving-one-out method.* It consists in omitting one object of the sample, and adapting the rule on the basis of the remaining $n-1$ objects. Next, the performance of the adapted rule is verified on the excluded element. In this way, we obtain successively n pairs consisting of the value of the predicted variable I for the excluded object and the prognosis $\hat{I}$ generated by the rule adapted on the basis of the remaining objects. From this sequence of pairs the estimate of the performance index is derived.

Another method frequently used is the *cross validation method.* It consists in dividing randomly a given learning sample into two parts and then adapting a given rule on the basis of the first part of the sample and verifying its performance on the basis of the second part of the sample. Next, we repeat the same procedure reversing the roles of the two parts of the sample. As a result we obtain two pairs, each consisting of the adapted rule and the estimate of its performance. These pairs together with the results derived from the whole sample give us an insight into performance of the adapted rule in question.

In estimating the performance of a rule it should be remembered that we are interested in the performance of the adapted rule and not of the solution δ_P of the corresponding probabilistic problem. Observe that for a parametric family $\mathscr{P}$ we could represent the probability of an incorrect decision for δ_P as a function of some parameters of the family $\mathscr{P}$ and then estimate those parameters by using a learning sample and, by substitution, obtain the estimation of the probability of an incorrect decision for δ_P. Such estimation is often wrongly regarded as estimating the probability of misclassification for the adapted rule.

3.4. Class separability measures

The vector Z of observable features is often composed of features chosen in a more or less arbitrary way. Now, it is desirable to prefer features of which each in itself discriminates strongly the two classes of objects, i.e., features whose distributions in the two classes are essentially different. At the same time, the features considered should be only weakly dependent since the use of a set of strongly interdependent features would be not much better than the use of a single feature, while the cost of collecting and processing data increases with the number of features. Hence, it is important to make a preliminary data reduction consisting in specifying a small set of weakly dependent features which are functions of crude data and discriminate the two classes in an appropriate way.

In some cases we content ourselves with selecting some of the primary features. Otherwise, new features are introduced by transforming the primary ones, and then the researcher should look for their interpretation which may provide a new insight into the data. Seeking for features which best discriminate classes may be considered a problem in itself and is even sometimes regarded as the main topic of research in discriminant analysis. According to this viewpoint we now present an optimal linear data reduction.

The first step of the reduction is the choice of a class separability measure. The most frequently used measure compares the dispersion of values of the vector Z with that of the vectors Z_1 and Z_2. Denote by k the dimension of Z. If $k = 1$, we have

$$\operatorname{var} Z = \pi\sigma_1^2+(1-\pi)\sigma_2^2+\pi(1-\pi)(\mu_1-\mu_2)^2, \tag{3.4.1}$$

where μ_i, σ_i^2 are, respectively, the expected value and the variance of the random variable Z_i. The total variance $\operatorname{var} Z$ consists of the *intra-class variance*

$$\sigma^2 = \pi\sigma_1^2+(1-\pi)\sigma_2^2 \tag{3.4.2}$$

and the *inter-class variance*

$$\pi(1-\pi)(\mu_1-\mu_2)^2. \tag{3.4.3}$$

We assume that $\sigma_1^2+\sigma_2^2 > 0$ and define the measure ϱ_F, called the *Fisher index*, equal to the quotient of the inter-class variance and the intra-class variance

$$\varrho_F(I, Z) = \frac{\pi(1-\pi)(\mu_1-\mu_2)^2}{\sigma^2}. \tag{3.4.4}$$

If $k > 1$, the parameters $\operatorname{var} Z$, σ_1^2, σ_2^2 are replaced by the matrices $\operatorname{cov} Z$, Σ_1, Σ_2, respectively. We assume that the matrices Σ_1, Σ_2 are positive definite. The matrix $\operatorname{cov} Z$ is the sum of matrices

$$\Sigma = \pi\Sigma_1+(1-\pi)\Sigma_2, \tag{3.4.5}$$

called the *intra-class variation matrix*, and

$$M = \pi(1-\pi)(\mu_1-\mu_2)(\mu_1-\mu_2)^T, \tag{3.4.6}$$

called the *inter-class variation matrix*. This representation generalizes the formula (3.4.1). If $\mu_1 \neq \mu_2$ the matrix M is of rank 1, and then the equation

$$|M-\lambda\Sigma| = 0 \tag{3.4.7}$$

has exactly one nonzero solution

$$\lambda = \pi(1-\pi)(\mu_1-\mu_2)^T\Sigma^{-1}(\mu_1-\mu_2), \tag{3.4.8}$$

which can be regarded as a natural generalization of formula (3.4.4). Hence, for any k, we define

$$\varrho_F(I, Z) = \lambda.$$

If we extend the definition of the Mahalanobis distance (cf. formula (2.5.7)) to distributions of (I, Z) with arbitrary positive definite covariance matrices Σ_1, Σ_2 of random vectors Z_1, Z_2 by assuming that the matrix Σ appearing in (2.5.7) is of the form (3.4.5), then by (3.4.8) we obtain

$$\varrho_F(I, Z) = \pi(1-\pi)\gamma. \tag{3.4.9}$$

Having defined the class separability measure, we choose a family $\mathscr{G}$ of functions $g\colon \mathbf{Z} \to \bar{\mathbf{Z}}$. In this family we want to find a function for which the transformed distributions of the vectors Z_1, Z_2 are maximally differentiated with respect to the chosen class separability measure. Traditionally, for the Fisher index (or equivalently the Mahalanobis distance), the family $\mathscr{G}$ consists of linear transformations

$$g(z) = Vz \tag{3.4.10}$$

where V is a matrix of k columns and s rows, $1 \leqslant s \leqslant k$, and $V\Sigma V^T$ is a nonsingular diagonal matrix of dimension $s\times s$. The latter condition means that components of $g(Z)$ are noncorrelated. The value of the measure ϱ_F is well defined for any pair (I, VZ) and, moreover, each matrix with rows $V_1^T, \ldots, V_s^T$ satisfying

$$V_1^T = (\mu_1-\mu_2)^T\Sigma^{-1},$$

and

$$V_i^T\Sigma V_j = 0 \quad \text{for} \quad i \neq j, \quad i,j = 1, \ldots, s,$$

maximizes $\varrho_F(I, VZ)$ over the family $\mathscr{G}$. For any $s \in \{1, \ldots, k\}$ the value of the Fisher index for the vector Z thus transformed is equal to its value before the transformation. Hence, with respect to ϱ_F, we can restrict ourselves to the transformations (3.4.10) for $s = 1$, and consequently we obtain only one new feature

$$Z' = (\mu_1 - \mu_2)^T \Sigma^{-1} Z.$$

The random variable Z' differs from LDF$(Z, \mu_1, \mu_2, \Sigma)$ (defined by (3.2.10)) only by a constant. Such transformation is often applied because it is convenient to describe objects by means of one feature which can easily be computed from the values of the primary features.

If the distribution P is not known, the values of Z' are not observable but can be approximated by the estimates of μ_i, Σ_i $(i = 1, 2)$ and π derived from the learning sample.

*

We have described above a certain particular case of determining the new features which best differentiate the classes of objects in a given family of transformations with respect to a given class separability measure. What are the relations between this approach to discriminant analysis and the approach based on the prediction of I by using the vector Z (cf. Section 3.2)? Is it possible to define, for a given prediction problem, the class separability measure and the family of transformations so as to obtain a solution being a function of the new features? What is the influence of data reduction on the properties of the optimal predictor?

To answer this questions we shall define class separability measures which are natural for probabilistic problems of risk minimization.

If the set Δ_P contains constant rules we compare the performance of the optimal predictor δ^* with the performance of the optimal constant predictor δ^c in the set Δ_P. Thus, if $R(\delta^c, P) > 0$, then with each distribution P of (I, Z) one can associate a measure ψ defined as

$$\psi(I, Z) = \frac{R(\delta^c, P) - R(\delta^*, P)}{R(\delta^c P)}. \tag{3.4.11}$$

The measure ψ estimates the quality of the prediction of I on the basis of Z, and will be called the *prediction power*. The prediction power takes on values in the interval $[0, 1]$; it is equal to 0 if and only if $R(\delta^c, P) = R(\delta^*, P)$ (i.e., the observation of the vector Z does not contribute to the prediction); it is equal to 1 if and only if $R(\delta^*, P) = 0$. Hence, for $L_{12} > 0$ and $L_{21} > 0$, we have $\psi(I, Z) = 1$ if and only if δ^* is a faultless rule.

EXAMPLE 3.4.1. *Prediction power for the minimization of misclassification probability when deferring decision is excluded.*

The risk is minimized for $L_{12} = L_{21} = 1$ over the set $\Delta_P = \Delta_F$. By (3.2.24), for an arbitrary Z, we have

$$\psi(I, Z) = \frac{\min(\pi, 1-\pi) - \int_Z \min\big(\pi f_1(z), (1-\pi) f_2(z)\big)\, d\nu(z)}{\min(\pi, 1-\pi)} \tag{3.4.12}$$

since the optimal constant predictor δ^c is

$$\delta^c = \begin{cases} (0, 1, 0) & \text{if} \quad \pi > 1-\pi, \\ (0, p, 1-p) & \text{if} \quad \pi = 1-\pi, \\ (0, 0, 1) & \text{if} \quad \pi < 1-\pi, \end{cases}$$

for any $p \in [0, 1]$, and since the probability of misclassification for the optimal constant predictor is equal to $\min(\pi, 1-\pi)$. Hence, for $\pi = \frac{1}{2}$, we have

$$\psi(I, Z) = 1 - \int_Z \min\big(f_1(z), f_2(z)\big)\, d\nu(z). \tag{3.4.13}$$

When $Z_i \sim N_k(\boldsymbol{\mu}_i, \Sigma)$, $i = 1, 2$, and $\pi = \frac{1}{2}$, we obtain the prediction power in the form

$$\psi(I, Z) = 2F_{N(0,1)}(\tfrac{1}{2}\gamma) - 1. \tag{3.4.14}$$

For a discrete variable Z assuming values in the set $\boldsymbol{Z} = \{1, \ldots, r\}$, the distribution P of the pair (I, Z) is given by the probabilities

$$p_{ij} = P(I = i, Z = j), \quad i = 1, 2, \quad j = 1, \ldots, r. \tag{3.4.15}$$

Put

$$p_{i\cdot} = \sum_{j=1}^{r} p_{ij}, \quad i = 1, 2. \tag{3.4.16}$$

Here, π is equal to $p_{1\cdot}$, and $f_i(j)$ is equal to $p_{ij}/p_{i\cdot}$; consequently, the prediction power [1] is equal to

$$\psi(I, Z) = \frac{\sum_{j=1}^{r} \max(p_{1j}, p_{2j}) - \max(p_{1\cdot}, p_{2\cdot})}{1 - \max(p_{1\cdot}, p_{2\cdot})} \tag{3.4.17}$$

and for $\pi = \frac{1}{2}$ we have

$$\psi(I, Z) = 1 - \sum_{j=1}^{r} \min\big(f_1(j), f_2(j)\big) = \sum_{j=1}^{r} |f_1(j) - f_2(j)|. \tag{3.4.18}$$

This parameter is known in the literature as a measure of divergence between two discrete distributions. □

Generally, each standard class of risk minimization problems over the set Δ_P defines a certain class separability measure over the set of pairs (I, Z). Now, we want to compare the values of ψ (prediction powers) for the distribution of (I, Z) and for

[1] In this case the prediction power reduces to the commonly used class separability measure usually denoted by λ_a, cf. Goodman and Kruskal (1954).

the distribution of $(I, g(Z))$, the latter being obtained when Z is reduced by a transformation g of Z. For any function g we have

$$\psi(I, Z) \geqslant \psi(I, g(Z)), \tag{3.4.19}$$

and the equality holds if and only if each solution of the problem for the pair (I, Z) is equivalent to a rule which is measurable with respect to g (in particular, the equality holds when g is a prediction sufficient statistic for (I, Z).

Consider now a family $\mathscr{G}$ of transformations of Z. We want to find a function g_{opt}, belonging to $\mathscr{G}$, which maximizes $\psi(I, g(Z))$ for $g \in G$. This corresponds to the case where we solve the original problem restricting the set Δ_p to those rules contained in it which are measurable with respect to g, $g \in \mathscr{G}$. If $\mathscr{G}$ contains a prediction sufficient statistic for (I, Z), then this statistic preserves the prediction power and thus is an optimal transformation.

Suppose now that $\mathscr{G}$ consists of linear functions of the form (3.4.10). If a distribution P has the property that there exists an increasing function whose composition with the likelihood ratio h_P belongs to $\mathscr{G}$, then this composition is an optimal function both with respect to the Fisher index and with respect to the prediction power. In particular, this is true for distributions in which $Z_i \sim N_k(\mu_i, \Sigma)$. For distributions which do not possess this property, optimal linear functions diminish the prediction power.

Loss of prediction power could be useful for evaluating the negative impact of reduction which is optimal in a given family of transformations on the properties of the optimal predictor. In practice, however, this fact is of little use since it is difficult to determine explicitly either the optimal reduction in a given class $\mathscr{G}$ or the prediction power after reduction. For instance, it is so for the family of linear functions (3.4.10).

Thus, optimal reduction is not an easy task even if P is known. The difficulties grow when part of the information on P is obtained from learning samples. In practice, the reduction is carried out by various methods of selecting and transforming primary features. However, the effects of applying these methods are in general not sufficiently known and in practice it is often necessary to rely on tradition and intuition.

*

We have introduced the prediction power for problems of risk minimization over sets Δ_P which can be represented by constraints imposed on distributions P_δ. The next example shows a case where there is no such representation of Δ_P.

EXAMPLE 3.4.2. *Prediction power for the minimization of misclassification probability when only two rules are allowed: one defined by the prior distribution and the other defined by the posterior distribution.*

We minimize the risk for $L_{12} = L_{21} = 1$ over the set $\Delta_P = \{\delta, \delta'\}$, where

$$\delta(z) \equiv (0, \pi, 1-\pi),$$
$$\delta'(z) = (0, P(I = 1 \mid Z = z), P(I = 2 \mid Z = z)).$$

The rule δ assigns the object to class 1 or class 2 according to the prior distribution of I, while δ' classifies the object according to the posterior distribution (cf. formula (3.2.5)).

We obtain the prediction power in the form

$$\psi(I, Z) = 1 - \int_Z \frac{f_1(z) f_2(z)}{\pi f_1(z) + (1-\pi) f_2(z)} \, d\nu(z). \tag{3.4.20}$$

When Z is a discrete variable considered in Example 3.4.1, the prediction power [1] is equal to

$$\psi(I, Z) = 1 - \sum_{j=1}^{r} \frac{p_{1j} p_{2j}}{p_{1\cdot}\, p_{2\cdot}\, (p_{1j} + p_{2j})}. \tag{3.4.21}$$

□

Reduction by an arbitrary prediction sufficient statistic transforms each problem from the class discussed in Example 3.4.2 into another problem from this class, and the prediction power remains unchanged.

3.5. Final remarks

Some practical discrimination problems such as proving fatherhood and problems in population genetics will be discussed in Chapters 6 and 7. These problems confirm our contention that the specific features of a given problem are decisive in solving it and a direct use of standard theoretical procedures is not possible.

In practical problems we often have to refrain from assuming that a learning sample or its subsets taken from the individual classes form a simple random sample. This affects the quality of classifications. There exist also more complex observation schemes than those described here, e.g. if no data are available for some features and/or objects, if the observation of values of some features is possible only in the case when some other features assume priviledged values, if the sample unit consists not of one object but of a set of objects (possibly with a random number of elements) chosen from the same class, etc. Moreover, in various subsets of the learning sample various schemes of observation may be available. For instance, owing to lack of coordination in hospitals the data are recorded according to various schemes. Another example of nonstandardized recording can be found in archeological research, e.g. in establishing the age of the graves in prehistoric burial grounds by means of traditional methods involving different age scales with overlapping categories.

---

(1) In this case the prediction power reduces to the commonly used class separability measure usually denoted by η_a and called measure of proportional prediction. The name comes from the fact that the prognosis is made with chances proportional to the posterior probability.

In practice preliminary data reduction is of utmost importance. This is usually done quite arbitrarily because of large number and variety of observed features, and evidently may change the final classification.

Since in practice we encounter a large variety of cases a systematic discussion of all of them is impossible. What we can do is to discuss simple typical examples. This has been done in Sections 3.2 – 3.4 and we hope that the reader will find it helpful in formalizing and solving practical problems.

In the present chapter we have dealt first with probabilistic analysis for a pair (I, Z) and next with statistical analysis for a triple $(I, Z, E^{(n)})$. In our opinion the distinction between these two cases is not sufficiently stressed in the literature and this can be a cause of misunderstandings.

It was our concern to stress the fact that in probabilistic analysis we cannot restrict ourselves to the minimization of the risk over the set Δ_F which plays a primary role in the discrimination literature. Little attention, on the other hand, is generally paid to the problem of decision deferrence which, in our opinion, is of fundamental importance. Devijver and Kittler (1982) mention decision deferrence in unconstrained risk minimization (under the assumption $L_{10} = L_{20}$, $L_{12} = L_{21}$). Anderson (1969) minimizes the probability of decision deferrence with constraints imposed on the misclassification probabilities in each class (this is referred to in Example 3.2.2). Marshall and Olkin (1968) investigate various modifications of this problem. In a modern monograph on discriminant analysis written by Hand (1981) decision deferrence is only mentioned. In fact, the author confines himself to giving some references, in particular Fukunaga (1972) and Aitchison *et al.* (1977). Broffitt (1982) observes that imposing bounds on the probabilities of the first and the second error does not ensure adequacy of decision taken in problems which admit decision deferrence. This is a continuation of some earlier papers, in particular those of Broffitt *et al.* (1976), and of Beckman and Johnson (1981). Niewiadomska-Bugaj (1987) considers bounding of the ratios of the probabilities of incorrect and correct decisions in the case of two classes (Example 3.2.3).

The theorem (mentioned in Section 3.2) on the admissibility of threshold rules in the set of all rules as well as in the set of rules determined by linear constraints is proved in Bromek and Niewiadomska-Bugaj (1987).

The basic facts concerning the adaptation of discrimination rules in parametric and nonparametric models are to be found in the already mentioned monograph by Hand (Chapters 2, 3, 4). Among other topics, Hand discusses sequential estimation of distribution parameters. In the present chapter we omit all the methods in which the learning sample or the set of features is supplemented in a sequential way.

The adaptation of the function F_1+F_2 is related to the rank discriminant functions, discussed e.g. by Niewiadomska-Bugaj (1987), Broffitt *et al.* (1976), Conover and Iman (1980), Randles *et al.* (1978).

In Section 3.3 we have omitted many popular adaptation rules, e.g. logistic ones. For the reader's information let it be said here that in logistic adaptation the family $\mathscr{P}$ consists of distributions P for which $\ln h_P(Z)$ is a linear function of the form $\beta_0+\beta'z$. This choice of the family $\mathscr{P}$ directly determines the form of the discriminant function. The family $\mathscr{P}$ includes random vectors Z composed of continuous and discrete features. The estimation of parameters β_0 and β is fairly simple from the numerical point of view (Anderson, 1982). The merits and drawbacks of other adaptation methods, especially for families of discrete distributions are discussed by Hand in Section 5.5 of his book where parametric and nonparametric methods are considered. In families of discrete distributions difficulties in the adaptation of functions f_1 and f_2 (by frequencies taken directly from the contingency table) are due to the fact that if the number of the probabilities to be estimated is very large in relation to the size of the learning sample we have $\hat{f}_1(z) = \hat{f}_2(z) = 0$ for many cells, and then the ratio $\hat{f}_2(z)/\hat{f}_1(z)$ is not defined. Logistic parametrization and related methods allows us to by-pass this difficulty. Various parametrizations of families of multidimensional binary distributions are given by Goldstein and Dillon (1978).

The notion of consistency of a sequence of discriminant rules depending on $(Z, E^{(n)})$ with a family of rules depending on Z has been introduced by Kowalczyk and Mielniczuk (1987) and is useful for description of the asymptotic properties of the adapted rule. However, it should be stressed that of particular interest are investigations of the properties of discriminant rules for learning samples of finite size.

In problems excluding decision deferrence the rule δ is fully determined by its component δ_1, which can thus be identified with the discrimination rule. Let ζ denote the density of the sample $E^{(n)}$ with respect to a measure ξ. To a rule δ_1 which depends on $(Z, E^{(n)})$ we assign a rule $\bar{\delta}_1$ which depends on Z and is given by the formula

$$\bar{\delta}_1(z) = \int_{E^{(n)}} \delta_1(z, e^{(n)})\zeta(e^{(n)})\, d\xi(e^{(n)}).$$

Since Z and $E^{(n)}$ are independent and since

$$\begin{aligned} P_{\delta_1}(\hat{I}_{\delta_1} = 1 \mid I = i) &= \int_{Z\times E^{(n)}} \delta_1(z, e^{(n)})\zeta(e^{(n)})f_i(z)\, d\xi(e^{(n)})\, d\nu(z) \\ &= \int_Z \bar{\delta}_1(z)f_i(z)\, d\nu(z) = P_{\bar{\delta}_1}(\hat{I}_{\bar{\delta}_1} = 1 \mid I = i), \end{aligned}$$

the distributions of pairs $(I, \hat{I}_{\delta_1})$ and $(I, \hat{I}_{\bar{\delta}_1})$ are identical. Hence, every standard discrimination problem for $(I, Z, E^{(n)})$ can be reduced to the corresponding problem for (I, Z), and the investigation of the properties of a given rule δ_1 can be replaced by the investigation of the corresponding rule $\bar{\delta}_1$ which depends on Z. By using this fact, it has been proved, among others, that if Z_1 and Z_2 have discrete distributions and if $L_{12} \neq L_{21}$, then the Bayesian rule δ^{FB} adapted by sample frequencies does not belong to the class of rules which are essentially complete with respect to the risk (Gnot, 1977, 1978). The proof of this result makes use of some relationships between discrimination on a finite set Z and testing a certain special family of hypotheses for exponential distributions.

Numerous investigations aim at estimating the probability of misclassification when the sample size n is fixed. There exist different approaches to this problem. For example in the case of discrete distributions Glick (1973) gives a bound of the difference between the probability of a correct decision for the rule δ^{FB} and the expected value of this probability for an adapted rule. A similar result for normal distributions is obtained by Lachenbruch and Mickey (1968), Rejtő and Revesz (1973) investigate the theoretical properties of a Bayesian rule adapted by means of kernel density estimators in the case of continuous distributions. In practice, however, the estimation is usually done by means of the leaving-one-out method.

In the literature on discriminant analysis data reduction is widely discussed. A survey of that literature, especially useful for readers interested in applications, can be found in Hand's monograph, namely in Chapter 6 on the selection of variables. Hand's book contains also a survey of various class separability measures.

If there are more than two classes, linear data reduction based on Fisher's index ϱ_F becomes more complicated than in the case of two classes described in Section 3.4. Then equation (3.4.7) has generally more than one nonzero solution which necessitates introducing several new features (called *canonical variables*) linearly related to the crude data. In practice there arise problems of estimating canonical variables and establishing their number. Interesting remarks on uncritical application of popular methods of establishing the number of canonical variables can be found in Klecka (1981).

A survey of up-to-date trends in discriminant analysis is contained in the second volume of a giant handbook of statistics edited by Krishnaiah and Kanal (1982). The volume comprises also cluster analysis and pattern recognition. It reflects the considerable development of discriminant analysis and increasing interest in it. Several monographs have appeared in this field, e.g., Anderson (1958), Cacoullos (1973), Tou and Gonzales (1972), Lachenbruch (1975), Hand (1981), Devijver and Kittler (1982).

The development of discriminant analysis is hampered, among other things, by the lack of standardized terminology. This follows not only from the existence of different traditions but also from the lack of coordination in various branches of science. This is evident particularly in the theory of indicators developed in social sciences by Nowak (1970) and Lazarsfeld (1967) and transferred to historical sciences and archeology (Fritz, 1975).

In the sequel we shall briefly present relations between theory of indicators and discriminant analysis.

Let us observe that if decisions are not allowed to be deferred or randomized then in the case of two classes a discrimination rule can be regarded as a binary observable feature. This feature is sometimes called an indicator. In the theory of indicators we consider a pair (A, B) of events, where A is observable and B is not. On the basis of the occurrence of event A the conclusion is drawn that event B has also occurred. The indicator is the characteristic function of event A.

The quality of an indicator is measured by the so-called *inclusion power*, defined as $P(A|\ B)$, and the *rejection power* defined as $P(B \mid A)$. The term "inclusion power" is motivated by the fact that if $B \subset A$ (i.e., the set of objects for which A holds includes the set of objects for which B holds), then the power of inclusion assumes its greatest value which is equal to 1. The term "rejection power" is motivated by the fact that if $A \subset B$ then among the objects about which conclusion A has been drawn there are no objects to which conclusion B does not apply. Roughly speaking, if $A \subset B$ then the indicator "faultlessly rejects objects for which the decision would be wrong".

For illustration, let B be an event consisting in examining by a medical board a person being a carrier of a contagious disease, and let A denote the result of a test on the basis of which the person examined is recognized as a carrier and is quarantined. Clearly, we want the indicator A to possess a great power of inclusion since the consequences of releasing a carrier from quarantine would be more serious than quarantining a healthy person.

On the other hand, if B means that a person brought to court is guilty, and A denotes the verdict "guilty", the jury should possess a great power of rejection. The consequences of condemning an innocent person are more serious than the consequences of exculpating a law-breaker.

Using our notation, we write the power of inclusion in the form

$$P_\delta(\hat{I}_\delta = 1 \mid I = 1) = 1 - a_{12}(\delta).$$

This is the probability of a correct decision in class 1. The power of rejection is equal to

$$P_\delta(I = 1 \mid \hat{I}_\delta = 1),$$

i.e., to the posterior probability of the fact that the object belongs to class 1 under the condition that it is assigned to this class.

In the next chapter we consider screening problems in which it is required that the solutions should have a sufficiently large power of rejection.

Chapter 4

Screening problems

Key words: *atomic screening, statistical quality control, acceptable object, screening function, lot, sublot, threshold rule.*

4.1. Introduction

In practice we often have to do with selecting certain sets S from a population. In statistical quality control a population is called a *lot*, the selected set of objects is called a *sublot*, and the process of selection is called *screening*. In what follows we use this terminology.

The aim of screening is to obtain a sublot which is, in some sense, better than the lot. Let W be a feature whose values contain all the information that is needed for the study. Usually W is not observable. Thus, another observable feature Z must be used to form a sublot.

The whole population is divided into two groups according to the values of the feature W; the objects from the first group are regarded as acceptable. Now, the aim of screening is to form a sublot in which the fraction of acceptable objects is greater than in the lot. For instance, in statistical quality control of products those products are acceptable which conform to the current standards. In periodical health examinations we select persons suspected of being ill; thus, in this screening process it is the sick persons that are "acceptable"; university entrance examinations aim at selecting sufficiently gifted persons, etc.

In the above examples the features W and Z are defined in different ways according to the character of the investigation. In the simplest case the value of W informs us only whether the object in question is acceptable or not, and therefore W is a binary feature. The distribution P of (W, Z) may be known either exactly or only with an accuracy up to a certain family $\mathcal{P}$ of distributions. In the latter case it is advisable to have a learning sample consisting of objects for which the values of W are fully or partially observable.

Among the different screening methods we distinguish those in which each object from the lot is treated in exactly the same way and the decision of including (or not) an element in a sublot is taken independently of the values of Z assumed by the remaining elements of the lot. The size of the sublot is then a random variable. Such procedures are called *atomic screening*. Nonatomic screening takes place in situations where the size of the sublot is fixed, for example, in university entrance examinations or a competion for a fixed number of posts. In this chapter we restrict ourselves to atomic screening.

In atomic screening we take into account only one element chosen randomly from the lot (population), and hence, the lot itself does not enter explicitly into our considerations. The decisions "include in the sublot" and "exclude from the sublot" can be reformulated in terms of discriminant analysis, i.e., "the object belongs to class 1 (is acceptable)" and "the object belongs to class 2 (is not acceptable)". If W is a binary feature, and either the distribution of (W, Z) is known or a learning sample is available, then atomic screening may be identified with two-class discriminant analysis. Therefore, the examples of discriminant problems given in Section 3.2 may also be treated as examples of atomic screening provided that W is a binary feature and the distribution of (W, Z) is known.

However, practical atomic screening problems require that the screening rules should satisfy other conditions than those considered in Section 3.2. For instance. the size of the sublot may be required to be as large as possible under the condition that the object included in the sublot should be in fact acceptable with sufficiently large probability. Screening problems with this requirement are considered in Section 4.2 – 4.4.

In Section 4.2 screening problems are solved for an arbitrary but known dis tribution of (W, Z). The solution is not given in an analytic form but if

$$(W, Z) \sim N_2(\mu_W, \mu_Z, \sigma_W^2, \sigma_Z^2, \varrho),$$

the solution can easily be obtained by using special tables given by Owen *et al*, (1975).

In Section 4.3 we assume that

$$(W, Z) \sim N_2(\mu_W, \mu_Z, \sigma_W^2, \sigma_Z^2, \varrho)$$

where some (or all) parameters of this distributions are unknown, and a learning sample is given. The proposed solutions are modified adaptations of the solutions of the related probabilistic problem. The modifications is aimed at controlling the expected quality of the sublot better than it would be possible by direct adaptation.

Next, in Section 4.4 the assumption concerning the distribution of (W, Z) used in Section 4.3 is considerably weakened. Namely, we only require a certain monotone dependence of W on Z. For this model we propose a screening method based on an estimator of isotonic regression. On the basis of simulation results the proposed method is compared with the solutions obtained in probabilistic and statistical models with normal distributions.

The construction of the present chapter is close to that of Chapter 3. This helps to show the relationships between discriminant analysis and atomic screening. We also use some notions and symbols introduced in Sections 3.2 and 3.3. Moreover, we refer to the notions of a standard class of problems and of a prediction sufficient statistic defined in Chapter 2.

4.2. PROBABILISTIC SCREENING PROBLEMS

In Section 3.2 we considered a population of objects described by a pair of features (I, Z), where I is assumed to be an unobservable feature with values 1 and 2 indicating to which of the two classes constituting the population a given object belongs. Classification of object is based on an observable feature Z.

Now we assume more generally that a population of objects is described by the distribution of a pair (W, Z), where W is an unobservable feature and takes on values in a certain set $\boldsymbol{W} \subset \boldsymbol{R}$. We split the set $\boldsymbol{W}$ into two disjoint subsets $\boldsymbol{W}_1$ and $\boldsymbol{W}_2$ and introduce an indicator I which indicates the class membership of an object, i.e.,

$$I = i \quad \Leftrightarrow \quad W \in \boldsymbol{W}_i, \quad i = 1, 2.$$

We denote by π the probability that $I = 1$.

Objects from class 1 are called *acceptable*. Evidently, object classification depends exclusively on the unobservable feature W. As in Section 3.2, we want to classify a given object by using the observable values of the feature Z, assuming additionally that the distribution P of (W, Z) is known. More precisely, we have a lot, consisting of N_0 elements, which is a simple random sample from (W, Z), and we select a sublot by applying a certain classification rule to each object. The required sublot should contain mainly objects from class 1 and hence the decision "include the object in the sublot" can be interpreted as "the object belongs to class 1", and the decision "do not include the object in the sublot" can be interpreted as "the object belongs to class 2".

Therefore, we deal with randomized decision rules $\delta = (\delta_0, \delta_1, \delta_2)$ defined on the set $\boldsymbol{Z}$ of values of Z (cf. the definition at the beginning of Section 3.2).

Let f denote the density of (W, Z) with respect to a certain measure which is the product of a measure λ on $\boldsymbol{W}$ and a measure ν on $\boldsymbol{Z}$. Denote by f_i, $i = 1, 2$, the density of the distribution of the random variable $(Z \mid I = i)$ (i.e., $(Z \mid W \in \boldsymbol{W}_i)$).

We have

$$f_i(z) = \frac{\int_{\boldsymbol{W}_i} f(w, z)\, \mathrm{d}\lambda}{\int_{\boldsymbol{W}_i} \int_{\boldsymbol{Z}} f(w, z)\, \mathrm{d}\nu\, \mathrm{d}\lambda.}\,.$$

We take into consideration only pairs (W, Z) such that $f_1 \neq f_2$. This excludes, in particular, the case of independent W and Z. The likelihood ratio of f_2 to f_1 is denoted by h_P:

$$h_P(z) = \frac{f_2(z)}{f_1(z)}, \qquad z \in \boldsymbol{Z}.$$

As in Section 3.2 (p. 57) we assign to each distribution P of (W, Z) and to each rule δ a pair of random variables $(I, \hat{I}_\delta)$ with the distribution P_δ. Hence, $\hat{I}_\delta = 1$ if an object is included in the sublot and $\hat{I}_\delta = 2$, otherwise. We do not take into account the possibility of decision deferment and thus we have $\delta_0(z) \equiv 0$ and $P_\delta(\hat{I}_\delta = 0) = 0$. Therefore, rule δ is uniquely determined by its component δ_1 (the probability of including an object in the sublot), called the *screening function*,

For an arbitrary rule δ we denote by $\theta_1(\delta)$ the conditional probability that an object belongs to class 1 under the condition that it has been acknowledged as an object of class 1, i.e.,

$$\theta_1(\delta) = P(I = 1 \mid \hat{I}_\delta = 1) = \frac{\int\limits_{W_1}\int\limits_{Z} \delta_1(z) f(w, z)\, \mathrm{d}\nu\, \mathrm{d}\lambda}{\int\limits_{W}\int\limits_{Z} \delta_1(z) f(w, z)\, \mathrm{d}\nu\, \mathrm{d}\lambda}. \tag{4.2.1}$$

Hence, $\theta_1(\delta)$ is the expected value of the fraction of objects from class 1 in the sublot obtained by means of rule δ, whereas the expected fraction of objects from class 1 in the lot $P(I = 1)$ is equal to π.

Let θ_1^* be a number chosen from the interval $(\pi, 1)$. This number defines the smallest admissible value of the probability $\theta_1(\delta)$, i.e., we consider only rules δ for which

$$\theta_1(\delta) \geqslant \theta_1^*. \tag{4.2.2}$$

In the set of rules satisfying (4.2.2) we seek those rules for which the size of the sublot is maximal. Denote by $\theta_0(\delta)$ the probability that an arbitrary element of the lot (i.e., an element randomly selected from the lot) is included in the sublot. We have

$$\theta_0(\delta) = P_\delta(\hat{I}_\delta = 1) = \int\limits_{W}\int\limits_{Z} \delta_1(z) f(w, z)\, \mathrm{d}\nu\, \mathrm{d}\lambda.$$

A rule δ is the solution of the problem

$$\theta_0(\delta) = \max\{\theta_0(\delta') \colon \theta_1(\delta') \geqslant \theta_1^*\}. \tag{4.2.3}$$

Condition (4.2.3) is equivalent to the condition

$$\theta_0(\delta) = \max\{\theta_0(\delta') \colon \theta_1(\delta') = \theta_1^*\}. \tag{4.2.4}$$

The class of screening problems here considered is similar to the classes of risk minimization problems which were treated in Section 3.2. There the risk was minimized over a given set Δ_P. Here, we seek the maximum of the function $P_\delta(\hat{I}_\delta = 1)$ on the set

$$\Delta_P = \{\delta = (0, \delta_1, 1-\delta_1) \colon P_\delta(I = 1 \mid \hat{I}_\delta = 1) \geqslant \theta_1^*\}.$$

In practice, this formulation of the problem means that we want to include in the sublot the largest possible number of elements provided that the expected value of the fraction of elements of class 1 is sufficiently large in the sublot.

This class of screening problems is standard in the sense of the definition given in Section 2.3. As for the problems presented in Section 3.2, any solution of a screening problem is a threshold rule with respect to the ratio h_P. Namely, the screening function is of the form

$$\delta_1(z) = \begin{cases} 0 & \text{if} \quad h_P(z) > \varkappa, \\ p & \text{if} \quad h_P(z) = \varkappa, \\ 1 & \text{if} \quad h_P(z) < \varkappa, \end{cases}$$

for $\varkappa \in \boldsymbol{R}^+$ and $p \in [0, 1]$ satisfying the condition

$$\frac{P(I = 1, h_P(Z) < \varkappa) + pP(I = 1, h_P(Z) = \varkappa)}{P(h_P(Z) < \varkappa) + pP(h_P(Z) = \varkappa)} = \theta_1^* \tag{4.2.5}$$

If

(i) W_1 *is an interval* (L, ∞), *where* $-\infty < L < \infty$;

(ii) *for some constants* $\varrho_1, \ldots, \varrho_k$ *and a certain positive constant* σ^2, *the density of* W *under the condition that* $Z = (z_1, z_2, \ldots, z_n)$ *is equal to*

$$f(w \mid z_1, \ldots, z_k) = \frac{1}{\sqrt{2\pi}} \exp\left(-\frac{1}{2\sigma^2}\left(w - \sum_{i=1}^{k} \varrho_i z_i\right)^2\right),$$

then, for any distribution of vector Z, *we have*

$$h_P(z) = \frac{\dfrac{\pi}{1-\pi} F_{N(0,1)}\left(\dfrac{1}{\sigma}\left(L - \sum_{i=1}^{k} \varrho_i z_i\right)\right)}{1 - F_{N(0,1)}\left(\dfrac{1}{\sigma}\left(L - \sum_{i=1}^{k} \varrho_i z_i\right)\right)}.$$

Under the assumptions (i), (ii) h_P is a decreasing function of the regression function of W on Z, which is, in turn, equal to the linear combination $\sum_{i=1}^{k} \varrho_i Z_i$. Therefore, solutions of screening problems can be represented as threshold rules with respect to that combination.

If Z is a univariate random variable, the screening function δ_1 is a threshold function with respect to Z, provided h_P is monotone. The ratio h_P is decreasing if and only if, for any $z, z' \in \boldsymbol{Z} \subset \boldsymbol{R}$, we have

$$z > z' \Rightarrow P(I = 1 \mid Z = z) > P(I = 1 \mid Z = z'), \tag{4.2.6}$$

and is increasing if and only if the implication

$$z > z' \Rightarrow P(I = 1 \mid Z = z) < P(I = 1 \mid Z = z').$$

holds for any $z, z' \in \boldsymbol{Z} \subset \boldsymbol{R}$.

In the sequel we shall consider problems in which an object is said to belong to class 1 if the corresponding value of W belongs to the interval (L, ∞), for a given L. Moreover, it will be assumed that Z is a continuous univariate random variable and h_P is decreasing. Then, by the continuity of Z, we can choose a solution with the screening function of the form

$$\delta_1(z) = \begin{cases} 0 & \text{if} \quad z < t, \\ 1 & \text{if} \quad z \geqslant t. \end{cases}$$

Hence, the rule $\delta = (0, \delta_1, 1-\delta_1)$ is determined by the threshold t for the screening function δ_1. For such rules we use the symbols $\theta_0(t)$, $\theta_1(t)$ instead of $\theta_0(\delta)$ and $\theta_1(\delta)$, respectively:

$$\begin{aligned} \theta_0(t) &= P(Z \geqslant t), \quad t \in \boldsymbol{Z}, \\ \theta_1(t) &= P(W > L \mid Z \geqslant t). \end{aligned} \tag{4.2.7}$$

By (4.2.5), the threshold t_0 which is the solution of the optimization problem under consideration can be determined from the equality

$$\theta_1(t_0) = P(W > L \mid Z \geqslant t_0) = \theta_1^*. \tag{4.2.8}$$

Observe that if h_P were increasing (instead of decreasing) then, for a univariate continuous variable Z, the optimal screening function would be of the form

$$\delta_1(z) = \begin{cases} 0 & \text{if} \quad z > t_0', \\ 1 & \text{if} \quad z \leqslant t_0', \end{cases}$$

where t_0' satisfies (4.2.8) for the pair $(W, -Z)$.

*

If $(W, Z) \sim N_2(\mu_W, \mu_Z, \sigma_W^2, \sigma_Z^2, \varrho)$, the likelihood ratio h_P is decreasing for $\varrho > 0$ and increasing for $\varrho < 0$. The problem is solvable for any L and θ_1^*. We present an algorithm for finding the optimal screening threshold t_0 for the distribution parameters μ_W, μ_Z, σ_W^2, σ_Z^2, δ and the problem parameters L, θ_1^*.

Let $P^{(\varrho)}$ denote the distribution $N_2(0, 0, 1, 1, \varrho)$ and let $(W', Z') \sim P^{(\varrho)}$. Since $((W-\mu_W)/\sigma_W,\ (Z-\mu_Z/\sigma_Z) \sim P^{(\varrho)}$, we have

$$\theta_1(t_0) = P(W > L \mid Z \geqslant t_0) = P^{(\varrho)}(W' > -q_\pi \mid Z' \geqslant -q_\beta),$$

where $\pi = P(W > L)$, $\beta = \theta_0(t_0)$, and q_π, q_β denote the quantiles $q_{N(0,1)}(\pi)$, $q_{N(0,1)}(\beta)$, respectively. We proceed as follows:

(i) for given μ_W, σ_W, L we determine

$$\pi = P(W > L) = 1 - F_{N(0,1)}\left(\frac{L-\mu_W}{\sigma_W}\right); \tag{4.2.9}$$

(ii) for given ϱ, π, θ_1^* we determine β from the condition

$$P^{(\varrho)}(W' > -q_\pi \mid Z' \geqslant -q_\beta) = \theta_1^*; \tag{4.2.10}$$

(iii) for given β, μ_Z, σ_Z we calculate

$$t_0 = \mu_Z - q_\beta \sigma_Z. \tag{4.2.11}$$

It follows from (i) – (iii) that the threshold t_0 depends on (μ_W, σ_W, L) only through π, which is determined in step (i). In step (ii) we can use a tabulated function g (cf. Owen *et al.*, 1975), which assigns to the triple ϱ, π, θ_1^* a number β satisfying (4.2.10). Function g has the following properties:

1° for a fixed pair (ϱ, π) the function $g(\varrho, \pi, \cdot)$ is a decreasing function of θ_1^* (i.e., the smaller is the lower bound θ_1^* of the probability that a given object included in the sublot belongs to class 1 the larger the probability β that the object is included in the sublot);

2° for a fixed pair (ϱ, θ_1^*) the function $g(\varrho, \cdot, \theta_1^*)$ is an increasing function of π;

3° for a fixed pair (π, θ_1^*) the function $g(\cdot, \pi, \theta_1^*)$ is an increasing function of ϱ.

The algorithm (i) – (iii) of finding t_0 can also be applied if

$$(G(W), G(Z)) \sim N_2(\mu_1, \mu_2, \sigma_1^2, \sigma_2^2, \varrho)$$

for a certain increasing function G. L is then replaced by $G(L)$, the numbers π and β are determined according to (i) and (ii) and, finally, the screening threshold is calculated from the formula

$$t_0 = G^{-1}(\mu_2 - q_\beta \sigma_2).$$

4.3. Statistical screening problems in a normal model

Assume now that $(W, Z) \sim N_2(\mu_W, \mu_Z, \sigma_W^2, \sigma_Z^2, \varrho)$ and not all distribution parameters are known. Moreover, we have at our disposal a learning sample given in the form of a simple random sample of size n, i.e., in the form of a sequence $E^{(n)} = (W^{(1)}, Z^{(1)}), \ldots, (W^{(n)}, Z^{(n)})$ of independent pairs of random variables, the distribution of $(W^{(i)}, Z^{(i)})$ coinciding for any i with that of (W, Z).

As stated in Section 4.2, if the likelihood ratio is monotone and the distribution of Z is continuous, the solution of a probabilistic screening problem is defined by a threshold t_0. In statistical problems we adapt the solution by finding an estimator of the threshold, say $\hat{t}_0^{(n)}$, which is a function of sample $E^{(n)}$. We require the sequence of adapted threshold $\hat{t}_0^{(n)}$ to converge to t_0 with probability 1. Then the sequence of adapted solutions of the statistical screening problems is consistent with the family of the probabilistic solutions (here consistency is understood in the sense of the definition given in Section 3.3).

We start with the adaptation of the algorithm (i) – (iii). Observe that the optimal threshold (4.2.11) depends on μ_W and σ_W only through the parameter π. If any of the parameters π, μ_Z, σ_Z^2, ϱ is not known, we determine the corresponding estimate from the sample. Next, by using the algorithm (i) – (iii) we calculate the adapted threshold $\hat{t}_0^{(n)}$. If the sequences of estimators of the unknown parameters converge to those parameters with probability 1, then the sequence of adapted thresholds $(\hat{t}_0^{(n)})$ converges to t_0 with probability 1.

The probability $\theta_1(\hat{t}_0^{(n)})$ that an element from the sublot is acceptable is a random variable which depends on the learning sample. Clearly, if $n \to \infty$ the sequence $\theta_1(\hat{t}_0^{(n)})$ converges to $\theta_1(t_0)$ and therefore condition (4.2.3) is asymptotically satisfied. For an arbitrary fixed sample size n, on the other hand, condition (4.2.4) takes the form

$$E\big(\theta_1(\hat{t}_0^{(n)})\big) = \theta_1^* \tag{4.3.1}$$

and is not satisfied in our case. Now the question arises: whether we can find a sequence of rules which satisfies condition (4.3.1) for an *arbitrary* n, and which is consistent with the family of probabilistic solutions with the threshold (4.2.11). Such a sequence of rules can easily be constructed in the following three cases: the only unknown parameter is μ_Z; the only unknown parameter is σ_Z; the unknown parameters are both μ_Z and σ_Z.

In the first case we adapt the algorithm (i) – (iii) by replacing μ_Z with the value $\bar{z}$ of the variable

$$\bar{Z} = \frac{1}{n} \sum_{i=1}^{n} Z^{(i)}.$$

Thus, the screening threshold is of the form

$$\bar{Z}-q_{\beta}\sigma_Z. \tag{4.3.2}$$

To satisfy condition (4.3.1) it is necessary to modify the algorithm through replacing q_β by

$$q = \sqrt{\frac{n+1}{n}}\; q_{\bar{\beta}} \tag{4.3.3}$$

where

$$\bar{\beta} = g\left(\varrho\sqrt{\frac{n}{n+1}}, \pi, \theta_1^*\right).$$

This follows from the form of the algorithm (i) – (iii) and the fact that

$$\left(\frac{W-\mu_W}{\sigma_W}, \sqrt{\frac{n}{n+1}}\,\frac{Z-\bar{Z}}{\sigma_Z}\right) \sim N_2\left(0, 0, 1, 1, \varrho\sqrt{\frac{n}{n+1}}\right).$$

In the second case, where σ_Z is the only unknown parameter, we can estimate it by means of the statistic

$$S_0 = \left(\frac{1}{n}\sum_{i=1}^{n}(Z^{(i)}-\mu_Z)^2\right)^{1/2}.$$

Next, in the family of rules with the random threshold $\big((\mu_Z-qS_0),\ q\in \boldsymbol{R}\big)$ we seek a rule for which

$$E(\theta_1(\mu_Z-qS_0)) = \theta_1^*.$$

Doing this, we exploit the fact that the distribution of the pair

$$(W'', Z'') = \left(\frac{W-\mu_W}{\sigma_W}, \frac{Z-\mu_Z}{S_0}\right)$$

depends only on ϱ and n since

$$Z'' \sim St(n, 0, 1),$$

$$(W'' \mid Z'' = z'') \sim N\big(z'', \sqrt{1-\varrho^2}\big).$$

Denote the distribution of (W'', Z'') by $P^{(\varrho, n)}$ and let g' be a function assigning to any quadruple $(\varrho, \pi, \theta_1^*, n)$ a number q satisfying an equality which is similar to (4.2.10), namely

$$P^{(\varrho,n)}(W'' > -q_\pi \mid Z'' \geqslant -q) = \theta_1^*. \tag{4.3.4}$$

The function g' is tabulated in Owen and Haas (1978). The required threshold is then of the form $\mu_Z - g'(\varrho, \pi, \theta_1^*, n)S_0$.

In the case where μ_Z and σ_Z are unknown, they are estimated by means of $\bar{Z}$ and S, where

$$S = \left(\frac{1}{n-1}\sum_{i=1}^{n}(Z^{(i)}-\bar{Z})^2\right)^{1/2}.$$

Then the rule with the threshold $\bar{Z}-qS$ satisfies the condition

$$E(\theta_1(\bar{Z}-qS)) = \theta_1^*$$

for

$$q = \sqrt{\frac{n+1}{n}}\, g'\left(\varrho\sqrt{\frac{n}{n+1}}\,, \pi, \theta_1^*, n-1\right).$$

This follows from the fact that

$$Z'' = \sqrt{\frac{n}{n+1}}\frac{Z-\bar{Z}}{S} \sim St(n-1, 0, 1),$$

$$(W'' \mid Z'' = z'') \sim N\left(z'', \sqrt{1-\varrho^2\frac{n}{n+1}}\right).$$

Rules satisfying condition (4.3.1) can also be constructed in cases where other parameters are unknown. Then the formulae are more complicated and we shall not present them here. We shall define for those cases, however, rules with thresholds $t^{(n)}$ depending on the learning sample where $\theta_1(t^{(n)})$ is not less than θ_1^* with probability equal to a given number η. More precisely, for a given confidence level η and arbitrary $\mu_W, \mu_Z, \sigma_W, \sigma_Z, \varrho$ a random threshold $t^{(n)}$ satisfies the equality:

$$P^n(\theta_1(t^{(n)}) \geqslant \theta_1^*) = \eta, \tag{4.3.5}$$

where P^n is the n-th power of the distribution $N_2(\mu_W, \mu_Z, \sigma_W^2, \sigma_Z^2, \varrho)$. We shall present sequences of rules with the threshold $t^{(n)}$ which are consistent with the family of probabilistic solutions with the threshold (4.2.11) and which satisfy (4.3.5).

To begin with, assume that ϱ is the only unknown parameter. By property (iii) of the function g (as stated in Section 4.2), condition (4.3.5) is satisfied whenever $q = q_{\bar{\beta}}$ with $\bar{\beta} = g(\hat{\varrho}_0, \pi, \theta_1^*)$ and $\hat{\varrho}_0$ is the lower confidence bound for ϱ at level η. Hence

$$P^n(\hat{\varrho}_0 \leqslant \varrho) = \eta.$$

Thus we adapt the probabilistic rule with threshold (4.2.11). This adaptation consists in replacing ϱ by $\hat{\varrho}_0$.

We proceed in a similar way if the parameters μ_Z, σ_Z, ϱ are known, and π is unknown. We then use the lower confidence bound $\hat{\pi}_0$ of the confidence interval for π. This bound is determined in different ways according to which of the parameters μ_W, σ_W is unknown. If, for instance, μ_W is unknown, then $\hat{\pi}_0$ is given by the formula

$$q_{\hat{\pi}_0} = -\frac{q_\eta}{\sqrt{n}} + \frac{\bar{W}-L}{\sigma_W}, \tag{4.3.6}$$

where

$$\bar{W} = \frac{1}{n}\sum_{i=1}^{n} W^{(i)}.$$

In the case where either ϱ or π is unknown and at the same time μ_Z is unknown, we use the rule derived in the case of μ_Z unknown by substituting for ϱ or π the lower bounds of the confidence intervals. For example, when (μ_W, μ_Z) are unknown, the random threshold is of the form

$$\bar{Z}-\sqrt{\frac{n+1}{n}}\, q_{\bar{\beta}}\,\delta_Z, \tag{4.3.7}$$

where

$$\bar{\beta} = g\left(\varrho\sqrt{\frac{n}{n+1}}, \hat{\pi}_0, \theta_1^*\right),$$

and $\hat{\pi}_0$ is defined by (4.3.6).

It is clear that as far as condition (4.3.1) is concerned the proposed rules differ only a little from the adapted rule. On the other hand, the rules satisfying (4.3.5) differ considerably from the adapted rule. It is true that condition (4.3.5) enforces the required quality of the sublot, but it is done at the cost of restricting the number of elements assigned to it.

4.4. Screening in a nonparametric model

Consider a nonparametric model in which W and Z are continuous and satisfy condition (4.2.6), and $Z = (a, b)$ where $a \geqslant -\infty$, $b \leqslant +\infty$. Let us assume that the likelihood ratio h_P is decreasing (if h_P is increasing, the pair $(W, -Z)$ should be considered). Moreover, we assume $I = 1$ if and only if $W > L$. It follows from these assumptions that the function θ_1 given by formula (4.2.7) is nondecreasing and the optimal threshold t_0 appearing in the probabilistic model is determined by the equality $\theta_1(t_0) = \theta_1^*$.

As previously, we assume that the learning sample $E^{(n)} = ((W^{(1)}, Z^{(1)}), \ldots$ $\ldots, (W^{(n)}, Z^{(n)}))$ is a simple random sample from the population and is fully observable.

We seek an estimator $\hat{t}_0^{(n)}$ of t_0 derived from the learning sample $E^{(n)}$ such that the sequence $\hat{t}_0^{(n)}$ converges to t_0 with probability 1. To this aim we construct a certain estimator $\theta_1^{(n)}$ of θ_1 whose realizations are nondecreasing step functions continuous on the left with possible jumps appearing only at points which are values of the order statistics $Z_{(1)}^n, \ldots, Z_{(n)}^n$ for the sequence $(Z^{(1)}, \ldots, Z^{(n)})$. Next we find the smallest jump point z_0 for which $\theta_1^{(n)}(z_0) \geqslant \theta_1^*$. If $\theta_1^{(n)}(z_0) = 1$ we put $t_0^{(n)} = z_0$, and if $\theta_1^{(n)}(z_0) < 1$ we put $t_0^{(n)}$ to be equal to the arithmetic mean of the values of those order statistics for which the function $\theta_1^{(n)}$ takes on the value $\theta_1^{(n)}(z_0)$. In Fig. 4.4.1 the function $\theta_1^{(n)}$ is presented for a certain sample for $n = 50$ and the corresponding value of the estimator $t_0^{(n)}$ for $\theta_1^* = 0.90$ and $\theta_1^* = 0.95$.

The formal definition of $t_0^{(n)}$ is as follows. First we introduce the sequence

$$u_{n,j} = \frac{\sum_{i=1}^{n} \chi_{\{W^{(i)}>L,\, Z^{(i)}>Z_{(j)}^n\}}}{\sum_{i=1}^{n} \chi_{\{Z^{(i)}\geqslant Z_{(j)}^n\}}}, \qquad j = 1, \ldots, n,$$

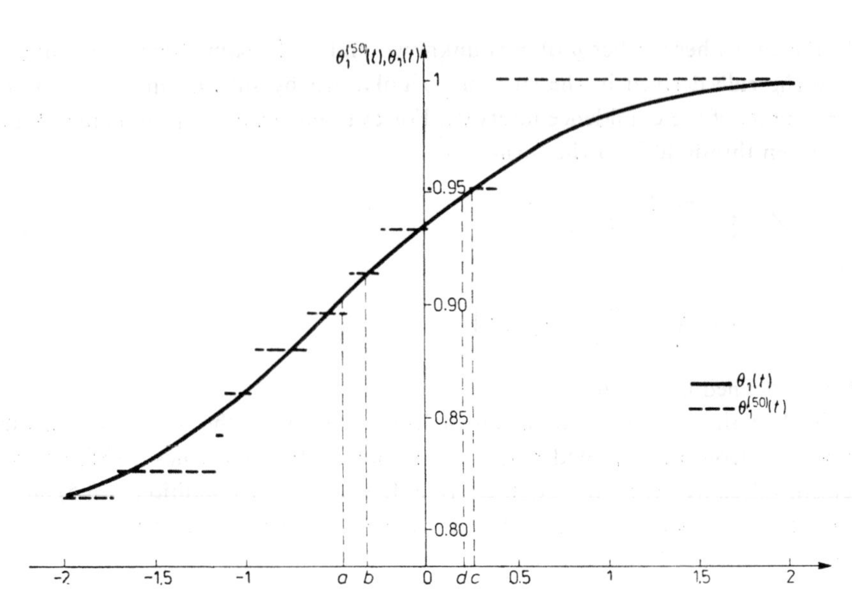

FIG. 4.4.1. The function θ_1, for $(W, Z) \sim N_2(0, 0, 1, 1, 0.6)$, and its estimator $\theta_1^{(50)}$ for sample $E^{(50)}$ and the corresponding screening thresholds: $a = t_0$ for $\theta_1^* = 0.90$, $b = t_0^{(50)}$ for $\theta_1^* = 0.90$, $c = t_0$ for $\theta_1^* = 0.95$, $d = t_0^{(50)}$ for $\theta_1^* = 0.95$.

and hence the sequence

$$u_{n,j}^{\circ} = \min_{w \geqslant j} \max_{s \leqslant j} \left(\sum_{r=s}^{w} \frac{u_{n,r}}{w-s+1} \right), \quad j = 1, \ldots, n.$$

The second sequence is nondecreasing and satisfies the condition

$$\sum_{j=1}^{n} (u_{n,j} - u_{n,j}^{\circ})^2 = \min_{\{\tilde{u}_{n,j}\}} \sum_{j=1}^{n} (u_{n,j} - \tilde{u}_{n,j})^2.$$

where $\{\tilde{u}_{n,j}\}$ is an arbitrary nondecreasing sequence. If $\{u_{n,j}\}$ is monotone, then

$$u_{n,j}^{\circ} = u_{n,j}, \quad j = 1, \ldots, n.$$

The sequence $\{u_{n,j}^{\circ}\}$ is called the *isotonic regression of the sequence* $\{u_{n,j}\}$ (the properties of the isotonic regression of an arbitrary finite sequence can be found in Barlow *et al.*, 1972).

The estimator $\theta_1^{(n)}$ is defined as

$$\theta_1^{(n)}(t) = \begin{cases} u_{n,1}^{\circ} & \text{if} \quad a < t \leqslant Z_{(1)}^n, \\ u_{n,j}^{\circ} & \text{if} \quad Z_{(j-1)}^n < t \leqslant Z_{(j)}^n, \quad j = 2, \ldots, n \\ u_{n,n}^{\circ} & \text{if} \quad Z_{(n)}^n < t \leqslant b. \end{cases}$$

Let $r(n)$ and $l(n)$ be statistics taking on the values $1, \ldots, n$ and $0, 1, \ldots, n-r(n)$, respectively, and let

$$r(n) = \begin{cases} 1 & \text{if} \quad u_{n,1}^{\circ} \geqslant \theta_1^*, \\ j & \text{if} \quad u_{n,j-1}^{\circ} < \theta_1^* \leqslant u_{n,j}^{\circ}, \\ n & \text{if} \quad u_{n,n}^{\circ} < \theta_1^*, \end{cases}$$

$$l(n) = \begin{cases} j & \text{if} \quad u^{\circ}_{n,r(n)+j} = u^{\circ}_{n,r(n)} \\ & \quad \text{and } (u^{\circ}_{n,r(n)+j+1} > u^{\circ}_{n,r(n)}), \\ n-r(n) & \text{if} \quad u^{\circ}_{n,r(n)} = u^{\circ}_{n,n} < 1, \\ 0 & \text{if} \quad u^{\circ}_{n,r(n)} = u^{\circ}_{n,n} = 1. \end{cases}$$

We now define the estimator $t_0^{(n)}$ by the formula

$$t_0^{(n)} = \frac{1}{l(n)+1} \sum_{s=r(n)}^{r(n)+l(n)} Z^n_{(s)}. \tag{4.4.1}$$

Hence, $r(n)$ is the index of the order statistic whose value is the required smallest jump point z_0 and $l(n)+1$ is the number of order statistics whose arithmetic mean is equal to $t_0^{(n)}$.

For example, let $L = -0.8411$ and consider, for $(W, Z) \sim N_2(0, 0, 1, 1, 0.6)$, the following simple random sample $E^{(50)}$:

$$\begin{aligned} E^0 = \; & (-0.062, \; 0.372), && (0.437, -0.011), && (-0.678, -1.614), \\ & (-1.141, -1.132), && (0.951, -0.464), && (-0.153, -0.377), \\ & (0.425, \; 0.563), && (0.898, \; 0.616), && (1.314, \; 0.347), \\ & (-1.618, -2.626), && (-1.650, -1.077), && (-0.778, -0.494), \\ & (-0.311, \; 0.789), && (1.468, \; 0.401), && (-0.797, \; 0.131), \\ & (-0.953, -2.043), && (-0.769, -0.433), && (0.465, \; 0.609), \\ & (2.076, \; 0.414), && (0.514, \; 0.734), && (-0.417, \; 0.404), \\ & (-0.314, -0.916), && (0.238, -0.364), && (0.161, \; 0.486), \\ & (-1.297, -1.808), && (-0.591, \; 0.931), && (0.526, \; 2.816), \\ & (2.544, \; 1.314), && (1.083, -0.107), && (-0.847, -0.044), \\ & (0.688, -0.210), && (-1.437, -0.410), && (0.812, \; 1.048), \\ & (0.155, -1.237), && (-1.361, -1.126), && (-1.418, -0.632), \\ & (0.413, \; 1.484), && (2.385, \; 0.869), && (1.019, \; 0.759), \\ & (0.734, \; 0.666), && (1.165, \; 0.637), && (-0.909, \; 0.379), \\ & (0.538, \; 0.364), && (0.907, -0.543), && (-1.045, -0.340), \\ & (-0.455, \; 0.037), && (-0.064, -0.074), && (0.003, -0.111), \\ & (0.882, -0.290), && (-0.339, -0.038). \end{aligned}$$

In Table 4.4.1 we present the results of estimation of the threshold t_0 by $t_0^{(50)}$ for this sample, where θ_1^* is equal to 0.90 and to 0.95. The threshold t_0 was determined from the formula (4.2.11) for $\varrho = 0.6$ and $\pi = 0.8$.

Figure 4.4.1 shows, for sample $E^{(50)}$ where θ_1^* is equal to 0.90 and to 0.95, the graphs of the function θ_1 for $(W, Z) \sim N_2(0, 0, 1, 1, 0.6)$ and of its estimate $\theta_1^{(50)}$ and also the thresholds t_0 and its estimate $t_0^{(50)}$.

To compare the parameters characterizing screening performed by the rule with threshold (4.4.1) with those corresponding to selected rules described in Section 4.2 – 4.3 simulation methods were used.

TABLE 4.4.1.
Sequences $\{u_{50,j}\}$, $\{u^{\circ}_{50,j}\}$, thresholds t_0 and $t_0^{(50)}$, for $\theta_1^* = 0.90$ and 0.95, and sample $E^{(50)}$.

i	$u_{50,j}$	$u^{\circ}_{50,j}$	θ_1^*	t_0	$r(50)$	$l(50)$	$t_0^{(50)}$
1	0.780	0.780					
2	0.796	0.896					
3	0.813	0.813					
4	0.830	0.826	0.90	−0.498	16	2	$\frac{1}{3}(Z^{50}_{(16)}+Z^{50}_{(17)}+Z^{50}_{(18)}) = -0.36$
5	0.826	0.826					
6	0.822	0.826					
7	0.841	0.841					
8	0.860	0.860	0.95	0.23	25	7	$\frac{1}{8}(Z^{50}_{(25)}+\ \dots\ +Z^{50}_{(32)}) = 0.198$
9	0.881	0.880					
10	0.878	0.880					
11	0.900	0.894					
12	0.897	0.894					
13	0.895	0.894					
14	0.892	0.894					
15	0.889	0.894					
16	0.914	0.912					
17	0.912	0.912					
18	0.909	0.912					
19	0.938	0.932					
20	0.935	0.932					
21	0.933	0.932					
22	0.931	0.932					
23	0.929	0.932					
24	0.926	0.932					
25	0.962	0.955					
26	0.960	0.955					
27	0.958	0.955					
28	0.956	0.955					
29	0.954	0.955					
30	0.952	0.955					
31	0.950	0.955					
32	0.947	0.955					
33 – 50	1	1					

The investigations comprised

(i) the rule with the threshold given by (4.2.11) (i.e., the optimal rule for normal distribution with known parameters); this rule will be denoted by N;

(ii) the rule with the random threshold $\bar{Z}-q\sigma_Z$ and the coefficient q defined by (4.3.3) (i.e., the optimal rule for the normal distribution with unknown mean value μ_Z); this rule will be denoted by $\mathrm{N}|\mu_Z$;

(iii) the rule with the random threshold defined by (4.4.1) (i.e., the optimal rule for a continuous distribution of (W, Z) with the monotone likelihood ratio); this rule will be denoted by MI.

For any rule $\delta = \mathrm{N}, \mathrm{N}|\mu_Z$, MI the threshold will be denoted by $t_0[\delta]$.

The behaviour of these rules was investigated in the case of (W, Z) having a normal distribution $N_2(\mu_W, \mu_Z, \sigma_W^2, \sigma_Z^2, \varrho)$ and in the case of (W, Z) having a certain two-dimensional gamma distribution with six parameters: $\alpha, \mu_W, \mu_Z, \sigma_W^2, \sigma_Z^2, \varrho$, where $\alpha > 0$, $0 < \varrho \leqslant 1$. This gamma distribution will be denoted by $\Gamma_2(\alpha, \mu_W, \mu_Z, \sigma_W^2, \sigma_Z^2, \varrho)$, or shortly Γ_2. We define it as follows: (W, Z) has the distribution $\Gamma_2(\alpha, \mu_W, \mu_Z, \sigma_W^2, \sigma_Z^2, \varrho)$ if

$$\begin{aligned} W &= \frac{\sigma_W}{\sqrt{\alpha}}(X_0+X_1)+\mu_W-\sigma_W\sqrt{\alpha}\,, \\ Z &= \frac{\sigma_Z}{\sqrt{\alpha}}(X_0+X_2)+\mu_Z-\sigma_Z\sqrt{\alpha}\,, \end{aligned} \tag{4.4.2}$$

where (X_0, X_1, X_2) are independent random variables with distributions $\Gamma(\alpha\varrho)$, $\Gamma(\alpha(1-\varrho))$, $\Gamma(\alpha(1-\varrho))$, respectively. The parameters $\mu_W, \mu_Z, \sigma_W^2, \sigma_Z^2, \varrho$ denote the means and the variance of the variables W, Z and their correlation coefficient, respectively. The parameter α defines how close Γ_2 is to the corresponding distribution N_2, since for $\alpha \to \infty$ the sequence of distributions $\Gamma_2(\alpha, \mu_W, \mu_Z, \sigma_W^2, \sigma_Z^2, \varrho)$ weakly converges to $N_2(\mu_W, \mu_Z, \sigma_W^2, \sigma_Z^2, \varrho)$. The marginal distributions in Γ_2 are linear transformations of the distribution Γ:

$$\left(\sqrt{\alpha}\,\frac{W-\mu_W}{\sigma_W}+\alpha\right) \sim \left(\sqrt{\alpha}\,\frac{Z-\mu_Z}{\sigma_Z}+\alpha\right) \sim \Gamma(\alpha).$$

The distribution of the pair $\big((W-\mu_W)/\sigma_W, (Z-\mu_Z)/\sigma_Z\big)$ is symmetric. The regression function $E(W \mid Z = z)$ is a linear function of z. For an arbitrary L, the likelihood ratio of the variable $Z_2 = (Z \mid W \leqslant L)$ and the variable $Z_1 = (Z \mid W > L)$ is a decreasing function, i.e., distribution Γ_2 satisfies condition (4.2.6) for each L. Thus, the distribution Γ_2 exemplifies situations in which the variables W and Z are right-skew and positive dependent, and satisfy condition (4.2.6). Such distributions often appear in statistical quality control.

When generating realizations of (W, Z) according to Γ_2 we exploit the fact that, for natural γ, $\Gamma(\gamma)$ is a distribution of the sum of γ independent random variables with the exponential distribution with parameter 1. Hence, if $\alpha\varrho$ and $\alpha(1-\varrho)$ are natural numbers, then X_0, X_1, X_2 appearing in formula (4.4.2) can be generated as the corresponding sums of independent random variables with the exponential distribution with parameter 1.

For each of the screening rules N, N$|\mu_Z$, MI we are interested in the values of the probabilisties

$$\begin{aligned} \theta_0(\delta) &= P_\delta(\hat{I}_\delta = 1), \\ \theta_1(\delta) &= P_\delta(I = 1 \mid \hat{I}_\delta = 1). \end{aligned}$$

Clearly, these probabilities depend on sample size and are different for different rules, and, in general, do not coincide with probabilities θ_0, θ_1 for the optimal solution of the probabilistic problem.

To estimate $\theta_0(\delta)$ and $\theta_1(\delta)$ training lots were generated as simple random samples from the distribution of (W, Z), independently of the learning sample used previously to construct rule δ. As estimators of probabilities $\theta_0(\delta)$ and $\theta_1(\delta)$, their sample counterparts were obtained from the training lot. Each training lot contained 300 elements.

Let (W_i', Z_i'), $i = 1, \ldots, 300$ denote the training lot. For any rule $\delta = \mathrm{N}, \mathrm{N}|\mu_Z, \mathrm{MI}$ and any training lot, the sublot obtained by means of δ was formed. Let $\hat{\theta}_0[\delta]$ and $\hat{\theta}_1[\delta]$ be, respectively, the fraction of elements included in the sublot, and the fraction of acceptable elements in the sublot. Then

$$\hat{\theta}_0[\delta] = \frac{1}{300}\sum_{i=1}^{300} \chi_{\{Z_i' \geq t_0[\delta]\}}, \tag{4.4.3}$$

$$\hat{\theta}_1[\delta] = \frac{\sum_{i=1}^{300} \chi_{\{W_i' > L, Z_i' \geq t_0[\delta]\}}}{\sum_{i=1}^{300} \chi_{\{Z_i' \geq t_0[\delta]\}}}. \tag{4.4.4}$$

The estimators $\hat{\theta}_0[\delta]$ and $\hat{\theta}_1[\delta]$ are invariant under linear transformations with positive coefficients of variables W and Z.

In view of this property simulation was performed for $\mu_W = \mu_Z = 0$ and $\sigma_W^2 = \sigma_Z^2 = 1$ since

$$(W, Z) \sim N_2(\mu_W, \mu_Z, \sigma_W^2, \sigma_Z^2, \varrho)$$

$$\Rightarrow \left(\frac{W-\mu_W}{\sigma_W}, \frac{Z-\mu_Z}{\sigma_Z}\right) \sim N_2(0, 0, 1, 1, \varrho),$$

$$(W, Z) \sim \Gamma_2(\alpha, \mu_W, \mu_Z, \sigma_W^2, \sigma_Z^2, \varrho)$$

$$\Rightarrow \frac{Z-\mu_Z}{\sigma_Z} \sim \Gamma_2(\alpha, 0, 0, 1, 1, \varrho).$$

Hence, it remained to fix the parameter ϱ in the case of the normal distribution and the parameters α and ϱ in the case of gamma distribution, and also the parameters θ_1^*, L related directly to the given screening problem. In the simulation experiment we assumed $\alpha = 5$, $\varrho = 0.6$ (consequently, $\alpha\varrho = 3$, $\alpha(1-\varrho) = 2$) and $\theta_1^* = 0.95$. The parameter L was chosen so as to ensure the same value of the probability $\pi = P(W > L)$ for the normal and the gamma distributions. For $\alpha = 5$ this was achieved for $\pi = 0.8$, i.e.,

$$\frac{L-\mu_W}{\sigma_W} = q_{N(0,1)}(0.8) = -0.8416.$$

For these values of the parameters of the problem the threshold (4.2.11) in the rule N ($\varrho = 0.6$) is equal to

$$-q_\beta = 0.231148.$$

The simulation proceeded as follows:

(i) for each of the distributions $N_2(0, 0, 1, 1, 0.6)$ and $\Gamma_2(5, 0, 0, 1, 1, 0.6)$ a training lot and a learning sample were generated, each consisting of 300 elements;

(ii) by using the learning sample the threshold $t_0[\delta]$, for $\delta = \mathrm{N}|\mu_Z$, MI was calculated;

(iii) on the basis of the training lot the quantities $\hat{\theta}_0[\delta]$ and $\hat{\theta}_1[\delta]$ for $\delta = \mathrm{N}$, $\mathrm{N}|\mu_Z$, MI were calculated according to formulae (4.4.3), (4.4.4).

(iv) steps (i) – (iii) were repeated 24 times.

The results are presented in Table 4.4.2 and Figs. 4.4.2 – 4.4.3 (cf. p. 102).

TABLE 4.4.2.

Results of simulation experiments for rules N, $\mathrm{N}|\mu_Z$ and MI, for distributions $N_2(0, 0, 1, 1, 0.6)$ $\Gamma_2(5, 0, 0, 1, 1, 0.6)$.

distribution	rule	Percentage of lots satisfying the condition		
		$\hat{\theta}_0[\delta] \geqslant 0.38$ (fractions of elements included in the sublot $\geqslant 0.38$)	$\hat{\theta}_1[\delta] \geqslant 0.95$ (fraction of elements preferred in the sublot $\geqslant 0.95$)	$\hat{\theta}_0[\delta] \geqslant 0.38$ $\hat{\theta}_1[\delta] \geqslant 0.95$
$N_2(0, 0, 1, 1, 0.6)$	N	70.8	54.17	45.8
	$\mathrm{N}\|\mu_Z$	66.7	54.17	41.7
	MI	45.8	66.7	20.8
$\Gamma_2(5, 0, 0, 1, 1, 0.6)$	N	41.7	62.5	29.17
	$\mathrm{N}\|\mu_Z$	25	50	12.5
	MI	50	91.7	45.8

In Table 4.4.2 the values $\hat{\theta}_0$ and $\hat{\theta}_1$ are compared with the respective critical values θ_0^* and θ_1^*. We put $\theta_1^* = 0.95$, $\theta_0^* = 0.38$. The choice of θ_0^* can be motivated as follows. For the distribution $N_2(0, 0, 1, 1, 0.6)$ and $\pi = 0.8$, the probability of including an element in the sublot (provided the optimal rule is applied) is equal to $g(0.6, 0.8, 0.95) = 0.4086$. On the other hand, the standard deviation of the fraction of elements included in the sublot and selected from the lot consisting of 300 elements is equal to $(0.4086 \cdot 0.5914)^{1/2}/300^{1/2} = 0.0284$. By assuming $\hat{\theta}_0^*$ to be equal to the difference $0.4086 - 0.0284 \approx 0.38$ we find $\hat{\theta}_0[\delta]$ to be less than the mean fraction (for the optimal rule) by one standard deviation. In Table 4.4.2 we give, for each distribution and each rule separately, the percentage of simulated lots (the total number is 24) satisfying the conditions:

(a) $\hat{\theta}_0[\delta] \geqslant \theta_0^*$;

(b) $\hat{\theta}_1[\delta] \geqslant \theta_1^*$;

(c) $\hat{\theta}_0[\delta] \geqslant \theta_0^*$ and $\hat{\theta}_1[\delta] \geqslant \theta_1^*$.

As can be seen, for the distribution N_2, the rule N is slightly better than $\mathrm{N}|\mu_Z$ and is definitely better than MI with respect to $\hat{\theta}_0$ (70.8% versus 15.8%), but slightly worse than MI with respect to $\hat{\theta}_1$ (54.17% versus 66.7%).

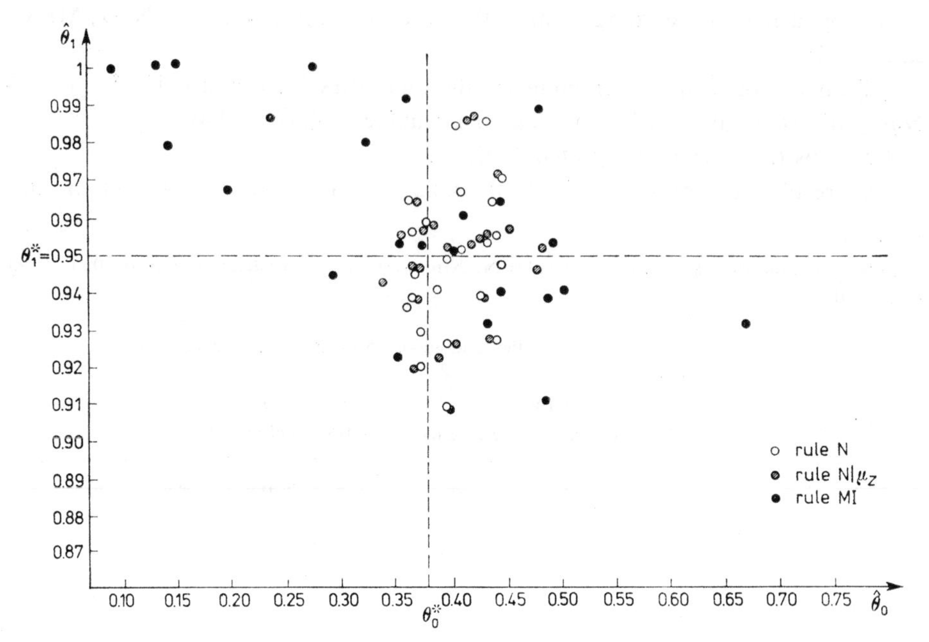

FIG. 4.4.2. Results of simulation experiments for rules N, N|μ_Z, MI for the distribution $N_2(0,0,1,1,0.6)$.

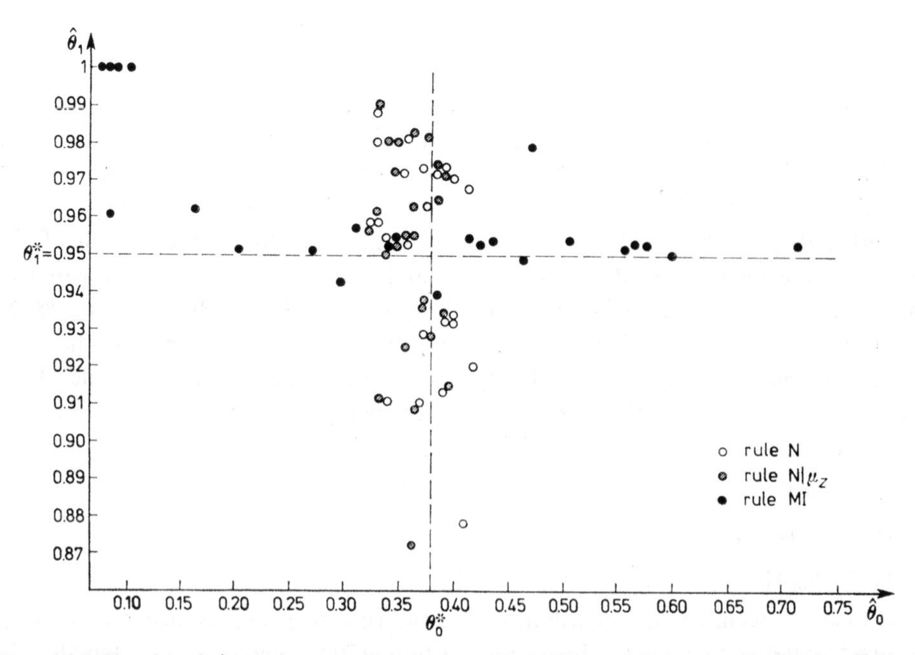

FIG. 4.4.3. Results of simulation experiments for rules N, N|μ_Z, MI for the distribution $\Gamma_2(5,0,0,1,1,0.6)$.

The number of lots with both critical values exceeded is, for rule N, two times greater than that for the rule MI (45.8% versus 20.8%). For the distribution Γ_2 the rule MI is the best one. It dominates both other rules, especially with respect to $\hat{\theta}_1$ (91,7% versus 62.5% and 50%). The rule $N|\mu_Z$ is decisively worse than N.

Figures 4.4.2 and 4.4.3 present in a more detailed way the results obtained. For each rule, the point $(\hat{\theta}_0, \hat{\theta}_1)$ is shown separately for N_2 and Γ_2.

It can be seen in Fig. 4.4.2 that in the case of N_2 the dispersion of each component is much larger for rule MI than in the case of both other rules. For one of the training lots which were generated the sublot obtained contained 65% of the lot elements 95% of which were acceptable elements. There are 9 lots for which the percentage of acceptable objects in the sublot exceeds 98, but in as many as 5 cases the percentage of objects included in the sublot is less than 20. Thus, the cost of the rule MI measured by the number of good elements which are neglected (not included in the sublot) is high. The rules N and $N|\mu_Z$ are quite similar. In both cases the dispersion of θ_1 is relatively large and the dispersion of θ_0 is rather small.

Figure 4.4.3 shows that for the rule MI the dispersion of $\hat{\theta}_0$ is as large for Γ_2 as for N_2, while the dispersion of $\hat{\theta}_1$ is much smaller for Γ_2 than for N_2. In most cases $\hat{\theta}_1$ exceeds θ_1^*. For rules N and $N|\mu_Z$ the dispersion of $\hat{\theta}_0$ is for Γ_2 as small as for N_2 while the dispersion of $\hat{\theta}_1$ is much greater for Γ_2 than for N_2.

Figures 4.4.2 and 4.3.3 suggest that the dispersion of $\hat{\theta}_0$ and the expected value of $\hat{\theta}_1$ are not influenced by the change of Γ_2 to N_2 (for $\alpha = 5$). On the other hand, the expected value of $\hat{\theta}_0$ and the dispersion of $\hat{\theta}_1$ seem to be very sensitive to such a change.

This study should only be regarded as a pilot study. The cost of the simulation allows us to consider only 24 lots and a limited number of values of the parameters $\alpha, \varrho, \pi, \theta_1^*$. Moreover, we did not consider other rules from among those proposed in Section 4.3 for the normal distribution with unknown parameters. We restricted ourselves to $N|\mu_Z$ since our main aim was to compare rule MI with rule N and to gain some indications for further modifications and investigations of MI.

Simulation permits estimating the joint distribution of $\hat{\theta}_0[\delta]$ and $\hat{\theta}_1[\delta]$ for different screening rules and different distributions of (W, Z). The distribution of $(\hat{\theta}_0, \hat{\theta}_1)$ describes those screening effects which are essential for practical purposes. In this respect it would certainly be better to formulate screening problems by imposing requirements directly on the joint distribution of $(\hat{\theta}_0, \hat{\theta}_1)$. We are unable, however, to solve such difficult problems. The usual approach (which we follow here) is to begin with the investigation of rules solving some simple probabilistic problems (as those considered in Section 4.2). Next, after adaptation, the properties of adapted rules are examined analytically and by simulation methods.

4.5. Final remarks

Among the first authors who dealt with probabilistic screening problems were Birnbaum and Chapman (1950) who maximized $\theta_0(\delta)$ under the constraint $\theta_1(\delta) \geqslant \theta_1^*$ for the pair (W, Z), where the random vector Z has an arbitrary continuous distribution, and the distribution of the random variable $(W \mid Z = z)$ is normal for $z \in Z$. A more general case of an arbitrary multidimensional continuous distribution of (W, Z) has been considered in Section 4.2 of this chapter. Birnbaum and Chapman (1950) solved also the following probabilistic screening problem (under the assumption that $(W \mid Z = z)$ is normal): find a rule δ maximizing $\theta_0(\delta)$ under the condition $E(W \mid \hat{I}_\delta = 1) \geqslant E^*$, where E^* is a given number. A problem of this kind is not standard but is an example of a class of screening problems in which requirements are imposed on the joint distribution of the pair $(W, \hat{I}_\delta)$.

As far as we know, the relationships, presented in Section 4.2, between standard probabilistic screening problems and discriminant analysis problems have not been discussed in the literature.

Practical screening problems are discussed in the literature concerning statistical quality control.

*

One of the notions in statistical quality control is the *average outgoing quality* (*AOQ*), i.e. $\theta_1(\delta)$. Different screening algorithms are considered according to what is observable in the lot and in the learning sample if there is one. The assumption usually made is that the distribution of (W, Z) is normal and that its parameters are totally or partially unknown.

Among the papers on statistical quality control which assume the normality of (W, Z) an important role is played by those papers which were globally treated in Section 4.3. They deal with the adaptation of screening algorithms in the problem of maximizing the probability $\theta_0(\delta)$ under the constraint $\theta_1(\delta) \geqslant \theta_1^*$. Owen and Boddie (1976) considered various screening procedures according to which of the parameters $\mu_W, \sigma_W, \mu_Z, \sigma_Z$ were unknown and what additional requirements were imposed on solutions. Owen and Su (1977) analysed the case where π, ϱ, and possibly also μ_Z, σ_Z are unknown. In Owen and Haas (1978) the unknown parameters are μ_Z and σ_Z; the authors presented tables to be used in this case. In Owen and Li (1979) the condition $W > L$ characterizing acceptable elements was replaced by the condition $L < W < U$, and both probabilistic and statistical cases were considered for various combinations of unknown parameters.

A nonparametric model in which the normality of the distribution of (W, Z) was replaced by the monotonicity of the likelihood ratio of (I, Z) was considered by Kowalczyk and Mielniczuk (1987). Their algorithm is not identical with that proposed in Section 4.4 since they use the estimator $\theta_1^{(n)}(t)$ restricted to $(-\infty, b_n)$, where b_n is determined from the learning sample and converges to $\sup(t\colon F_Z(t) < 1)$ with probability 1 if $n \to \infty$.

Owing to this modification the sequence of thresholds converges with probability 1 (under a certain additional assumption) to the optimal threshold.

Screening problems taking into account costs of investigation were considered, e.g., by Yahav (1979), who proposed a screening procedure for choosing from an infinite population (i.e., from a lot with infinitely many elements) a sublot of a given size N and the expected fraction of acceptable elements not smaller than a given bound π^*. Yahav assumes that the fraction π of acceptable objects in the population is known, and chooses π^* greater than π. Each element can be tested repeatedly and there are only two possible realizations of the test, called *success* and *failure*. It is assumed also that the sequence of test repetitions is a sequence of independent random variables. Moreover, the probability of success is equal to a given number α_0 for all acceptable elements and to α_1 ($\alpha_1 < \alpha_0$) for the remaining elements. The cost of a single test is constant and the screening should be performed in such a way as to minimize the cost of the whole operation. The problem is solved by a sequential procedure in which both the number of tests performed for elements tested and the number of tested elements are random variables.

A totally different approach to screening is presented in papers dealing with the elimination of outliers. These are problems in which a given population is contaminated by a small number of alien elements and it is not possible to decide in a direct way whether a given element is alien or not. The conditional distribution of an observable feature Z in the original population (before contamination) is assumed to be a member of a given family of distributions (e.g. normal). We know nothing about the conditional distribution of Z in the class of alien elements except that it is "distinctly different" from the distribution of Z in the noncontaminated population. On the basis of these assumptions we can expect that there are only a few alien elements in the lot and that the values of Z for those elements differ considerably from the values of Z in the noncontaminated population. Consequently, the elimination of alien elements inevitably reduces to the elimination of elements with outlying values of Z.

An elementary guide to those problems is contained in Chapter 14 of Dunn and Clark (1972). A more advanced presentation is given by Neyman and Scott (1971).

Chapter 5

Evaluation of stochastic dependence

Key words: *monotone dependence, quadrant dependence, Frechet distribution, symmetrized distribution, codependence, monotone dependence function, intraclass correlation coefficient.*

5.1. Introduction

The term "correlation" often appears in colloquial use without any reference to statistics. In Huxley's book *Eye-less in Gaza* the following correlation between civilization and sexuality is noted: the more advanced civilization the more intensive sexuality. This is an excellent appeal to the reader's intuition. But how can we define correlation? How can we measure it? Should we start by measuring "civilization" and "sexuality" on scales permitting the gradation indicated by Huxley? What would happen when features being examined have values which cannot be ordered in a natural way?

Partial answers to the above questions are contained in numerous papers devoted to stochastic dependence and its evaluation. However, the problem turns out to be very complex and commonly accepted methods of evaluation do not exist. This gives rise to a great variety of different approaches and methods which confuses practicians. We aim at clarifying the basic relationships between the concepts of stochastic dependence according to a scheme which seems to conform to the intensions of most authors.

In different families of distributions different types of dependence can be evaluated. Roughly speaking, the more homogeneous the family the more precise, in some sense, the evaluation can be. We begin our considerations with very simple families of distributions, namely, the family consisting of all bivariate distributions of binary random variables, and the families which are its subsets. In Section 5.2 we formulate postulates concerning the numerical measures of dependence for binary variables and we give a fairly exhausted survey of measures which are used for such pairs.

In Section 5.3 we extend our considerations to monotone dependent pairs of random variables. We begin with presenting different families of pairs with monotone dependence for variables measured on at least ordinal scale. Among these families we distinguish families of pairs with linear dependence for variables measured on at least interval scale. Next, we formulate postulates concerning nu merical measures of monotone dependence and analyze some examples.

In Section 5.4 we present the monotone dependence function which can be regarded as a function-valued measure of dependence. In Section 5.5 we discuss the evaluation of dependence for couples of random variables regarded as "unordered pair".

In the literature the number of papers devoted to statistical inference concerning measures of dependence considerably exceeds the number of papers devoted to describing types of dependence and to selecting parameters measuring its strength. Various estimators of measures of dependence are defined and investigated in detail but usually without any deeper justification of the choice of a particular parameter in a particular study.

A great deal of attention is also paid to testing independence in different models. This is motivated by the fact that in practice we often want to verify the hypothesis that the values of a given feature are not influenced by the values of other features. However, we are not interested in idealized absolute lack of influence but in a practically meaningless influence. Any hypothesis of absolute lack of influence, when expressed formally as a hypothesis of independence, will be almost always rejected by any test of power asymptotically equal to 1 if applied to a sample of sufficiently large size. This follows from the fact that in empirical investigations some deviation from independence always exists. In particular, the independence hypothesis will be almost always rejected if between the features in question there exists a practically meaningless dependence. This undesirable situation is avoided if the hypothesis of independence is replaced by the hypothesis stating that suitably formalized deviations from independence do not exceed a given threshold. Measures of strength of dependence are used to formulate such a kind of hypotheses of "almost independence".

Families of distributions with dependence of a given type are indispensable for describing empirical phenomena and especially for formulating alternative hypotheses in testing independence and almost independence. Also, measures of dependence play an important role in stochastic modelling.

5.2. Dependence between two binary random variables

In the present section we consider dependence between two binary features describing objects from a certain population. We assume that both features are nondegenerate, i.e., none of them is constant on the whole population. The population can be formally represented by a nondegenerate 2×2 distribution, i.e., by a distribution P of a pair of binary random variables (X, Y) with values $1, 2$. This probability distribution is given by a matrix of probabilities $p_{ij} = P(X = i, Y = j)$ for $i, j = 1, 2$, where $0 < p_{12}+p_{22} < 1$, $0 < p_{21}+p_{22} < 1$.

The family of all nondegenerated 2×2 distributions defined on $\{1, 2\}^2$ will be denoted by $\mathscr{P}_{2\times 2}$. This family can be indexed in a one-to-one manner by triples

$(\gamma_0, \gamma_1, \gamma_2)$ of numbers from the interval $[0, 1]$ satisfying the inequalities $0 < \gamma_0 < 1$, $0 < \gamma_1+\gamma_2 < 2$, where

$$\gamma_0 = P(X = 1) = p_{11}+p_{12},$$

$$\gamma_1 = P(Y = 1 \mid X = 1) = \frac{p_{11}}{p_{11}+p_{12}},$$

$$\gamma_2 = P(Y = 1 \mid X = 2) = \frac{p_{21}}{p_{21}+p_{22}}.$$

The probabilities p_{ij}, expressed by $(\gamma_0, \gamma_1, \gamma_2)$, can be rewritten in the following form:

$i \backslash j$	1	2	$P(X = i)$
1	$p_{11} = \gamma_0\gamma_1$	$p_{12} = \gamma_0(1-\gamma_1)$	$p_{1.} = p_{11}+p_{12}$
2	$p_{21} = (1-\gamma_0)\gamma_2$	$p_{22} = (1-\gamma_0)(1-\gamma_2)$	$p_{2.} = p_{21}+p_{22}$
$P(Y = j)$	$p_{.1} = p_{11}+p_{21}$	$p_{.2} = p_{12}+p_{22}$	

Hence, γ_0 defines the distribution of the random variable X, and γ_1, γ_2 define the conditional distributions of Y under the conditions $X = 1$, $X = 2$, respectively The distribution corresponding to a triple $(\gamma_0, \gamma_1, \gamma_2)$ will be denoted by $P_{\gamma_0,\gamma_1,\gamma_2}$.

To each triple $(\gamma_0, \gamma_1, \gamma_2)$ defining the distribution of (X, Y) we assign another triple $(\theta_0, \theta_1, \theta_2)$ defining the distribution of (Y, X) in the following way

$$\theta_0 = P(Y = 1) = \gamma_0\gamma_1+(1-\gamma_0)\gamma_2,$$

$$\theta_1 = P(X = 1 \mid Y = 1) = \frac{\gamma_0\gamma_1}{\gamma_0\gamma_1+(1-\gamma_0)\gamma_2},$$

$$\theta_2 = P(X = 1 \mid Y = 2) = \frac{\gamma_0(1-\gamma_1)}{\gamma_0(1-\gamma_1)+(1-\gamma_0)(1-\gamma_2)}.$$

The set $\mathscr{P}_{2\times 2}$ is the union of two sets

$$\mathscr{P}^+ = \{P_{\gamma_0,\gamma_1,\gamma_2}: \gamma_0 \in (0, 1), \gamma_1 \geqslant \gamma_2\},$$

$$\mathscr{P}^- = \{P_{\gamma_0,\gamma_1,\gamma_2}: \gamma_0 \in (0, 1), \gamma_1 \leqslant \gamma_2\}.$$

A distribution P belongs to the common part

$$\mathscr{P}^0 = \mathscr{P}^+ \cap \mathscr{P}^- = \{P_{\gamma_0,\gamma_1,\gamma_2}: \gamma_0 \in (0, 1), \gamma_1 = \gamma_2\}$$

if and only if

$$P(Y = i \mid X = j) = P(Y = i), \quad i, j = 1, 2,$$

Thus, $P \in \mathscr{P}^0$ if and only if X and Y are independent.

In any $P \in \mathscr{P}^+$ there is a tendency to the simultaneous appearance of identical values of the two variables. This is expressed by the inequalities

$$\begin{aligned} P(Y = 1 \mid X = 1) &\geqslant P(Y = 1), \\ P(Y = 2 \mid X = 2) &\geqslant P(Y = 2). \end{aligned} \tag{5.2.1}$$

In confront to this, in any $P \in \mathscr{P}^-$ there is a tendency to the simultaneous appearance of different values and, thus, in (5.2.1) opposite inequalities occur. We say that X is *positive dependent* on Y if $(X, Y) \in \mathscr{P}^+$, and is *negative dependent* if $(X, Y) \in \mathscr{P}^-$.

Positive or negative dependence of X on Y is equivalent to that of Y on X since

$$(X, Y) \in \mathscr{P}^+ \ (\mathscr{P}^-) \quad \Leftrightarrow \quad (Y, X) \in \mathscr{P}^+ \ (\mathscr{P}^-).$$

Among positive dependent distributions we distinguish a family of those for which $\gamma_1 = 1$, $\gamma_2 = 0$, i.e., $P(X = Y) = 1$, and we denote it by $\mathscr{P}^{c+}$. In the family of negative dependent distributions we distinguish a family of distributions for which $\gamma_1 = 0$, $\gamma_2 = 1$, i.e., $P(X = 3-Y) = 1$, and we denote it by $\mathscr{P}^{c-}$. The sum of $\mathscr{P}^{c+}$ and $\mathscr{P}^{c-}$ will be denoted by $\mathscr{P}^c$. Hence, $(X, Y) = \mathscr{P}^c$ iff X is a function of Y (with probability 1).

So, in $\mathscr{P}^0$ the variable X is not dependent on Y, while in $\mathscr{P}^c$ the dependence is complete. Now we want to evaluate the departure from independence (or in other words, the approach to complete dependence) of an arbitrary distribution P of (X, Y), while preserving the information whether P belongs to $\mathscr{P}^+$ or to $\mathscr{P}^-$. Informally speaking, we aim at evaluating the strength of monotone dependence, i.e., the *strength of positive dependence* if P belongs to $\mathscr{P}^+$, and the *strength of negative dependence* if P belongs to $\mathscr{P}^-$. To compare the strength of *monotone dependence* we need some ordering in the set $P_{2\times 2}$.

Let P denote a distribution of (X, Y) indexed by a triple $(\gamma_0, \gamma_1, \gamma_2)$ with the matrix $[p_{ij}]$ and let P' be a distribution of (X', Y') indexed by $(\gamma_0', \gamma_1', \gamma_2')$ with the matrix $[p_{ij}']$. It is intuitively clear that the strength of dependence increases in $\mathscr{P}^+$ with the tendency to the co-occurrence of identical values, and in $\mathscr{P}^-$ with the tendency to co-occurrence of different values of the two variables. Thus we define an ordering $\preccurlyeq$ in the set $\mathscr{P}_{2\times 2}$ in the following way:

$$\begin{aligned} &\text{for} \quad P, P' \in \mathscr{P}^+ \\ &P \preccurlyeq P' \quad \Leftrightarrow \quad p_{11} \leqslant p_{11}' \ \wedge \ p_{22} \leqslant p_{22}' \\ &\qquad\qquad\qquad \wedge \ p_{12} \geqslant p_{12}' \ \wedge \ p_{21} \geqslant p_{21}'; \\ &\text{for} \ P, P' \in \mathscr{P}^- \\ &P \preccurlyeq P' \quad \Leftrightarrow \quad p_{11} \geqslant p_{11}' \ \wedge \ p_{22} \geqslant p_{22}' \\ &\qquad\qquad\qquad \wedge \ p_{12} \leqslant p_{12}' \ \wedge \ p_{21} \leqslant p_{21}'. \end{aligned} \tag{5.2.2}$$

The elements of $\mathscr{P}^c$ are maximal elements of this relation.

For arbitrary $P, P' \in \mathscr{P}^+$ such that $\gamma_0 = \gamma_0'$ we have

$$P \preccurlyeq P' \quad \Leftrightarrow \quad \gamma_1 \leqslant \gamma_1' \ \wedge \ \gamma_2 > \gamma_2'.$$

A similar formula holds for distributions from $\mathscr{P}^-$.

To evaluate numerically the strength of monotone dependence by taking into account independence, positive dependence and negative dependence, and the ordering relation defined above, we use the function

$$\varkappa\colon \mathscr{P}_{2\times 2} \to [-1, 1]$$

satisfying the following conditions:

$$
\begin{aligned}
&\text{(i)} \quad \varkappa(P) = 0 \;\Leftrightarrow\; P \in \mathscr{P}^0; \\
&\text{(ii)} \quad \varkappa(P) = 1 \;\Leftrightarrow\; P \in \mathscr{P}^{c+}; \\
&\text{(iii)} \quad \varkappa(P) = -1 \;\Leftrightarrow\; P \in \mathscr{P}^{c-}; \\
&\text{(iv)} \quad P \preccurlyeq P',\ P, P' \in \mathscr{P}^+,\ P \neq P' \;\Rightarrow\; \varkappa(P) < \varkappa(P'); \\
&\text{(v)} \quad P \preccurlyeq P',\ P, P' \in \mathscr{P}^-,\ P \neq P' \;\Rightarrow\; \varkappa(P) > \varkappa(P'); \\
&\text{(vi)} \quad \text{if } P \in \mathscr{P}_{2\times 2} \text{ and } (P_n,\ n = 1, 2, \ldots) \text{ is a sequence of} \\
&\qquad \text{elements of } \mathscr{P}_{2\times 2} \text{ such that } P_n \xrightarrow{\mathscr{L}} P, \text{ then } \varkappa(P_n) \to \varkappa(P).
\end{aligned}
\tag{5.2.3}
$$

We have

$$
\begin{aligned}
P \in \mathscr{P}^+ \setminus \mathscr{P}^0 &\;\Leftrightarrow\; \varkappa(P) > 0. \\
P \in \mathscr{P}^- \setminus \mathscr{P}^0 &\;\Leftrightarrow\; \varkappa(P) < 0.
\end{aligned}
\tag{5.2.4}
$$

The composition of the parameter $\varkappa$ satisfying (5.2.3) with an arbitrary continuous strictly monotone function ϕ, $\phi(0) = 0$, which transforms the interval $[-1, 1]$ onto itself, also satisfies conditions (5.2.3). Conditions (5.2.3) are also satisfied by a parameter $\varkappa'$ defined as

$$\varkappa'(X, Y) = \varkappa(Y, X).$$

This follows from the symmetry of (5.2.3). As examples of parameters satisfying (5.2.3) we can quote

$$K(X, Y) = \gamma_1 - \gamma_2, \tag{5.2.5}$$

$$K'(X, Y) = \theta_1 - \theta_2, \tag{5.2.6}$$

and the correlation coefficient

$$\operatorname{cor}(X, Y) = \frac{p_{11}p_{22} - p_{12}p_{21}}{\sqrt{p_{1\cdot}p_{2\cdot}p_{\cdot 1}p_{\cdot 2}}}. \tag{5.2.7}$$

Parameter K satisfies in addition the following condition: for any $\gamma_0, \gamma_0' \in (0, 1)$, $\gamma_1, \gamma_2 \in [0, 1]$.

$$K(P_{\gamma_0, \gamma_1, \gamma_2}) = K(P_{\gamma_0', \gamma_1, \gamma_2}). \tag{5.2.8}$$

Hence, K depends upon the distribution of Y and does not depend upon the distribution of X; for K' the opposite holds. The correlation coefficient depends upon the distributions of X and Y. The parameters K, K', and cor are strictly related to each other. We have

$$\operatorname{cor}^2(X, Y) = K \cdot K', \tag{5.2.9}$$

$$\operatorname{cor}(X, Y) = K \cdot \sqrt{\alpha}, \tag{5.2.10}$$

$$K' = K \cdot \alpha, \tag{5.2.11}$$

where

$$\alpha(P) = \frac{\gamma_0(1-\gamma_0)}{\theta_0(1-\theta_0)}. \tag{5.2.12}$$

The ordering defined by (5.2.2) is not linear in $\mathscr{P}_{2\times 2}$ but it is linear in each subset of $\mathscr{P}_{2\times 2}$ consisting of distributions with fixed $p_{1\cdot}, p_{\cdot 1}$.

In the literature bivariate distributions are usually compared only if their marginal distributions are equal. Then we have the following ordering

$$\begin{aligned} &\text{for } P, P' \in \mathscr{P}^+ \\ &P \preccurlyeq P' \quad \Leftrightarrow \quad p_{1\cdot} = p'_{1\cdot} \ \wedge \ p_{\cdot 1} = p'_{\cdot 1} \\ &\qquad\qquad\qquad \wedge \ p_{ii} \leqslant p'_{ii}, \ i = 1, 2; \\ &\text{for } P, P' \in \mathscr{P}^-. \\ &P \preccurlyeq P' \quad \Leftrightarrow \quad p_{1\cdot} = p'_{1\cdot} \ \wedge \ p_{\cdot 1} = p'_{\cdot 1} \\ &\qquad\qquad\qquad \wedge \ p_{ii} \geqslant p'_{ii}, \ i = 1, 2. \end{aligned} \tag{5.2.13}$$

Evidently, the ordering defined by (5.2.2) is an extension of the ordering defined by (5.2.13).

To each distribution P the following distributions are uniquelly assigned: the independent distribution $P^0 \in \mathscr{P}^0$:

$$p^0_{ij} = p_{i\cdot} p_{\cdot j},$$

and distributions P^{q+}, P^{q-} representing the strongest positive and negative dependence in the set of distributions with fixed parameters γ_0, θ_0:

$$p^+_{12} p^+_{21} = 0 \quad \text{for} \quad P^{q+},$$

$$p^-_{11} p^-_{22} = 0 \quad \text{for} \quad P^{q-}.$$

Distributions P^{q+}, P^{q-} will be called *distributions of quasi-complete positive* and *negative dependence* (they are called *upper* and *lower Fréchet bounds*, respectively). For these distributions we have

$$P \in \mathscr{P}^+ \quad \Rightarrow \quad P^0 \preccurlyeq P \preccurlyeq P^{q+},$$

$$P \in \mathscr{P}^- \quad \Rightarrow \quad P^0 \preccurlyeq P \preccurlyeq P^{q-}.$$

Write

$$\mathscr{P}^{q+} = \{P^{q+}: \ P \in \mathscr{P}^+ \setminus \mathscr{P}^0\},$$

$$\mathscr{P}^{q-} = \{P^{q-}: \ P \in \mathscr{P}^- \setminus \mathscr{P}^0\}$$

and put

$$\mathscr{P}^q = \mathscr{P}^{q+} \cup \mathscr{P}^{q-}.$$

The set $\mathscr{P}^q$ is the set of maximal elements of the ordering relation (5.2.13). In postulates (5.2.3) we can replace $\mathscr{P}^{c+}$ and $\mathscr{P}^{c-}$ by $\mathscr{P}^{q+}$ and $\mathscr{P}^{q-}$, respectively, and the ordering relation (5.2.2) by (5.2.13). Parameters satisfying the postulates thus modified are compositions of the *cross product ratio* defined by

$$q(X, Y) = \begin{cases} \dfrac{p_{11} p_{22}}{p_{12} p_{21}}, & \text{if} \quad p_{12} p_{21} \neq 0, \\ +\infty, & \text{if} \quad p_{12} p_{21} = 0, \end{cases}$$

with some strictly monotone function mapping $[0, +\infty]$ onto $[-1, 1]$. Such compositions are the only symmetric parameters satisfying postulates (5.2.3) in the modified form and not depending on γ_0. In particular, consider the following parameters,

$$Q(X,Y)=\begin{cases}\dfrac{q(X,Y)-1}{q(X,Y)+1}, & \text{if} \quad p_{12}p_{21}\neq 0,\\ 1 & \text{otherwise},\end{cases}$$
$$L(X,Y)=\begin{cases}\dfrac{\sqrt{q(X,Y)}-1}{\sqrt{q(X,Y)}+1}, & \text{if} \quad p_{12}p_{21}\neq 0,\\ 1 & \text{otherwise}.\end{cases} \tag{5.2.14}$$

These parameters satisfy the equality

$$Q=f\circ L.$$

where f is a strictly increasing function of the form

$$f(t)=\frac{2t}{1+t^2}, \quad t\in[-1,1].$$

The values of parameters considered in this section may differ significantly from each other. For instance, for the distribution

$$\begin{bmatrix}0.1 & 0\\ 0.4 & 0.5\end{bmatrix} \tag{5.2.15}$$

we have

$$(\gamma_0,\gamma_1,\gamma_2)=(0.1,1,0.4(4)), \quad (\theta_0,\theta_1,\theta_2)=(0.5,0.2,0),$$

Hence, $K=0.5(5)$, $K'=0.2$, $\text{cor}=0.3(3)$, $Q=L=1$.

Therefore, the family $\mathscr{P}_{2\times 2}$ is differentiated so much that it is impossible to define a natural dependence measure in it. However, this can be done in some of its subsets. For instance, for a distribution with uniform marginal distributions the matrix $[p_{ij}]$ takes the form

$$\begin{bmatrix}\frac{1}{2}\gamma_1 & \frac{1}{2}(1-\gamma_1)\\ \frac{1}{2}(1-\gamma_1) & \frac{1}{2}\gamma_1\end{bmatrix} \tag{5.2.16}$$

and we obtain $\gamma_0=\theta_0=0.5$, $\gamma_1=\theta_1$, $\gamma_2=\theta_2=1-\theta_1$. Hence,

$$\begin{aligned}K(X,Y)&=K'(X,Y)=\text{cor}(X,Y)\\ &=L(X,Y)=2\gamma_1-1 \quad \text{for} \quad (X,Y)\in\mathscr{P}^{\text{u}},\end{aligned} \tag{5.2.17}$$

where $\mathscr{P}^{\text{u}}$ denotes the family of 2×2 distributions with uniform marginal distributions. Consequently, in the family $\mathscr{P}^{\text{u}}$ the parameter $2\gamma_1-1$ is the only function (up to suitable compositions) satisfying all the postulates (5.2.3) and can be accepted as a "natural" measure of dependence in $\mathscr{P}^{\text{u}}$.

*

Another idea of evaluating the dependence of a random variable X on a random variable Y is related to the prediction of an unobservable variable X by an observable variable Y. Such problems have been considered in Section 3.2. In Section

3.4. we introduce the prediction power $\psi(I, Z)$ for different prediction problems given by losses (L_{ij}) and set Δ_P of admissible decision rules (formula (3.4.11)). Intuitively, we expect the dependence of X on Y to increase with our ability to predict the values of X on the basis of the values of Y; thus it is possible to use as a measure of dependence the prediction power $\psi(X, Y)$ ((X, Y) replaces the pair (I, Z) considered in Section 3.4). The parameter thus defined has a natural interpretation related directly to the prediction problem defined by (L_{ij}) and Δ_P. In particular, the problem considered in Example 3.4.1 leads to the definition of the parameter λ_a by means of (3.4.17) for $r = 2$. The problem from Example 3.4.2 leads to the definition of the parameter η_a by means of (3.4.21) for $r = 2$. We find that

$$\eta_a(X, Y) = \operatorname{cor}^2(X, Y). \tag{5.2.18}$$

The prediction power is a parameter with values in $[0, 1]$. It measures the overall dependence of the random variable X on the random variable Y. This means that it does not preserve the information on whether the distribution of (X, Y) belongs to $\mathscr{P}^+$ or $\mathscr{P}^-$.

By (5.2.18), the parameter η_a satisfies only (i), (vi) of (5.2.3), and moreover

$$\eta_a(P) = 1 \quad \Leftrightarrow \quad P \in \mathscr{P}^c \tag{5.2.19}$$

$$P \preccurlyeq P' \ \wedge\ (P, P' \in \mathscr{P}^+ \ \vee\ P, P' \in \mathscr{P}^-) \ \wedge\ P \neq P'$$
$$\Rightarrow \quad \eta_a(P) < \eta_a(P'). \tag{5.2.20}$$

The parameter λ_a satisfies (vi) of (5.2.3) and (5.2.19) (with η_a replaced by λ_a), but does not satisfy (5.2.20). Instead of (i) in (5.2.3), λ_a satisfied the weaker condition

$$(X, Y) \in \mathscr{P}^0 \quad \Rightarrow \quad \lambda_a(X, Y) = 0.$$

If the marginal distribution of X is uniform, then

$$\lambda_a(X, Y) = |\gamma_1 - \gamma_2| = |K(X, Y)|.$$

The prediction power is not always symmetric. However, one can postulate symmetry by putting equal stress on the prognosis of X by means of Y and the prognosis of Y by means of X. For a given set of rules Δ_P and losses (L_{ij}) we can determine the arithmetic mean of risks for the optimal prognosis of X by means of Y and for the optimal prognosis of Y by X and the arithmetic mean of the risks for the optimal prognoses by means of constant rules belonging to Δ_P. Replacing in (3.4.11) the risk values by the corresponding arithmetic means of the risks we obtain a new measure, called the *symmetric prediction power* for (X, Y). This modification does not influence the symmetric parameter η_a, but instead of λ_a we obtain

$$\lambda(X, Y) = \frac{\frac{1}{2}\sum_{i=1}^{2} (\max(p_{i1}, p_{i2}) + \max(p_{1i}, p_{2i}))}{1 - \frac{1}{2}(\max(p_{\cdot 1}, p_{\cdot 2}) + \max(p_{1\cdot}, p_{2\cdot}))} - \frac{\frac{1}{2}(\max(p_{\cdot 1}, p_{\cdot 2}) + \max(p_{1\cdot}, p_{2\cdot}))}{1 - \frac{1}{2}(\max(p_{\cdot 1}, p_{\cdot 2}) + \max(p_{1\cdot}, p_{2\cdot}))}.$$

*

Clearly, the results of this section can be generalized to binary random variables X, Y taking on arbitrary real values, say, a_1, a_2 and $b_1, b_2, a_1 < a_2, b_1 < b_2$. The joint distribution of these variables X, Y is given by

$$p_{ij} = P(X = a_i, Y = b_j), \quad i, j = 1, 2.$$

The family of all nondegenerate distributions on $\{a_1, a_2\} \times \{b_1, b_2\}$, for any $a_1, a_2, b_1, b_2 \in \boldsymbol{R}$, $a_1 < a_2, b_1 < b_2$, will be denoted by $\tilde{\mathscr{P}}_{2\times 2}$. Analogously, we define $\tilde{\mathscr{P}}^+$ instead of $\mathscr{P}^+$, etc. In the class $\tilde{\mathscr{P}}_{2\times 2}$ we define the relation of equivalence of pairs of random variables as the equality of their probability matrices. This permits introducing in $\tilde{\mathscr{P}}_{2\times 2}$ the same notions as those introduced in $\mathscr{P}_{2\times 2}$. Hence, in $\tilde{\mathscr{P}}_{2\times 2}$ there exist orderings and dependence measures which are counterparts of respective notions in $\mathscr{P}_{2\times 2}$.

5.3. Dependence in case of bivariate distributions

In the preceding section we discussed various aspects of evaluation of dependence in family $\mathscr{P}_{2\times 2}$. Now we shall show that in more complex families of bivariate distributions the situation is more difficult.

In the sequel we assume that the families in question do not contain bivariate degenerate distributions, i.e., distributions with at least one marginal distribution concentrated at a point.

In Section 5.2 we considered monotone dependence between binary random variables; now we are going to extend this notion to the case of pairs of univariate random variables.

Intuitively, a random variable X is *positive dependent* on a random variable Y if greater values of X tend to co-appear with greater values of Y, and smaller values of X tend to co-appear with smaller values of Y. If the converse effect is observed, i.e., greater values of one variable tend to co-appear with smaller values of the other variables, we say that the dependence is *negative*.

This vague statement can be formalized in various ways. One of the most widely known formalizations is quadrant dependence.

A random variable X is *positive quadrant dependent* on a random variable Y if, for each pair $(x, y) \in \boldsymbol{R}^2$

$$P(X \leqslant x, Y \leqslant y) \geqslant P(X \leqslant x)P(Y \leqslant y), \quad (x, y) \in \boldsymbol{R}^2. \tag{5.3.1}$$

Analogously we define *negative quadrant dependence* by reversing the inequality sign in (5.3.1).

These definitions are symmetric with respect to both variables, i.e., if X is positive (negative) quadrant dependent on Y, then Y is positive (negative) quadrant dependent on X.

The set of all two-dimensional positive (negative) dependent distributions is denoted by QD^+ (QD^-), the union of the sets QD^+ and QD^- is denoted by QD. The common part $\mathrm{QD}^+ \cap \mathrm{QD}^-$ consists of independent distributions.

The term "quadrant dependence" derives from the fact that masses concentrated in quadrants $(-\infty, x] \times (-\infty, y]$, $(x, \infty) \times (y, \infty)$, $(x, \infty) \times (-\infty, y)$,

$(-\infty, x]\times(y, \infty)$ are compared with the masses concentrated in these quadrants and corresponding to independent variables with the same marginal distributions. Let $x_p, y_p, p \in (0, 1)$, denote arbitrary quantiles of order p of the random variables X, Y, respectively. The stronger the dependence the greater the mass concentrated in the quadrants $(-\infty, x_p]\times(-\infty, y_p]$, $[x_p, \infty)\times[y_p, \infty)$ if the dependence is positive (cf. Fig. 5.4.1), and in the quadrants $(-\infty, x_{1-p}]\times[y_p, \infty)$, $[x_{1-p}, \infty)\times$ $\times(-\infty, y_p]$ if the dependence is negative.

There exist many equivalent definitions of quadrant dependence. For example, in the case of positive dependence we have

$$P(X\leqslant x, Y \leqslant y)P(X > x, Y > y)$$
$$\geqslant P(X \leqslant x, Y > y)P(X > x, Y \leqslant y), \quad (x, y) \in \boldsymbol{R}^2,$$

or

$$P(X \leqslant x \mid Y \leqslant y) \geqslant P(X \leqslant x) \tag{5.3.2}$$

for $(x, y) \in \boldsymbol{R}^2$ such that $F_Y(y) > 0$, where F_Y is the distribution function of Y.

This conditions can be rewritten in the following way: for each $y \in \boldsymbol{R}$ such that $F_Y(y) > 0$, we have

$$(X \mid Y \leqslant y) \leqslant_{st} X.$$

Other formalizations of monotone dependence of X and Y are also known. Sometimes the resulting dependence is, in a sense, more regular than the quadrant dependence. Consider, for instance, the following families contained in QD^+:

Family $\widetilde{\mathrm{QD}}^+$: $(X, Y) \in \widetilde{\mathrm{QD}}^+$ if, for any y_1, y_2 satisfying $F_Y(y_j) > 0 \quad (j=1, 2)$,

$$y_1 < y_2 \quad\Rightarrow\quad (X \mid Y \leqslant y_1) \leqslant_{st} (X \mid Y \leqslant y_2). \tag{5.3.3}$$

Family RD^+ (regression dependence): $(X, Y) \in \mathrm{RD}^+$ if, for any y_1, y_2 satisfying $0 < F_Y(y_j) < 1$ $(i = 1, 2)$,

$$y_1 < y_2 \quad\Rightarrow\quad (X \mid Y = y_1) \leqslant_{st} (X \mid Y = y_2). \tag{5.3.4}$$

Family LRD^+ (likelihood ratio dependence): $(X, Y) \in \mathrm{LRD}^+$ if for any x_i, y_i $(i = 1, 2, 3)$ such that $x_1 < x_2 < x_3, y_1 < y_2 < y_3$,

$$P(x_1 < X \leqslant x_2, y_1 < Y \leqslant y_2)P(x_2 < X \leqslant x_3, y_2 < Y \leqslant y_3)$$
$$\geqslant P(x_1 < X \leqslant x_2, y_2 < Y \leqslant y_3)P(x_2 < X \leqslant x_3, y_1 < Y \leqslant y_2). \tag{5.3.5}$$

Note that for a pair (X, Y) with density f condition (5.3.5) is equivalent to the condition

$$f(x_1, y_1)f(x_2, y_2) \geqslant f(x_1, y_2)f(x_2, y_1). \tag{5.3.6}$$

which holds for any x_1, x_2, y_1, y_2 such that $x_1 < x_2, y_1 < y_2$.

It can easily be shown that

$$\mathrm{LRD}^+ \subset \mathrm{RD}^+ \subset \widetilde{\mathrm{QD}}^+ \subset \mathrm{QD}^+. \tag{5.3.7}$$

Analogously, we define families $\widetilde{\mathrm{QR}}^-$, RD^-, LRD^- and $\widetilde{\mathrm{QD}}$, RD, LRD. For these families we have the inclusions similar to (5.3.7).

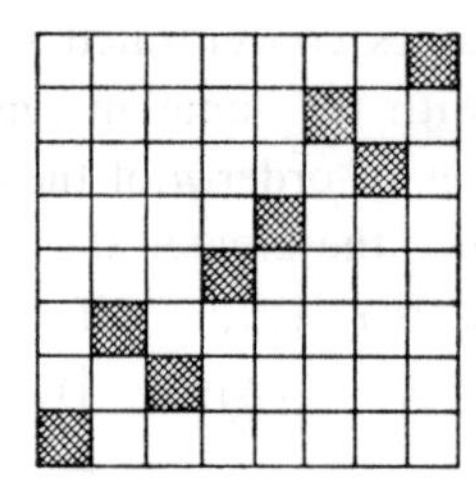
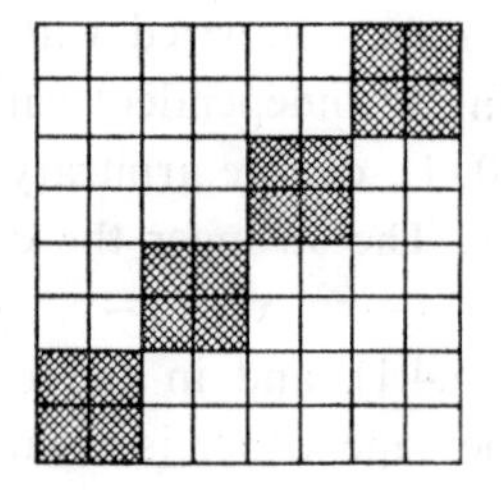

FIG. 5.3.1. Supports of uniform distributions $P_1 \in \mathrm{QD}^+ \setminus \mathrm{RD}^+$ and $P_2 \in \mathrm{RD}^+$.

Consider two distributions P_1, P_2, which are uniform on the corresponding shaded areas in Fig. 5.3.1. The distribution P_1 belongs to $\mathrm{QD}^+ \setminus \mathrm{RD}^+$ while P_2 belongs to RD^+. This example illustrates the intuitive meaning of the "greater regularity" of positive stochastic dependence in the class RD^+ than in the class QD^+.

Unlike the families QD and LRD the families $\widetilde{\mathrm{QD}}$ and RD are not symmetric with respect to the ordering of the variables.

We now give examples of families contained in LRD.

1. Family $\tilde{\mathscr{P}}_{2\times 2}$. Condition (5.3.5) takes the form

$$p_{11}p_{22} \geqslant p_{12}p_{21}$$

and is satisfied for any $P \in \tilde{\mathscr{P}}^+_{2\times 2}$.

2. The family of normal distributions

$$\mathscr{N}_2 = \{N_2(\mu_1, \mu_2, \sigma_1^2, \sigma_2^2, \varrho): \\ (\mu_1, \mu_2) \in \boldsymbol{R}^2, (\sigma_1^2, \sigma_2^2) \in (\boldsymbol{R}^+)^2, \varrho \in [-1, 1]\}.$$

3. For any number $\alpha > 0$, the family of bivariate gamma distributions

$$\mathscr{T}_2^{(\alpha)} = \{\Gamma_2(\alpha, \mu_1, \mu_2, \sigma_1^2, \sigma_2^2, \varrho): \\ (\mu_1, \mu_2) \in \boldsymbol{R}^2, (\sigma_1^2, \sigma_2^2) \in (\boldsymbol{R}^+)^2, \varrho \in [-1,1]\};$$

For $0 \leqslant \varrho \leqslant 1$ the distribution $\Gamma_2(\alpha, \mu_1, \mu_2, \sigma_1^2, \sigma_2^2, \varrho)$ (cf. Section 4.4) is defined as the distribution of the pair $(X, Y) = (\sigma_1 W + \mu_1, \sigma_2 Z + \mu_2)$, where $W = (V_0 + V_1 - \alpha)/\sqrt{\alpha}$, $Z = (V_0 + V_2 - \alpha)/\sqrt{\alpha}$, and V_0, V_1, V_2 are independent random variables with the distributions $\tilde{\Gamma}(\alpha\varrho), \tilde{\Gamma}(\alpha(1-\varrho)), \tilde{\Gamma}(\alpha(1-\varrho))$, respectively. For $\gamma > 0$ the distribution $\tilde{\Gamma}(\gamma)$ is equal to $\Gamma(\gamma)$, and for $\gamma = 0$ it is concentrated at zero. For $-1 \leqslant \varrho \leqslant 0$ the distribution $\Gamma_2(\alpha, \mu_1, \mu_2, \sigma_1^2, \sigma_2^2, \varrho)$ is defined as the distribution of a pair (X, Y) such that $(X, -Y) \sim \Gamma_2(\alpha, \mu_1, -\mu_2, \sigma_1^2, \sigma_2^2, -\varrho)$.

4. The family of distributions of pairs (X, Y) such that X takes on values 1 or 2, and the densities f_i of $(Y \mid X = 1)$, $i = 1, 2$, with respect to a given measure ν are positive on a given interval of $\boldsymbol{R}$, and the likelihood ratio (restricted to this interval) is monotone. Such families of distributions were considered in Chapters 3 and 4.

If we extend family 2 to family in which X has a continuous distribution and the conditional distribution of $(Y \mid X = x)$ remains unchanged, then the resulting family is also contained in LRD.

In these examples, each family $\mathscr{P}$ is the union of two families: $\mathscr{P}^+$ consisting of positive dependent distributions, and $\mathscr{P}^-$ consisting of negative dependent distributions. In each case the intersection $\mathscr{P}^0 = \mathscr{P}^+ \cap \mathscr{P}^-$ consists of independent distrbutions.

Now we introduce a family ED such that $\mathrm{ED} = \mathrm{ED}^+ \cup \mathrm{ED}^-$, where

$$(X, Y) \in \mathrm{ED}^+ \quad \text{iff} \quad E(X \mid Y \leqslant y) \leqslant EX \text{ for almost all } y$$

and

$$(X, Y) \in \mathrm{ED}^- \quad \text{iff} \quad E(X \mid Y \leqslant y) \geqslant EX \text{ for almost all } y.$$

Clearly, $(X, Y) \in \mathrm{ED}$ if $(X, Y) \in \mathrm{QD}$ and the expected value of X is finite. The converse implication is not true. Family ED contains family MR ("monotone regression") of all distributions with monotone regression of X on Y. The family MR is composed of sets MR^+ and MR^- of distributions with nondecreasing and nonincreasing regression functions, respectively. The common part MR^0 contains all distributions with constant regression, and in particular distributions of independent random variables with finite expected values.

The sets $\mathrm{QD} \setminus \mathrm{MR}$ and $\mathrm{MR} \setminus \mathrm{QD}$ are nonempty, e.g. if the regression of X on Y is constant and X, Y are not independent, then $(X, Y) \in \mathrm{MR} \setminus \mathrm{QD}$.

Family LR of distributions with linear regression ($\mathrm{LR} \subset \mathrm{MR}$) is also considered.

*

As in family $\mathscr{P}_{2\times 2}$, to each bivariate distribution P of (X, Y) we assign uniquely the distributions P^{q+}, P^{q-}, P^0 with the distribution functions F^{q+}, F^{q-}, F^0:

$$F^{q+}(x, y) = \min\big(F_X(x), F_Y(y)\big), \quad (x, y) \in \mathbf{R}^2,$$
$$F^{q-}(x, y) = \max\big(F_X(x)+F_Y(y)-1, 0\big), \quad (x, y) \in \mathbf{R}^2,$$
$$F^0(x, y) = F_X(x) F_Y(y).$$

The distributions P, P^{q+}, P^{q-}, P^0 have marginal distributions respectively equal and the distribution function F of P satisfies the inequality

$$F^{q-}(x, y) \leqslant F(x, y) \leqslant F^{q+}(x, y), \quad (x, y) \in \mathbf{R}^2.$$

The distributions P^{q+} and P^{q-} are called the *Fréchet distributions* (or *upper* and *lower Fréchet bounds*) of P. In the family $\mathscr{P}_{2\times 2}$ these distributions were defined as distributions with quasi-complete dependence, positive or negative. The Frechet distributions describe the strongest monotone dependence of the first random variable on the second in all pairs with given marginal distributions. In particular, if $X \sim f(Y)$ for a certain monotone function f, then P^{q+} (P^{q-}) is concentrated on the graph of f provided f is increasing (decreasing).

Distribution P^0 is the distribution of a pair of independent variables with given marginal distributions.

As in family $\mathscr{P}_{2\times 2}$, we denote by $\mathscr{P}^{q+}$ the set of distributions P^{q+} for $P \in \mathscr{P}^+ \setminus \mathscr{P}^0$ and by P^{q-} the set of distributions P^{q-} for $P \in \mathscr{P}^- \setminus \mathscr{P}^0$ The union $\mathscr{P}^{q+} \cup \mathscr{P}^{q-}$ will be denoted by $\mathscr{P}^q$.

Let $\preccurlyeq$ denote an ordering in $\mathscr{P}$. The distribution $P \in \mathscr{P}$ is called a *maximal* (*minimal*) *element* of the set $(\mathscr{P}, \preccurlyeq)$ if, for each $P' \in \mathscr{P}$, we have $P \preccurlyeq P'$ $(P' \preccurlyeq P)$ $\Rightarrow$ $P' \preccurlyeq P$ $(P \preccurlyeq P')$. In the sequel we consider orderings which formalize the idea of the strength of monotone dependence. We postulate the following conditions to be satisfied by these orderings

$$\begin{aligned} P^0 \preccurlyeq P \preccurlyeq P^{\mathrm{q}+} \quad &\text{for} \quad P \in \mathscr{P}^+, \\ P^0 \preccurlyeq P \preccurlyeq P^{\mathrm{q}-} \quad &\text{for} \quad P \in \mathscr{P}^-; \end{aligned} \tag{5.3.8}$$

if $(X, Y) \in \mathscr{P}$ and X, Y are independent, then (X, Y) is a minimal element of the set $(\mathscr{P}, \preccurlyeq)$; if $(X, Y) \in \mathscr{P}$ and $X = f(Y)$ a.e. for a monotone function f, then (X, Y) is a maximal element of the set $(\mathscr{P}, \preccurlyeq)$. (5.3.9)

The set $\mathscr{P}^{\mathrm{m}+}$ of maximal elements of the ordered set $(\mathscr{P}^+, \preccurlyeq)$ is a subset of $\mathscr{P}^{\mathrm{q}+}$, and the set $\mathscr{P}^{\mathrm{m}-}$ of maximal elements of $(\mathscr{P}^-, \preccurlyeq)$ is a subset of $\mathscr{P}^{\mathrm{q}-}$. The union $\mathscr{P}^{\mathrm{m}+} \cup \mathscr{P}^{\mathrm{m}-}$ will be denoted by $\mathscr{P}^{\mathrm{m}}$. For example, in the family $\tilde{\mathscr{P}}_{2\times 2}$ ordered by (5.2.2), this union is equal to $\tilde{\mathscr{P}}^{\mathrm{c}}_{2\times 2}$, and in $\tilde{\mathscr{P}}_{2\times 2}$ ordered by (5.2.13) this union is equal to $\tilde{\mathscr{P}}^{\mathrm{q}}_{2\times 2}$.

If $\mathscr{P}^{\mathrm{m}}$ contains only distributions concentrated on straight lines, it seems reasonable to investigate the deviation of the distributions belonging to $\mathscr{P}$ from linear functional dependence. For example, this can be done in the families $\mathscr{N}_2$, $\mathscr{T}_2^{(\alpha)}$ for any ordering satisfying (5.3.8) since in these families each distribution from $\mathscr{P}^{\mathrm{q}}$ is concentrated on a straight line.

In evaluating monotone or linear dependence we should take into account the measurement scales of X and Y (cf. page 25).

We distinguish two situations. In the first, both variables X, Y can be measured on the same scale: nominal, ordinal, interval, ratio, or absolute. We denote by $\mathfrak{f}$ the set of admissible mappings (for a given scale type cf. page 25). For (X, Y) this set is of the form

$$\{(f, f) \colon f \in \mathfrak{f}\}. \tag{5.3.10}$$

In the second, the two variables are measured on independently chosen scales. Therefore, the set of admissible transformations for (X, Y) is a Cartesian product of the form

$$\{(f, g) \colon f \in \mathfrak{f}, g \in \mathfrak{g}\}, \tag{5.3.11}$$

where $\mathfrak{f}$, $\mathfrak{g}$, are sets of admissible transformations for measurement scales of X and Y, respectively.

Denote by Φ_0 the set of injections of $\boldsymbol{R}$ (into $\boldsymbol{R}$), by Φ_1 the set of increasing mappings, by Φ_2 the set of increasing linear mappings, by Φ_3 the set of increasing linear homogeneous mappings, and by Φ_4 the one-element set containing only the identity mapping. The set Φ_0 is the set of admissible mappings for the nominal scale and Φ_1, Φ_2, Φ_3 are sets of admissible mappings for the ordinal, interval, and absolute scales, respectively. These scales are ordered by inclusion of the corresponding sets of admissible mappings. Clearly, if X and Y are measured on a scale which is at least ordinal, then $\mathfrak{f}, \mathfrak{g} \in \{\Phi_1, \Phi_2, \Phi_3, \Phi_4\}$, and if X and Y are measured on a scale which is at least interval, then $\mathfrak{f}, \mathfrak{g} \in \{\Phi_2, \Phi_3, \Phi_4\}$.

Let $\mathscr{S}_{\mathrm{M}}$ be the class of all admissible mappings if X and Y are measured on a scale which is at least ordinal:

$$\mathscr{S}_{\mathrm{M}} = \bigcup_{i,j=1}^{4} \{(f, g), f \in \Phi_i, g \in \Phi_j)\} \cup \bigcup_{i=1}^{4} \{(f, f), f \in \Phi_i\}.$$

In the same way we define the following classes: the class $\mathscr{S}_{\mathrm{L}}$ if X and Y are measured on scales which are at least interval (in the first union $i, j = 2, 3, 4$, and in the second union $i = 2, 3, 4$) and the class $\mathscr{S}_{\mathrm{LM}}$ if X is measured on a scale which is at least interval and Y on a scale which is at least ordinal (in the first union $i = 2, 3, 4, j = 1, 2, 3, 4$, and in the second union $i = 2, 3, 4$). Clearly, $\mathscr{S}_{\mathrm{L}} \subset \mathscr{S}_{\mathrm{LM}} \subset \mathscr{S}_{\mathrm{M}}$.

We consider it evident that the monotone dependence of X on Y can be evaluated only if X and Y are measured on scales which are at least ordinal and that the evaluation of the linear dependence of X and Y is possible only if the measurement scales of X and Y are at least interval. Otherwise, both monotone and linear dependence lose their intuitive sense.

Now we define ordered families of pairs (X, Y) in which it is possible to take into account measurement scales of X and Y in evaluating the monotone or linear dependence of X on Y. They are families of type M (for monotone dependence) and of type L (for linear dependence).

For any $S \in \mathscr{S}_{\mathrm{M}}$, an ordered family $(\mathscr{P}, \preccurlyeq)$ is of *type* M with respect to S if for any $(f, g) \in S$,

$$(X, Y) \in \mathscr{P} \quad \Rightarrow \quad (f(X), g(Y)) \in \mathscr{P}, \tag{5.3.12}$$

$$\mathscr{P} = \mathscr{P}^{+} \cup \mathscr{P}^{-}, \qquad \mathscr{P}^{\mathfrak{q}+} \cup \mathscr{P}^{0} \subset \mathscr{P}^{+}, \qquad \mathscr{P}^{\mathfrak{q}-} \cup \mathscr{P}^{0} \subset \mathscr{P}^{-}, \tag{5.3.13}$$

for any $(f_i, g_i) \in S \quad (i = 1, 2)$ and $(X, Y), (X', Y') \in \mathscr{P}$

$$(X, Y) \preccurlyeq (X', Y') \quad \Rightarrow \quad (f_1(X), g_1(Y)) \preccurlyeq (f_2(X'), g_2(Y')). \tag{5.3.14}$$

For any $S \in \mathscr{S}_{\mathrm{L}}$, an ordered family $(\mathscr{P}, \preccurlyeq)$ is of *type* L with respect to S if the conditions (5.3.12) – (5.3.14) are satisfied and moreover

each distribution $P \in \mathscr{P}^{\mathfrak{q}}$ is concentrated on a line. (5.3.15)

A family $(\mathscr{P}, \preccurlyeq)$ of type L with respect to $S \in \mathscr{S}_{\mathrm{L}}$ is of type M with respect to S. The largest subset $\mathscr{P}$ of a family of type M with respect to $S \in \mathscr{S}_{\mathrm{L}}$, that satisfies the condition

$$(X, Y) \in \mathscr{P}^{\mathfrak{q}}$$
$$\Rightarrow (X = aY + b \text{ a.e. for some } a, b \in \mathbf{R},\ a \neq 0), \tag{5.3.16}$$

is of type L with respect to S.

Families QD, $\widetilde{\mathrm{QD}}$, RD, LRD with an arbitrary ordering satisfying (5.3.14) are of type M with respect to any $S \in \mathscr{S}_{\mathrm{M}}$. For any fixed $S \in \mathscr{S}_{\mathrm{M}}$, natural ordering in QD is defined in the following way:

> $(X, Y) \preccurlyeq (X', Y')$ if and only if there exists a pair $(f, g) \in \boldsymbol{S}$ such that $f(X) \sim X'$, $g(Y) \sim Y'$ and the distribution function F, F' of the pairs $(f(X), g(Y)), (X', Y')$, respectively, satisfy the following condition: for any $x, y \in \boldsymbol{R}$
>
> $F(x, y) \leqslant F'(x, y)$ for $(X, Y), (X', Y') \in \mathrm{QD}^+$,
>
> $F(x, y) \geqslant F'(x, y)$ for $(X, Y), (X', Y') \in \mathrm{QD}^-$.

For any $\boldsymbol{S} \in \mathscr{S}_{\mathrm{M}}$, this ordering relation satisfies the condition (5.3.14).

Families ED and MR with an arbitrary ordering satisfying (5.3.14) are of type M with respect to any $\boldsymbol{S} \in \mathscr{S}_{\mathrm{LM}}$. For any fixed $\boldsymbol{S} \in \mathscr{S}_{\mathrm{LM}}$, the natural ordering in ED is defined as follows:

> $(X, Y) \preccurlyeq (X', Y')$ if and only if there exists a pair $(f, g) \in \boldsymbol{S}$ such that $X' \sim f(X)$, $Y' \sim g(Y)$ and
>
> $E\big(f(X) \mid g(Y) \leqslant y\big) \geqslant E(X' \mid Y' \leqslant y)$ a.e. for $(X, Y), (X', Y') \in \mathrm{ED}^+$,
>
> $E\big(f(X) \mid g(Y) \leqslant y\big) \leqslant E(X' \mid Y' \leqslant y)$ a.e. for $(X, Y), (X', Y') \in \mathrm{ED}^-$.

For any $\boldsymbol{S} \in \mathscr{S}_{\mathrm{LM}}$ this ordering satisfies (5.3.14) and can be regarded as an ordering in ED or MR. The family LR (defined as the set of pairs (X, Y) with the linear regression of X on Y) is not of type M (and hence not of type L) since it does not satisfy the condition (5.3.13).

The reason is that there are distributions with linearly increasing regression such that the mass of the distribution $P^{\mathfrak{q}+}$ does not lie in a straight line. Hence $P^{\mathfrak{q}+}$ is not a distribution with linear regression, and $P^{\mathfrak{q}+} \notin \mathrm{LR}^+$. The same holds for distributions with linearly decreasing regression, for which $P^{\mathfrak{q}-} \notin \mathrm{LR}^-$. It is worthwhile to eliminate these distributions from LR as irregular. Consequently, the following distributions remain: distributions from $\mathrm{LR}^0 = \mathrm{LR}^+ \cap \mathrm{LR}^-$, distributions from $\mathrm{LR}^+ \setminus \mathrm{LR}^0$ in which the mass of $P^{\mathfrak{q}+}$ is concentrated on an upward straight line, and distributions from $\mathrm{LR}^- \setminus \mathrm{LR}^0$ in which the mass of $P^{\mathfrak{q}-}$ is concentrated on a downward straight line. This family will be denoted by LRLF (linear regression and Fréchet distribution concentrated on a line). It contains the families $\mathscr{N}_2$ and $\mathscr{T}_2^{(\alpha)}$ for $\alpha > 0$.

For any $\boldsymbol{S} \in \mathscr{S}_{\mathrm{L}}$, family LRLF with an arbitrary ordering satisfying (5.3.14) is of type L with respect to $\boldsymbol{S}$. Note that if $(X', Y') \in \mathrm{LRLF}^+$ and $X \sim X'$, $Y \sim Y'$, then for both pairs the distributions $P^{\mathfrak{q}+}$ are concentrated on the same upward straight line which intersects the increasing regression lines of X on Y, and of X' on Y' at the same point (EX, EY). Hence, we say that the linear dependence of X on Y is weaker than the linear dependence of X' on Y' if the straight line on which the distribution $P^{\mathfrak{q}+}$ is concentrated lies closer to the line of regression of X' on Y' than to the line of regression of X on Y. A similar definition can be given for distributions from LRLF^-. Extending this natural ordering to pairs (X, Y), (X', Y') such that $X' \sim f(X)$, $Y' \sim g(Y)$ for a certain pair of mappings $(f, g) \in \boldsymbol{S} \in \mathscr{S}_{\mathrm{L}}$, we obtain an ordering in LRLF which is equivalent to the ordering in ED considered above

The set LRLF can be split into disjoint classes in such a way that (X, Y) and (X', Y') are in the same class if and only if, for some increasing linear functions f, g, we have $X' \sim f(X)$, $Y' \sim g(Y)$. Evidently, the ordering previously defined on LRLF is linear if restricted to any of these classes. Observe that this is the coarsest partition of LRLF with this property. Each of the families $\mathcal{N}_2, \mathcal{T}_2^{(\alpha)}$ is contained in a different class, and is linearly ordered.

It remains to discuss families of 2×2 distributions.

Clearly, any measurement of a binary random variable on the ordinal scale can be regarded as a measurement on the interval scale. Thus, it is meaningless to distinguish subsets $\mathcal{S}_{LM}$ and $\mathcal{S}_L$ of $\mathcal{S}_M$ with regard to families of 2×2 distributions.

Suppose that $S \in \mathcal{S}_M$. The family $\mathcal{P}_{2\times 2}$ with the ordering given by (5.2.2) or (5.2.13) is of type M but not of type L. The ordering (5.2.13) coincides with the natural ordering in QD restricted to $\mathcal{P}_{2\times 2}$. The intersection of $\tilde{\mathcal{P}}_{2\times 2}$ and LRLF, which contains distributions from $\mathcal{P}^+_{2\times 2}$ with equal marginal distributions ($\gamma_0 = \theta_0$), and distributions from $\mathcal{P}^-_{2\times 2}$ with one distribution being a mirror reflection of the second one ($\gamma_0 = 1-\theta_0$), with ordering (5.2.2) or (5.2.13), is of type L.

*

We will investigate the evaluation of monotone dependence only in families of type M and the evaluation of linear dependence only in families of type L. Moreover, we shall take into account only families $\mathcal{P}$ of the form $\mathcal{P} = \mathcal{P}^+ \cup \mathcal{P}^-$, where $\mathcal{P}^+$ and $\mathcal{P}^-$ are formalizations of positive and negative monotone dependence and the intersection $\mathcal{P}^0 = \mathcal{P}^+ \cap \mathcal{P}^-$ contains independent distributions from $\mathcal{P}$. In particular, we shall consider families contained in QD or ED.

For any ordered family $(\mathcal{P}, \preccurlyeq)$ of type M and any set $S \in \mathcal{S}_M$, the *measure of monotone dependence* in $(\mathcal{P}, \preccurlyeq)$ with respect to S will be defined as a function

$$\varkappa\colon \mathcal{P} \to [-1, 1]$$

satisfying the following conditions:

(i) $\varkappa(P) = 0 \;\Leftrightarrow\; P \in \mathcal{P}^0$;

(ii) $\varkappa(P) = 1 \;\Leftrightarrow\; P \in \mathcal{P}^{m+}$;

(iii) $\varkappa(P) = -1 \;\Leftrightarrow\; P \in \mathcal{P}^{m-}$;

(iv) $P \preccurlyeq P' \wedge P, P' \in \mathcal{P}^+ \;\Rightarrow\; \varkappa(P) \leqslant \varkappa(P')$;

(v) $P \preccurlyeq P' \wedge P, P' \in \mathcal{P}^- \;\Rightarrow\; \varkappa(P) \geqslant \varkappa(P')$;

(vi) if $P \in \mathcal{P}$ and $P_n \in \mathcal{P}$ $(n = 1, 2, \ldots)$ are distributions such that $P_n \xrightarrow{\mathcal{L}} P$, then $\varkappa(P_n) \to \varkappa(P)$.

A function $\varkappa\colon \mathcal{P} \to [-1, 1]$ satisfying conditions (i) – (vi) for a family of type L with respect to $S \in \mathcal{S}_L$ will be called the *measure of linear dependence* in $(\mathcal{P}, \preccurlyeq)$ with respect to S.

It follows from conditions (iv), (v) and (5.3.14) that for arbitrary $(X, Y) \in \mathcal{P}$ and $(f, g) \in S$ we have

$$\varkappa(X, Y) = \varkappa\big(f(X), g(Y)\big).$$

If $\varkappa$ is a measure of monotone or linear dependence, then so is the composition $\phi \circ \varkappa$, where ϕ is a continuous increasing function which maps $[-1, 1]$ onto $[-1, 1]$, and is such that $\phi(0) = 0$.

Let $S \in \mathscr{S}_M$. In the family QD with the natural ordering taking into account set S no measure of monotone dependence is known. However, if we apply the same ordering to the set of distributions from QD with continuous marginal distributions (thus a set forming a family of type M with respect to S), then we can use *Kendall's* τ and *Sperman's* ϱ_S as measures of monotone dependence. They are defined as follows:

$$\tau(X, Y) = 4 \int_{-\infty}^{+\infty} \int_{-\infty}^{+\infty} F(x, y) \, dF(x, y) - 1,$$

$$\varrho_S(X, Y) = 12 \int_{-\infty}^{+\infty} \int_{-\infty}^{+\infty} (F(x, y) - F_X(x) F_Y(y)) \, dF_X(x) \, dF_Y(y),$$

where F, F_X, F_Y are distribution function of (X, Y), X, Y, respectively.

If

$$(X, Y) \sim N_2(\mu_1, \mu_2, \sigma_1^2, \sigma_2^2, \varrho),$$

then

$$\tau(X, Y) = \frac{2}{\pi} \arcsin \varrho,$$

$$\varrho_S(X, Y) = \frac{6}{\pi} \arcsin \frac{\varrho}{2}.$$

The correlation coefficient

$$\operatorname{cor}(X, Y) = \frac{1}{(\operatorname{var}(X) \operatorname{var}(Y))^{1/2}} \int_{-\infty}^{+\infty} \int_{-\infty}^{+\infty} (F(x, y) - F_X(x) F_Y(y)) \, dx \, dy$$

is not a measure of monotone dependence in the family QD of distributions with continuous marginal distributions since it does not satisfy conditions (ii) and (iii). On the other hand, for $S \in \mathscr{S}_L$, the correlation coefficient is a measure of linear dependence in each subfamily of QD which is an ordered family of type L with finite second moments. In particular, the correlation coefficient is a measure of linear dependence in the family of all distributions from QD with uniform marginal distributions $U[0, 1]$. If $(X, Y) \in$ QD and the distributions X and Y are continuous, then transforming X and Y by means of their distribution functions we obtain the distribution of $(F_X(X), F_Y(Y))$ with uniform marginal distributions. Hence, the monotone dependence of $F_X(X)$ on $F_Y(Y)$ can be measured by the correlation coefficient $\operatorname{cor}(F_X(X), F_Y(Y))$. It can easily be verified that this correlation coefficient is equel to ϱ_S.

As in the family QD, in the family ED with the natural ordering corresponding to S (for any $S \in \mathscr{S}_{LM}$) no real-valued measure of monotone dependence is known. In the next section we shall show that in ED there exists a function-valued measure of monotone dependence with properties similar to those postulated for a real-valued measure of monotone dependence.

In the family LRLF we denote by $b_1(X, Y)$ the slope of the line of regression of X on Y, and by $b_2(X, Y)$ the slope of the line on which the distribution P^{q+} (if $(X, Y) \in \mathrm{LRLF}^+$) or the distribution P^{q-} (if $(X, Y) \in \mathrm{LRLF}^-$) is concentrated. Let

$$\varrho(X, Y) = \operatorname{sgn}\big(b_2(X, Y)\big)\frac{b_1(X, Y)}{b_2(X, Y)}. \tag{5.3.17}$$

The parameter ϱ is a measure of linear dependence in the family LRLF with the natural ordering corresponding to $\boldsymbol{S}$ (for any $\boldsymbol{S} \in \mathscr{S}_{\mathrm{L}}$). When the second moments of X and Y are finite, then

$$\varrho(X, Y) = \operatorname{cor}(X, Y).$$

In the family $\tilde{\mathscr{P}}_{2\times 2}$ measures of monotone dependence are:

— the parameters Q and L for the ordering (5.2.13);

— the parameters K, K', and the correlation coefficient for the ordering (5.2.2).

In the already mentioned intersection of $\tilde{\mathscr{P}}_{2\times 2}$ with LRLF, all the parameters Q, L, K, K', cor are measures of linear dependence for each of the orderings considered.

5.4. Monotone dependence function

A single number used to evaluate monotone dependence gives little information about the character of that dependence. To illustrate the kind of information which may be of interest, let us consider the pair $(X, Y) \in \mathrm{QD}^+$ with a distribution which is uniform in the shaded area shown in Fig. 5.4.1. It is intuitively clear that for

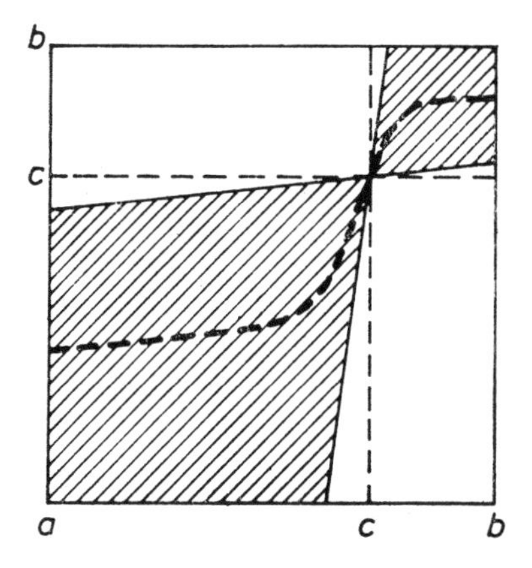

Fig. 5.4.1. Regression function of uniform distribution concentrated on the shaded area.

the set of smaller values of the two variables positive dependence for X and Y is weaker than for the set of larger values, and that the strength of dependence first increases and then decreases with the increase of the values of the variables.

A formalization of these intuitions is not easy and should be based on parameters taking values in more general sets than intervals of real numbers. In this section we define a parameter with values in the set of continuous functions h: $(0, 1) \to [-1, 1]$. This parameter is suitable for the formalization of intuitions relating to Fig. 5.4.1.

We assume that X and Y have finite expected values. We also assume for the moment that the distributions of X, Y are continuous. As previously, the quantiles of order p of X and Y will be denoted by x_p and y_p, respectively.

Consider the variables $(X \mid X < x_p)$, $(X \mid Y < y_p)$ and X. If $(X, Y) \in \mathrm{QD}^+$, then, for any $p \in (0, 1)$, these variables are stochastically ordered:

$$(X \mid X < x_p) \leqslant_{\mathrm{st}} (X \mid Y < y_p) \leqslant_{\mathrm{st}} X.$$

The distribution of $(X \mid Y < y_p)$ coincides with the distribution of $(X \mid X < x_p)$ if and only if the distribution of (X, Y) is concentrated on the union of quadrants $\{x < x_p, y < y_p\}$ and $\{x > x_p, y > y_p\}$. The variable X is an almost everywhere increasing function of Y if and only if, for each $p \in (0, 1)$, the distribution of (X, Y) is concentrated on the union of these quadrants. Finally, X is independent of Y if and only if $(X \mid Y < y_p) \sim X$ for each $p \in (0, 1)$.

Clearly, the family $\big((X \mid X < x_p), p \in (0, 1)\big)$ is stochastically increasing and the distributions of $(X \mid X < x_p)$ weakly converge to the distribution of X for $p \to 1$. If $(X, Y) \in \widetilde{\mathrm{QD}}^+$, the family $\big((X \mid Y < y_p), p \in (0, 1)\big)$ has the same property.

To any (X, Y) we assign a function $\mu^+_{X,Y}$, defined on $(0, 1)$ and taking on values in $[0, 1]$. It is given by the formula

$$\mu^+_{X,Y}(p) = \frac{EX - E(X \mid Y < y_p)}{EX - E(X \mid X < x_p)}, \quad p \in (0, 1). \tag{5.4.1}$$

For each p, the numerator of the expression $\mu^+_{X,Y}(p)$ measures the discrepancy between X and $(X \mid Y < y_p)$, while the denominator does the same for X and $(X \mid X < x_p)$.

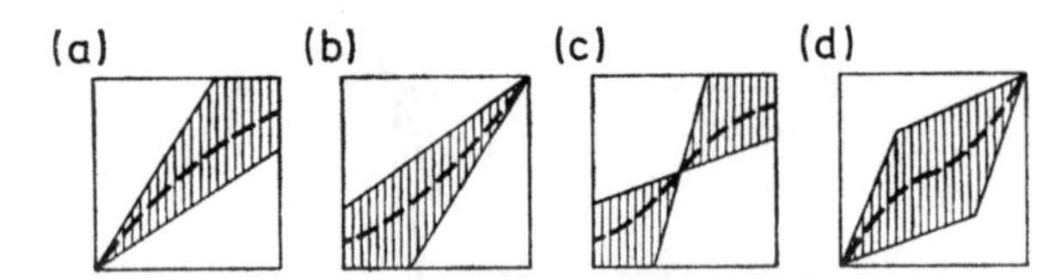

FIG. 5.4.2. Regression functions of uniform distributions concentrated on the shaded areas.

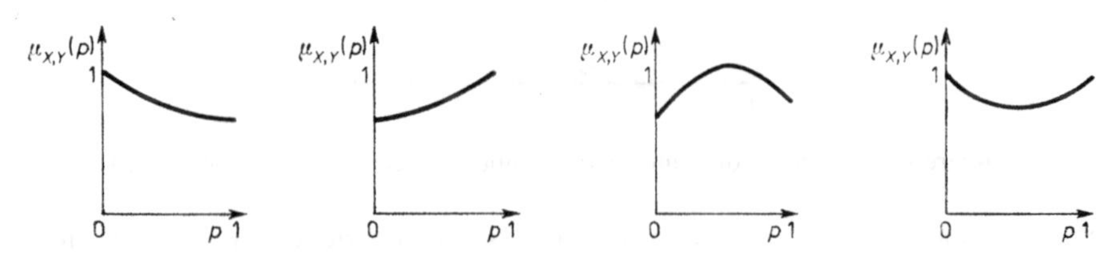

FIG. 5.4.3. Monotone dependence functions for uniform distributions with supports depicted in Fig. 5.4.2.

In Fig. 5.4.2 we present the supports of a few uniform distributions from QD^+. The graphs of the corresponding functions $\mu^+_{X,Y}$ are shown in Fig. 5.4.3. For the first distribution the function $\mu^+_{X,Y}$ is decreasing. This reflects the intuition that the greater the values of Y the weaker the dependence of X and Y. Interpretation of graphs of other functions $\mu^+_{X,Y}$ is similar.

For distributions from QD^- and for each $p \in (0, 1)$, we have

$$X \leqslant_{st} (X \mid Y < y_p) \leqslant_{st} (X \mid X > x_{1-p}).$$

As before, to any (X, Y) we assign a function $\mu^-_{X,Y}$ given by the formula

$$\mu^-_{X,Y}(p) = \frac{EX - E(X \mid Y < y_p)}{E(X \mid X > x_{1-p}) - EX}, \quad p \in (0, 1). \tag{5.4.2}$$

Now let us omit the assumption of continuity for X and Y. Definitions (5.4.1) and (5.4.2) will be extended to the set of all pairs of nondegenerate random variables X, Y with finite expected values. Taking into account that now the differences $p - P(X < x_p)$, $p - P(Y < y_p)$, $p - P(X > x_{1-p})$ may differ from zero, we get the following definitions:

$$\mu^+_{X,Y}(p) = \frac{pEX - E(X; Y < y_p) - E(X \mid Y = y_p)(p - P(Y < y_p))}{pEX - E(X; X < x_p) - x_p(p - P(X < x_p))}, \quad p \in (0, 1) \tag{5.4.3}$$

$$\mu^-_{X,Y}(p) = \frac{pEX - E(X; Y < y_p) - E(X \mid Y = y_p)(p - P(Y < y_p))}{E(X; X > x_{1-p}) - pEX + x_{1-p}(p - P(X > x_{1-p}))}, \quad p \in (0, 1) \tag{5.4.4}$$

where

$$E(X; Y < y_p) = \int_{-\infty}^{+\infty} \int_{-\infty}^{y_p} x \, dF(x, y),$$

$$E(X; X < x_p) = \int_{-\infty}^{x_p} \int_{-\infty}^{+\infty} x \, dF(x, y).$$

The expressions (5.4.3) and (5.4.4) do not depend on the choice of the quantiles x_p, y_p if these quantiles are not uniquely defined.

The functions $\mu^+_{X,Y}$ and $\mu^-_{X,Y}$ have the same signs; they are both positive for $(X, Y) \in QD^+ \cup MR^+$, and both negative for $(X, Y) \in QD^- \cup MR^-$. Moreover, $\mu^+_{X,Y} \leqslant 1$, $\mu^-_{X,Y} \geqslant -1$. By using these functions we define a parameter called the *monotone dependence function*. This parameter assigns to each pair of nondegenerated random variables X and Y with finite expected values a function $\mu_{X,Y}$: $(0, 1) \to [-1, 1]$ defined by the formula

$$\mu_{X,Y}(p) = \begin{cases} \mu^+_{X,Y}(p) & \text{if} \quad \mu^+_{X,Y}(p) \geqslant 0, \\ \mu^-_{X,Y}(p) & \text{if} \quad \mu^+_{X,Y}(p) < 0, \end{cases} \quad p \in (0, 1). \tag{5.4.5}$$

The monotone dependence function has the following properties:

(i) For any $p \in (0, 1)$, $-1 \leqslant \mu_{X,Y}(p) \leqslant 1$.

(ii) For any $p \in (0, 1)$, the equality $\mu_{X,Y}(p) = 1$ holds if and only if either

$$P(X < x_p, Y \geqslant y_p) = P(X > x_p, Y \leqslant y_p) = 0$$
$$\text{and} \quad P(Y < y_p) < p < P(Y \leqslant y_p)$$

or

$$P(X < x_p, Y > y_p) = P(X > x_p, Y \leqslant y_p) = 0$$
$$\text{and} \quad P(Y \leqslant y_p) = p$$

or

$$P(X < x_p, Y \geqslant y_p) = P(X > x_p, Y < y_p) = 0$$
$$\text{and } P(Y < y_p) = p < P(Y \leqslant y_p).$$

A similar condition can be formulated for $\mu_{X,Y}(p) = -1$.

(iii) $\mu_{X,Y}(p) \equiv 1$ if and only if there exists a nondecreasing function f: $\boldsymbol{R} \to \boldsymbol{R}$ such that $X = f(Y)$ P-a.e.,

$\mu_{X,Y}(p) = 0$ if and only if $E(X \mid Y = y) = E(X)$ P-a.e.,

$\mu_{X,Y}(p) = -1$ if and only if there exists a nonincreasing function f: $\boldsymbol{R} \to \boldsymbol{R}$ such that $X = f(Y)$ P-a.e..

(iv) $(X, Y) \in \mathrm{QD}^+$ (QD^-) if and only if for each nondecreasing function f such that $E(|f(X)|) < \infty$ we have $\mu_{f(X),Y} \geqslant 0$ $(\mu_{f(X),Y} \leqslant 0)$; hence X and Y are independent if and only if $\mu_{f(X),Y}(p) \equiv 0$ for any such function f.

(v) If (X, Y), $(X', Y') \in \mathrm{QD}$ and $X' \sim f(X)$, $Y' \sim g(Y)$, where $f \in \Phi_2$, $g \in \Phi_1$, then

$$(X, Y) \preccurlyeq_{\mathrm{QD}} (X', Y') \;\Rightarrow\; |\mu_{X,Y}| \leqslant |\mu_{X',Y'}|.$$

(vi) If a sequence $(P_n, n = 1, 2, \ldots)$ of distributions of pairs (X_n, Y_n) is weakly convergent to a distribution P of a pair (X, Y) such that

$$P(|X| \leqslant A) = 1 \tag{5.4.6}$$

for a positive constant A, then

$$\mu_{X_n, Y_n}(p) \xrightarrow[n \to \infty]{} \mu_{X,Y}(p).$$

(vii) For any $a, b \in \boldsymbol{R}$, $a \neq 0$, and an arbitrary function g: $\boldsymbol{R} \to \boldsymbol{R}$

$$\mu_{aX+b,g(Y)}(p) = \begin{cases} (\operatorname{sgn}(a))\mu_{X,Y}(p) & \text{if } g \text{ is increasing } P\text{-a.e.,} \\ (-\operatorname{sgn}(a))\mu_{X,Y}(1-p) & \text{if } g \text{ is decreasing } P\text{-a.e..} \end{cases}$$

(viii) If $X \sim X'$, $Y \sim Y'$, then

$$\mu_{X,Y} = \mu_{X',Y'} \;\Leftrightarrow\; E(X \mid Y = y) = E(X' \mid Y' = y) \;\; P\text{-a.e..}$$

(ix) Suppose that X, Y have continuous marginal distributions, f^+ is an increasing function such that $f^+(Y) \sim X$, f^- is an increasing function such that $f^-(Y) \sim (-X)$, m^+ is the regression function of X on $f^+(Y)$, and m^- is the regression function of X on $f^-(Y)$. Then

1° if m^+ is continuous, nonlinear, nondecreasing and convex (concave), then $\mu_{X,Y}$ is positive and increasing (decreasing);

2° if m^- is continuous, nonlinear, nondecreasing and convex (concave), then $\mu_{X,Y}$ is negative and increasing (decreasing).

(x) If there exists an increasing function f^+ such that $f^+(Y) \sim X$, then, for each $\varrho \in [0, 1]$, we have

$$\mu_{X,Y}(p) \equiv \varrho \quad \Leftrightarrow \quad m^+(y) = \varrho y + (1-|\varrho|)EX,$$

where m^+ is the regression function of X on $f^+(Y)$; if there exists an increasing function f^- such that $f^-(Y) \sim (-X)$, then for each $\varrho \in [-1, 0]$ we have

$$\mu_{X,Y}(p) \equiv \varrho \quad \Leftrightarrow \quad m^-(y) = \varrho y + (1-|\varrho|)EX,$$

where m^- is the regression function of X on $f^-(Y)$.

These properties show the differences existing between the monotone dependence function and the real-valued parameters. Let us note that the notion of a real-valued measure of monotone dependence with values in $[-1, 1]$ can be naturally extended to the notion of a function-valued measure of monotone dependence with values in the set of continuous functions h: $(0, 1) \to [-1, 1]$ ordered by the natural relation

$$h_1 \leqslant h_2 \quad \Leftrightarrow \quad \big(h_1(p) \leqslant h_2(p),\ p \in (0, 1)\big).$$

In particular, constant functions identically equal to $+1, 0, -1$ play the same role for a function-valued measure as the values $+1$, 0, -1 for a real-valued measure. It is easily seen that $\mu_{X,Y}$ is a function-valued measure of monotone dependence with respect to $S \in \mathscr{S}_{\mathrm{LM}}$ in the naturally ordered set ED restricted to distributions in which X has a bounded support (this restriction is necessary for maintaining the continuity of the measure). The same holds if ED is replaced by QD.

The shape of a monotone dependence function gives a great deal of information which could not be provided by any real-valued parameter. It follows from (iv) that if the function $\mu_{X,Y}$ changes the sign, then $(X, Y) \notin$ QD. By (ix), the convexity of m^+ and the concavity of m^- correspond to an "increasing strength of dependence" and a "decreasing strength of dependence" of X on Y. This is illustrated by distributions which are uniform on the supports presented in Fig. 5.4.2(a) and (b), where $X \sim Y$ and hence m^+ is the regression function of X on Y. In case (a), the strength of dependence "decreases", m^+ is concave, and $\mu_{X,Y}$ is decreasing; in case (b) the strength of dependence "increases", m^+ is convex, and $\mu_{X,Y}$ is increasing.

Of particular importance is the class of pairs (X, Y) for which the monotone dependence function is constant. Consider the set of pairs of variables X, Y which have finite second moments and are such that there exists an increasing function h

satisfying the conditions $h(Y) \sim X$ if $\mathrm{cor}(X, Y) > 0$ and $h(Y) \sim (-X)$ if $\mathrm{cor}(X, Y) < 0$. By (x), in this set of pairs (X, Y) the function $\mu_{X,Y}$ is constant if and only if $(X, h(Y)) \in \mathrm{LRLF}$. Moreover, $\mu_{X,Y}$ is equal to 0 if the regression of X on Y is constant, and otherwise, it is equal to the square root of the correlation ratio (cf. formula (5.3.21)) with the sign equal to that of $\mathrm{cor}(X, Y)$. Since the correlation ratio is equal to the squared correlation coefficient if and only if the regression of X on Y is linear, the following relation holds in the considered set of pairs:

$$\mu_{X,Y}(p) \equiv \mathrm{cor}(X, Y) \quad \Leftrightarrow \quad (X, Y) \in \mathrm{LRLF}. \tag{5.4.7}$$

In the above equivalence the correlation coefficient can be replaced by its extension defined in LRLF and denoted by ϱ (cf. formula (5.3.17)), if instead of the finiteness of the second moments of X and Y only the finiteness of the expected value of X is required.

The family LRLF formalizes the *regular linear dependence of X on Y*, and thus it is important in describing empirical phenomena. In this respect, the characterization of the family LRLF by means of (5.4.7) is very useful. Let us note that at the sample stage the sequence of sample correlation coefficients $\widehat{\mathrm{cor}}^{(n)}(X, Y)$ and the sequences of sample monotone dependence function $\hat{\mu}^{(n)}_{X,Y}(p)$, for any $p \in (0, 1)$, converge with probability 1 to the respective parameters of (X, Y) as $n \to \infty$. Hence, for sufficiently large samples, relation (5.4.7) is approximately satisfied for sample parameters.

In practice, the graph of a sample monotone dependence functions is compared with the value of the sample correlation coefficient, which gives better insight into the character of the dependence of X and Y. The application of this method to the investigation of the cell cycle will be presented in Section 8.4 (cf. Fig. 8.4.3 – 8.4.6, p. 208 – 211).

5.5. Codependence in a pair of random variables

Problems of evaluating the dependence of variables arise in practice when we deal with populations of objects described by a pair of features. These features are formally treated as random variables. Drawing an object from a given population regardless of the values of the features is formally equivalent to sampling the values according to the joint distribution of the features. Measures of dependence serve for evaluating the strength of the affinity between features for a randomly chosen object representing a given population.

Consider now a population of pairs of objects of a certain kind, e.g., a population of brothers and sisters, a population of married couples, or a population of sister-cells (i.e., cells originating from the division of the same mother-cell). For the sake of convenience we shall use for all pairs of a given kind a conventional name: family. Members of any family are elements of some primary population (for instance, the population of children from a given area, inhabitants of a given region, cells of a given type). In this population a certain feature V is defined, e.g., the weight of a child at the moment of birth, the average number of cigarettes smoked daily by a person during the year preceding the investigations, the length of life of a cell. Hence, to each family there correspond two values of the feature V. For a given ordering of the elements of a given family, let us denote the feature in question by X for the first element and by Y for the second element.

Let P denote a distribution of the pair (X, Y). Clearly, this distribution describes the simultaneous occurrence of the values of the feature in question in a given family for a given ordering of its elements. However, sometimes we are also interested in the simultaneous occurrence in the family of the values of the investigated feature with no regard to the ordering of elements. This means that we are treating the family as an "unordered pair". Roughly speaking, we want to choose randomly a family and to order it randomly by giving an equal chance to each permutation of the elements.

Denote by P^* a distribution defined on $\boldsymbol{R}^2$ such that, for any event $A \subset \boldsymbol{R}^2$,

$$P^*(A) = \tfrac{1}{2}(P(A)+P(A')), \tag{5.5.1}$$

where $A' = \{(x, y)\colon (y, x) \in A\}$. Let (X^*, Y^*) denote a pair of random variables with the distribution P^*. The pair (X^*, Y^*) represents the values of a certain feature V in a randomly chosen and randomly ordered family. This can be expressed formally by saying that a pair of random variables $(X, Y)\varepsilon+(Y, X)\cdot(1-\varepsilon)$, where ε is a random variable independent of (X, Y) and taking on values 0, 1 with probability $\frac{1}{2}$, has the distribution P^*. The distribution P^* is called a *symmetrized distribution* of (X, Y). It is symmetric and its marginal distribution is the distribution of the feature V in the primary population (i.e., in the one from which the population of families is formed).

*

Now let us investigate dependence between values of the feature V in a family regarded as an unordered pair, i.e. dependence between X^* and Y^*. It will be called the *codependence between* X and Y. Hence, definitions concerning dependence between X and Y lead to the corresponding definitions of codependence. In particular, the variables X, Y are called *coindependent* if (X^*, Y^*) are independent. Similarly, the variables X and Y are called *quadrant codependent* if X^* and Y^* are quadrant dependent. If random variables X and Y are quadrant dependent and $X \sim Y$, then X and Y are quadrant codependent. As a consequence of this fact, in the class of quadrant dependent random variables with the same marginal distributions, the independence of random variables is equivalent to the coindependence of those variables.

The assumption that X and Y are quadrant dependent is important here since there exist pairs (X, Y), $X \sim Y$, in which X and Y are coindependent but X and Y are not dependent. For example, it is so if the distribution of (X, Y) is uniform in the shaded area shown in Fig. 5.5.1 (the symmetrized distribution of (X, Y) is

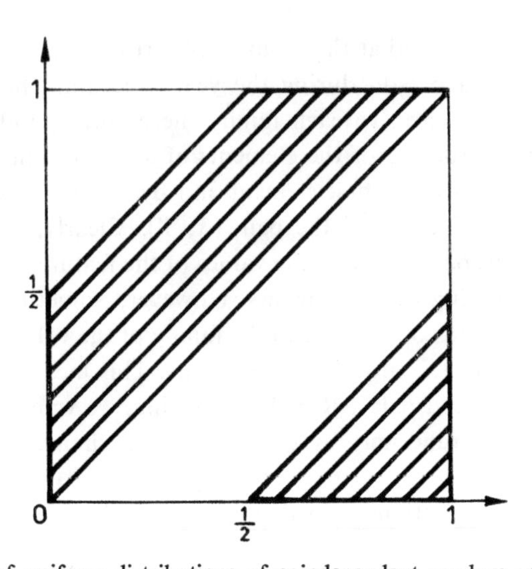

FIG. 5.5.1. Supports of uniform distributions of coindependent random variables X, Y.

uniform on $[0, 1]\times[0, 1]$) or if the distribution of (X, Y) is given by the matrix

$$\begin{bmatrix} \frac{1}{9} & \frac{1}{9}+\varepsilon & \frac{1}{9}-\varepsilon \\ \frac{1}{9}-\varepsilon & \frac{1}{9} & \frac{1}{9}+\varepsilon \\ \frac{1}{9}+\varepsilon & \frac{1}{9}-\varepsilon & \frac{1}{9} \end{bmatrix} \tag{5.5.2}$$

for any $\varepsilon \in (0, \frac{1}{9})$.

Quadrant codependence is preserved under change of the measurement scale of V if the scale is at least ordinal. Since X and Y represent the same feature V relating to the first and the second element, respectively, they must be measured on the same scale. The set $\boldsymbol{S}$ of admissible mappings is of the form $\boldsymbol{S} = \{(f,f)\colon f \in \mathfrak{f}\}$, where $\mathfrak{f}$ is the group of admissible transformations for V. If X and Y are quadrant codependent, then $f(X)$ and $f(Y)$ are also quadrant codependent for any $f \in \boldsymbol{\Phi}_1$. The coindependence of X and Y, in turn, is preserved under any group of transformations.

In the same way we can introduce codependence for other families of distributions defined in Section 5.3, e.g. for $\widetilde{\mathrm{QD}}$, RD, LRD, etc. To do this, we distinguish in a given family $\mathscr{P}$ a subset $\mathscr{P}^*$ of symmetric distributions and we consider pairs (X, Y) such that $(X^*, Y^*) \in \mathscr{P}^*$. For example, X and Y are called *regularly linearly codependent* if $(X^*, Y^*) \in \mathrm{LRLF}^*$, where LRLF* is the set of all symmetric distributions from LRLF. Regular linear codependence is preserved when the feature V is measured on at least interval scale, i.e., when $\mathfrak{f} \in \{\boldsymbol{\Phi}_2, \boldsymbol{\Phi}_3, \boldsymbol{\Phi}_4\}$.

The evaluation of codependence defined by a family of symmetric distributions $\mathscr{P}^*$ is obviously related to the evaluation of dependence in $\mathscr{P}^*$. Note that to each parameter $\varkappa$ we can assign a parameter $\varkappa^*$ by the formula

$$\varkappa^*(X, Y) = \varkappa(X^*, Y^*)$$

for (X, Y) such that (X^*, Y^*) belongs to the domain of $\varkappa$. The parameter $\varkappa^*$ will be called the *symmetrized parameter* $\varkappa$. For example, the *symmetrized correlation coefficient* is of the form

$$\operatorname{cor}^*(X, Y) = \frac{EXY - \frac{1}{4}(EX+EY)^2}{\frac{1}{2}(\operatorname{var} X + \operatorname{var} Y) + \frac{1}{4}(EX-EY)^2}$$

and the symmetrized parameter (5.3.20) is of the form

$$\eta^*(X, Y)$$

$$= \frac{1}{1-\frac{1}{4}\sum_{i=1}^{n}(p_{i.}+p_{.i})^2} \sum_{j=1}^{n}\sum_{i=1}^{n}\Big([\tfrac{1}{2}(p_{ij}+p_{ji}) -$$

$$-\tfrac{1}{4}(p_{.i}+p_{i.})]^2[(p_{i.}+p_{.i})^{-1}+(p_{.j}+p_{j.})^{-1}]\Big). \quad (5.5.3)$$

For $X \sim Y$ the symmetrized monotone dependence function is given by the formula

$$\mu^*_{X,Y} = \tfrac{1}{2}(\mu_{X,Y}+\mu_{Y,X}).$$

Evaluating codependence in a family

$$\{(X, Y)\colon (X^*, Y^*) \in \mathscr{P}^*\} \quad (5.5.4)$$

where $\mathscr{P}^*$ is a set of symmetric distributions we make use of the following fact: if the set $\mathscr{P}^*$ is closed under admissible transformations from a set $\boldsymbol{S}$ for a certain group $\boldsymbol{\Phi}$, then so is the set (5.5.4). Therefore, if $\varkappa$ is a measure of monotone or linear dependence in $\mathscr{P}^*$, then $\varkappa^*$ can be taken also as a measure of codependence in the set (5.5.4).

In particular, quadrant codependence between continuous variables X and Y can be measured by ϱ^*_S or τ^*, and the regular linear codependence by cor*.

In Section 5.4 we gave the characterization of the family LRLF (cf. (5.4.7)). Now we have the following analogous property of LRLF*:

$$\mu^*_{X,Y}(p) \equiv \operatorname{cor}^*(X, Y) \Leftrightarrow (X^*, Y^*) \in \mathrm{LRLF}^*. \quad (5.5.5)$$

*

A parameter $\varkappa$ satisfying the condition

$$\varkappa(X, Y) = \varkappa(X^*, Y^*) \quad (5.5.6)$$

is called *strongly symmetric*. Such parameters are used to evaluate codependence. Statistical inference concerning such parameters is usually based on strongly symmetric sample parameters.

Let (X_i, Y_i), $i = 1, \ldots, n$, be a simple random sample according to a distribution P. A statistic Z is called *strongly symmetric* if

$$Z\big((x_1, y_1), \ldots, (x_n, y_n)\big) =$$
$$= Z\big((x_1, y_1)\alpha_1 + (y_1, x_1)(1-\alpha_1), \ldots, (x_n, y_n)\alpha_n + (y_n, x_n)(1-\alpha_n)\big)$$

for arbitrary $\alpha_1, \ldots \alpha_n \in \{0, 1\}$. Every such statistic is of the form

$$Z\big((x_1, y_1), \ldots, (x_n, y_n)\big) = Z\big((v_1, w_1), \ldots, (v_n, w_n)\big), \quad (5.5.7)$$

where

$$v_i = \min(x_i, y_i), \quad w_i = \max(x_i, y_i), \quad i = 1, \ldots, n. \quad (5.5.8)$$

If the ordering of elements in a pair cannot be observed and, for each pair, we know only the two values of the feature in question, then Z is a maximal observable statistic. In this case a parameter of a distribution P is observably reducible if and only if it is strongly symmetric. For instance, it has been shown in Section 2.1 that for this type of observability the correlation coefficient is not observably reducible. Similarly, the hypothesis of the independence of X and Y is not observably reducible unless it coincides with the hypothesis of coindependence.

The estimation of parameters measuring the codependence is usually based on the estimation of the corresponding sample parameters. Let $P^{(n)}$ denote the empirical distribution for the sequence $\omega = ((x_1, y_1), \ldots, (x_n, y_n))$ and let $(P_\omega^{(n)})^*$ denote its symmetrization, i.e., the empirical distribution for the sequence (x_1, y_1), $(y_1, x_1), \ldots, (x_n, y_n)$, (y_n, x_n). By F_n^* we shall denote the symmetrized sample distribution function corresponding to $(P_\omega^{(n)})^*$,

$$F_n^*(x, y) = (P_\omega^{(n)})^* ((-\infty, x] \times (-\infty, y]).$$

If $\varkappa$ is a parameter whose domain contains all the empirical distributions, we can define the symmetrized sample parameter $\varkappa$:

$$\hat{\varkappa}_n^* = \varkappa((P_\omega^{(n)})^*). \tag{5.5.9}$$

For example, the symmetrized sample coefficient ϱ_S (i.e., the symmetrized rank correlation coefficient) is equal to

$$\begin{aligned} &\frac{48n}{4n^2-1} \sum_{i=1}^{n} F_n^*(x_i) F_n^*(y_i) - \frac{3(2n+1)}{2n-1} \\ &= 12 \sum_{i=1}^{n} \frac{r(i) r(n+i)}{n(4n^2-1)} - \frac{3(2n+1)}{2n-1}, \end{aligned} \tag{5.5.10}$$

where $r(i)$, $(r(n+i))$, $i = 1, \ldots, n$, denotes the rank x_i (y_i) of the sequence $x_1, \ldots, x_n, y_1, \ldots, y_n$.

Since for any continuous distribution P the sequence $\varrho_S(P_\omega^{(n)})$ is convergent to $\varrho_S(P)$ P-a.e., the sequence of expressions (5.5.10) is convergent to $\varrho_S^*(P)$. Clearly, this is true for any parameter $\varkappa$ for which the sequence $\varkappa(P_\omega^{(n)})$ is convergent to $\varkappa(P)$ P-a.e.

5.6. Final remarks

It follows from the considerations presented in this chapter that measures of dependence are sensible only in sufficiently "regular" families of distributions with natural linear orderings (e.g. in the family $\mathscr{P}^u$ of 2×2 distributions with uniform marginal distributions, in the family $\mathscr{N}_2$, in the family $\mathscr{T}_2^{(\alpha)}$ for any fixed $\alpha > 0$). For such "regular" families, measures of dependence satisfy some monotonicity conditions stronger than postulates (iv) and (v) appearing in the definition of a measure of monotone or linear dependence (cf. p. 121). Namely, for any two distributions from $\mathscr{P}^+$ ($\mathscr{P}^-$), the value of a measure for one distribution is greater (smaller) than the value of that measure for another distribution *if and only if* these distributions are ordered according to the strength of dependence.

In families in which a linear ordering is not defined it is rather not justified to compare the strength of dependence by means of numerical measures. Therefore, in those cases, measures of dependence are usually chosen on the basis of tradition rather than rational arguments.

In the statistical literature a great deal of attention has been devoted to defining classes of distributions of pairs (X, Y), where X and Y are monotonically dependent. One of the earliest and basic contributions to the problem has been presented in Lehmann (1966), where the definitions of the families QD, $\widetilde{\text{QD}}$, RD, LRD are given. Among further papers investigating this problem let us quote: Yanagimoto (1972), Esary and Proschan (1972), Alam and Wallenius (1976). Yanagimoto proposed classification of families of distributions with monotone dependence according to various criteria.

Papers devoted to axiomatic definitions of measures of stochastic dependence appear only occasionally.

The first axiomatic approach to the problem of evaluating stochastic dependence has been proposed by Renyi (1959). According to this approach stochastic dependence should be evaluated by symmetric measures which map the set of all pairs (X, Y) onto the interval $[0, 1]$. Moreover, Renyi postulated invariance of measures und er injective Borel functions. This means that an arbitrary scale, e.g. nominal, is admitted to measuring X and Y. A detailed discussion of Renyi's axioms is to be found in a paper by Schweitzer and Wolff (1981), where another set of axioms was also given. The authors considered symmetric measures which map the set of pairs (X, Y) with continuous marginals onto the interval $[0, 1]$. Invariance of measures under strictly monotone transformations of both variables was postulated; in other words, variables X, Y were assumed to be measured on ordinal scales. This invariance requirement means in fact that what the authors intend to evaluate is monotone dependence (but without the distinction between positive and negative dependence). However, if monotone dependence is to be dealt with, then, in our opinion, not *all* pairs (X, Y) with continuous marginals should be admitted to considerations. Moreover, the symmetry postulate may be debatable since it is evident that the strength of dependence of X on Y may be different from the strength of dependence of Y on X (e.g. if X and Y can be measured on different scales).

A detailed polemic with the approach presented in Schweitzer and Wolff (1981) is given in Bromek, Kowalczyk and Pleszczyńska (1988). The set of axioms proposed in their paper has been cited in Section 5.3 of the present book. Their axioms differ from those formulated by Schweitzer and Wolff mainly in considering an *ordered* set of distributions and postulating that a measure of dependence should be consistent with this ordering.

Schweitzer and Wolff limited postulates concerning ordering to the postulates that: 1° the value 0 corresponds to independence and value 1 to functional dependence of X on Y, 2° on the set of bivariate normal distributions the measure of dependence is an increasing function of the absolute value of the correlation coefficient. The latter postulate corresponds, in a way, to the requirement of consistency of the measure with the ordering in the family of normal distributions.

In Schweitzer and Wolff (1981) some new measures of dependence were proposed in the form of normalized distance measures between a given bivariate distribution with uniform marginals and the distribution of two independent random variables with uniform marginals. In this way Schweitzer and Wolff refer to measures based on grade transformations which were considered among others by Kimeldorf and Sampson (1975a, 1975b, 1978) and by Kowalczyk and Ledwina (1982). The following measure introduced by Schweitzer and Wolff,

$$\sigma(X, Y) = 12 \int_{-\infty}^{+\infty} \int_{-\infty}^{+\infty} |F(x, y) - F_X(x) F_Y(y)| \, \mathrm{d}F_X(x) \, \mathrm{d}F_Y(y)$$

satisfies all the postulates stated by these authors. On the other hand $|\varrho_S|$ and $|\tau|$ satisfy all these postulates excepted that the value 0 is assumed exclusively by independent distributions. Clearly, for $(X, Y) \in \mathrm{QD}$, $\sigma(X, Y) = |\varrho_S(X, Y)|$.

Neither Renyi nor Schweitzer and Wolff related the evaluation of dependence to the measurement scales for X and Y. In the paper by Dąbrowska (1985) measurement scales are not introduced in an explicit way either. Her paper deals with global and monotone dependence of X on Y. Global dependence is understood in a broad sense as a formalization of the influence of the realization of Y on the realization of X. Dąbrowska postulates that a measure should be invariant under linear increasing transformations of X and injective or increasing transformations of Y, where injective transformations correspond to global dependence, and increasing transformations to monotone dependence. This means that global dependence is evaluated only for X measured on the interval scale and Y measured on the nominal scale, while monotone dependence is evaluated only for X measured on the interval scale and Y measured on the ordinal scale. In our opinion, the evaluation of global dependence should be considered for any measurement scale for the pair (X, Y) and the evaluation of monotone (or linear) dependence should be possible for any scale such that $\boldsymbol{S} \in \mathscr{S}_{\mathrm{M}}$ ($\boldsymbol{S} \in \mathscr{S}_{\mathrm{L}}$) (cf. Section 5.3). Moreover, we are of the opinion that a measure of dependence should be invariant under all transformations admissible for the measurement of (X, Y). Hence, measures of global, monotone, or linear dependence should not be required to be invariant with respect to some particular group of transformations of (X, Y).

Dąbrowska postulates that in the case of monotone dependence the condition

$$(X, Y) \preccurlyeq (X', Y') \quad \Rightarrow \quad (f(X), g(Y)) \preccurlyeq (f(X'), g(Y'))$$

should be satisfied for any linear increasing function f and any increasing function g, while we admit in Section 5.3 various pairs (f_1, g_1) and (f_2, g_2) of such functions for (X, Y) and (X', Y'). Moreover, Dąbrowska deals with ordering such that (X, Y) and (X', Y') may be ordered only if $X' \sim X$ and $Y' \sim Y$. Observe that, for natural orderings in the sets QD, ED (cf. Section 5.3), these two pairs of variables can also be in an ordering relation if $X' \sim f(X)$, $Y' \sim g(Y)$ for a certain pair $(f, g) \in S$.

Dąbrowska's paper (1985) is related to earlier works on measures of location, dispersion, asymmetry, and kurtosis, for univariate distributions (Bickel and Lehmann, 1975a, 1975b, 1976). In the introduction to the above mentioned series of papers Bickel and Lehmann (1975a) pointed out the importance of an appropriate choice of the ordering relation in a family of distributions. They also mention restricting the measurement to sufficiently homogeneous families of distributions. Other requirements formulated by those authors concern the robustness of measures and some properties of their sample counterparts. In place of universally applicable measures Bickel and Lehmann recommend building measures suitable for particular needs and, above all, for given families of distributions.

Let us also mention the axiomatic approach to evaluation of dependence in the family $\mathscr{P}_{2\times 2}$ proposed by Edwards (1963). It is shown there that a measure of dependence in this family is symmetric and independent of marginal distributions, and, moreover, that it preserves the ordering defined by (5.2.13) if and only if it is an increasing function of the parameter $q = p_{11}p_{22}/p_{12}p_{21}$.

The evaluation of "global" dependence is closely related to the prognostic approach to the evaluation of dependence. This was one of the earliest approaches to the problem and it is still being continued (Goodman and Kruskal, 1954, 1959, 1963, 1972; De Groot, 1970; Bjørnstadt, 1975; Lissowski, 1978; Dąbrowska 1985). Bjørnstadt (1975) made an attempt to embody measurement scales in the prognostic approach. In general, global dependence could be regarded as a monotone dependence of X on a predictor $\hat{X}$ which solves a given prediction problem (taking into account measurement scales).

The definition and properties of monotone dependence function can be found e.g. in Kowalczyk (1977, 1982), Kowalczyk and Pleszczyńska (1977), Kowalczyk *et al.* (1979, 1980).

The concept of an "unordered pair" of random variables (cf. e.g. Hinkley, 1973) arises in a natural way in statistical inference with limited observability where the maximal observable statistic is of the form $(\min(X, Y), \max(X, Y))$. Various approaches to defining measure of codependence (i.e., a measure of dependence for unordered pairs of variables) are reported in the literature. The sample parameter cor*, called the *intra-class correlation coefficient*, was defined by Kendall and Stuart (1961). For testing coindependence the symmetrized sample rank correlation coefficient (see Shirahata, 1981) and the strongly symmetric statistic ψ^2 (see Haber, 1982) have been used. A definition of symmetrized distributions and properties of symmetrized measures cor*, μ^* have been presented by Ćwik *et al.* (1982). Relationships between dependence and codependence in the family QD have been presented by Bromek and Kowalczyk (1986).

The concept of codependence for a pair of random variables can easily be generalized to random vectors. Namely, for an arbitrary $k > 1$, we define a k-dimensional symmetrized distribution (as for $k = 2$), and we regard a vector $X^* = (X_1^*, \ldots, X_k^*)$ having this distribution as an unordered set of values of a certain feature

V in a family of size k chosen randomly from the population. Evidently, each pair of variables (X_i^*, X_j^*), $i \neq j$, has identic symmetric distribution and represents a pair of objects randomly chosen from a family. The codependence of components of X is defined as the dependence for the pair (X_1^*, X_2^*). Codependence can also be defined in a population of families differing in the number of elements but containing not less than two elements. In Ćwik (1985) the properties of the symmetrized correlation coefficient $\text{cor}^*(X_1, \ldots, X_k)$ were investigated and the relations between $\text{cor}^*(X_1, \ldots, X_k)$ and the Fisher measure ϱ_F (cf. Section 3.4) were found. Note that ϱ_F measures heterogeneity of a population composed of classes of objects.

There are also attempts to reduce the evaluation of dependence of random vectors to the bivariate case. The dependence of a random variable X on a random vector Y is often regarded as the dependence of X on $f(Y)$ for an appropriately chosen function f. Prediction approach can also be applied in a natural way to random vectors.

Part Three

SELECTED PRACTICAL PROBLEMS

Chapter 6

Statistical problems of population genetics

Key words: *chromosome, phenotype, gene, genotype, karyotype, Hardy–Weinberg hypothesis, random mating, multinomial model, coefficient of kinship, exponential family of distributions.*

6.1. Introduction

Population genetics deals with population studies of live organisms which transfer their genetic structure to their progeny according to certain stable laws. The genetic structure of an organism is not directly observable, however, some of the organism's traits, depending on it deterministically or stochastically, can be observed. The basic problems of population genetics concern investigation of the genetic structure of the population, its changes with time, comparisons of various populations, anticipation of their future development etc.

Problems of population genetics absorb statisticians because of their specificity resulting, among other things, from the complexity of the genetic structure of an organism, nontrivial observability, and interesting physical interpretation of population parameters. At the same time, application of statistical methods in solving these problem is particularly justified, since the assumptions of a random development of the population generally correspond well to reality.

Jacquard (1974) regards population genetics as the "field of study in which probabilistic models correspond most clearly to the reality of the phenomena being studied". The results of experiments inspired the research workers, whose intuition in creating models was then confirmed by discovery of concrete biological entities of hypothetical structures and the behaviour of the actual hereditary mechanism.

In Section 6.2 we introduce the basic concepts of genetic structure of a single organism (chromosome, gene, allele, locus, genotype, phenotype and karyotype) and we consider schemes of dependence of the phenotype on the genotype and the karyotype.

In Section 6.3 we present the simplest formalizations of the development of human populations and of the synthetic description of that development by means of selected parameters. We discuss a static model and a dynamic model (called primitive generation model) in which a single locus of an autosomal chromosome is considered and phenotype is assumed to depend deterministically on the genotype. In the static case we introduce, among others, the concept of gene distribution in a population of gametes, of the distribution of a "symmetrized genotype", of the independence and coindependence of genes in a genotype, and of indices describing deviations from independence. In the dynamic case we present—under simplifying assumptions— a scheme of dependence of the child's sex and genotype on the parental genotypes. This scheme implies stationarity of certain important parameters, which in turn makes it possible to formulate the hypothesis of (stationary) coindependence of genes in the genotype, called the *Hardy–Weinberg hypothesis.*

In Section 6.4 and Section 6.5 methods of estimating gene distribution in a population of gametes ("gene frequencies") and testing the Hardy–Weinberg hypothesis are presented.

6.2. The genetic structure of man and its relations with the phenotype

The nucleus of each human cell contains 23 pairs of chromosomes. The set of human cell chromosomes existing in identical copies in each of the cells is the only carrier of genetic information, and so it forms the basis of the genetic structure of the human organism. We can describe this structure in various ways, taking into account those of its elements which are significant from the point of view of a given study.

The pairs of chromosomes are ordered according to a certain key and numbered from 1 to 23. Chromosomes forming a pair are called *homologous.* Each pair consists of a copy of one of the two paternal chromosomes and a copy of one of the two maternal chromosomes which appear in father and mother in pairs bearing an identical number.

Using statistical methods, the geneticist usually considers a standard model in which each of the chromosomes is divided into a certain number of segments (the same for both homologous chromosomes and for all human individuals) in such a way that the DNA fragments forming a pair of homologous segments carry "the same" genetic information.

Each distinguished chromosome segment is called *gene* in the standard model, and its location in the chromosome is called *locus.* Both homologous segments have the same locus because of their identical location in the chromosome. Thus, a locus has two genes associated with it: paternal gene and maternal gene.

In the standard model it is assumed that for a given locus the paternal gene and the maternal gene may appear in various states called *alleles,* coming from the same set of states proper for the locus in each human organism, though the number of alleles in different loci varies from one to several dozen.

Let $\tilde{A}_1, \ldots, \tilde{A}_r$ be the set of alleles at a given locus. To each organism corresponds a pair of genes $(\tilde{A}_i, \tilde{A}_j)$ from this set, and we adopt the convention that the first element in the pair is the paternal gene and the second the maternal gene. In various studies we consider one or more loci, not necessarily for the same pair of homologous chromosomes. The set of gene pairs for the loci selected in a given investigation is called (in that investigation) *genotype of the organism*. Hence, in different research problems we consider different genotypes according to what loci are taken into consideration. (Geneticists also use the term *genotype* to define the set of all pairs of genes for a complete system of chromosomes).

The genotype is usually not observable, and information on it is based on a set of selected organism traits, accessible to observation in the particular study. The selected set of traits is called (in that study) *phenotype of the organism*. The modelling of stochastical relations between genotype G and phenotype F (in a given population and in a given study) is one of the basic stages of formalizing the problems of population genetics.

Most often it is assumed that the phenotype depends functionally on the genotype. Let u stand for the function mapping the set of all possible pairs of alleles from the set $\{\tilde{A}_1, \ldots, \tilde{A}_r\}$ into the set of values of a phenotypic trait connected with this locus. We assume that u is a symmetric function, i.e., paternal gene $\tilde{A}_i$ and maternal gene $\tilde{A}_j$ determine the same phenotype as paternal gene $\tilde{A}_j$ and maternal gene $\tilde{A}_i$. Let $G_{ij} = (\tilde{A}_i, \tilde{A}_j)$. Phenotypes corresponding to different genotypes G_{ii} are different, i.e.,

$$u(G_{ii}) \neq u(G_{jj}), \quad i \neq j,\ i, j = 1, \ldots, r. \tag{6.2.1}$$

With $i \neq j$ it can happen that one of the genes, for example $\tilde{A}_i$, dominates the other in such a sense that only $\tilde{A}_i$ influences the phenotype. This phenotype is then the same as in the case of the pair $(\tilde{A}_i, \tilde{A}_i)$:

$$u(G_{ij}) = u(G_{ji}) = u(G_{ii}), \quad i \neq j,\ i, j = 1, \ldots, r. \tag{6.2.2}$$

However, the influence of both genes may also be pronounced and then a phenotype different from those associated with genotypes G_{ii} or G_{jj} is obtained:

$$u(G_{ii}) \neq u(G_{ij}) \neq u(G_{jj}), \quad i \neq j,\ i, j = 1, \ldots, r. \tag{6.2.3}$$

In the first case we say that gene $\tilde{A}_i$ is *dominant* over $\tilde{A}_j$ (or in other words that gene $\tilde{A}_j$ is *recessive* to $\tilde{A}_i$), and in the second case, that gene $\tilde{A}_i$ is *codominant* with $\tilde{A}_j$ (and vice versa).

To give an illustration we shall find function u for the locus which causes appearance of the blood group system called AB0 *system*. Originally, four phenotypes were distinguished in this system, namely 0, A, B, and AB. At the same time it was assumed that there exist three alleles $\tilde{\mathrm{A}}$, $\tilde{\mathrm{B}}$, and $\tilde{0}$. Hence, $G_{11} = (\tilde{\mathrm{A}}, \tilde{\mathrm{A}})$, $G_{12} = (\tilde{\mathrm{A}}, \tilde{\mathrm{B}})$, $G_{22} = (\tilde{\mathrm{B}}, \tilde{\mathrm{B}})$, etc. It was further assumed that $\tilde{\mathrm{A}}$ codominates with $\tilde{\mathrm{B}}$, and $\tilde{0}$ is recessive to $\tilde{\mathrm{A}}$ and to $\tilde{\mathrm{B}}$. Thus, function u was defined as follows:

$$u(G_{11}) = u(G_{13}) = (G_{31}) = \mathrm{A},$$

$$u(G_{22}) = u(G_{23}) = (G_{32}) = \mathrm{B},$$

$$u(G_{33}) = 0, \tag{6.2.4}$$

$$u(G_{12}) = u(G_{21}) = \mathrm{AB}.$$

In the sequel six phenotypes 0, A_1, A_2, B, $\mathrm{A}_1\mathrm{B}$, $\mathrm{A}_2\mathrm{B}$ were distinguished instead of four, and instead of $\tilde{\mathrm{A}}$ two alleles $\tilde{\mathrm{A}}_1$ and $\tilde{\mathrm{A}}_2$ were introduced. It was found that gene $\tilde{0}$ is recessive to genes $\tilde{\mathrm{A}}_1$, $\tilde{\mathrm{A}}_2$ and $\tilde{\mathrm{B}}$, that $\tilde{\mathrm{A}}_1$ dominates $\tilde{\mathrm{A}}_2$ and codominates with $\tilde{\mathrm{B}}$, and finally, that $\tilde{\mathrm{A}}_2$ codominates with $\tilde{\mathrm{B}}$. A suitably modified function u is presented in Table 7.3.1 (cf. p. 169).

As a second example we shall consider the system Gm(1,2) with three alleles, denoted by serologists as $\mathrm{Gm}^{1,-2}$, $\mathrm{Gm}^{-1,-2}$ and $\mathrm{Gm}^{1,2}$. In this system we deal only with dominance of genes: $\mathrm{Gm}^{-1,-2}$ is recessive in regard to both the other genes, and $\mathrm{Gm}^{1,2}$ dominates $\mathrm{Gm}^{1,-2}$. Three phenotypes are present here denoted traditionally Gm(1, −2), Gm(−1, −2) and Gm(1, 2), where

$$\begin{aligned} u(G_{11}) &= u(G_{12}) = u(G_{21}) = \mathrm{Gm}(1, -2), \\ u(G_{22}) &= \mathrm{Gm}(-1, -2), \\ u(G_{33}) &= u(G_{13}) = u(G_{31}) = u(G_{23}) \\ &= u(G_{32}) = \mathrm{Gm}(1, 2). \end{aligned} \tag{6.2.5}$$

As seen, in the AB0 system with alleles $\tilde{\mathrm{A}}$, $\tilde{\mathrm{B}}$, $\tilde{0}$ we can ascertain by observation of the phenotype whether the genotype equals G_{33}, but this cannot be done for genotype G_{22} or genotype G_{11}. Similarly, in the Gm(1, 2) system it is possible to ascertain G_{22} but not G_{11} or G_{33}.

For any locus the codomination of two genes, say $\tilde{A}_i$ and $\tilde{A}_j$, causes that the events $G = \{G_{ii}\}$, $G \in \{G_{ij}, G_{ji}\}$, $G = \{G_{jj}\}$ are *atoms of observability*, i.e., observable events the subsets of which are not observable. The event $\{G = G_{ij}\}$ for $i \neq j$ is therefore not observable, thus, on the basis of the phenotype it is not possible to ascertain the origin of both the genes. If gene $\tilde{A}_i$ dominates gene $\tilde{A}_j$, then the event $G \in \{G_{ij}, G_{ji}, G_{ii}\}$ is the atom of observability, that is to say, on the basis of the phenotype we cannot distinguish the mixed genotype from the genotype of a pure dominating gene. However, if gene $\tilde{A}_i$ is recessive to all the remaining genes, then the event $\{G = G_{ii}\}$ is observable, and the phenotype allows to distinguish the pure genotype of the recessive gene.

A trait considered as a phenotype can depend functionally on two or more loci. This complicates the notation of function u, as well as the classification of the genotypes and the establishment which subsets of genotypes are observable. Let us consider a simple example. A selected phenotypic trait depends on two loci characterized by two-element sets of alleles $\{\tilde{A}_1, \tilde{A}_2\}$ and $\{\tilde{B}_1, \tilde{B}_2\}$, respectively. We assume that the trait values depend functionally only on the joint number of alleles $\tilde{\mathrm{A}}_1$ and $\tilde{\mathrm{B}}_1$ in the individual's genotype. Thus, the trait considered has five different values, corresponding to the joint number of alleles $\tilde{A}_1$ and $\tilde{B}_1$: 0, 1, 2, 3, 4. It happens frequently that these values are ordered in some natural and at the same time observable way (for instance lack of pigmentation, or poor, moderate, intense, very intense pigmentation), and that this ordering corresponds to the joint number of alleles $\tilde{A}_1$ and $\tilde{B}_1$.

Similar situations occur also when the number of loci is higher than two; moreover, it can be assumed that the influence of the number of alleles of a distinguished type on the "intensity" of the phenotypic trait differs in different loci. Genes with additive phenotypic effects are called *cumulative*.

Many phenotypic traits are partially shaped under the influence of the environment in which the organism grows, and then the dependence of the phenotype from the genotype is not deterministic but stochastic. This dependence is described by a family of conditional probability distributions of the phenotype given the genotype.

Phenotypic traits depending deterministically on a genotype have a finite set of values, while traits depending on it stochastically are not subject to such a limitation and may be discrete, continuous or mixed. Conditional observability of genotype on the basis of phenotype (cf. Sec. 1.5) usually does not appear in the case of traits conditioned stochastically, and for traits conditioned by cumulative genes the genotype can be recognized on the basis of phenotype only in exceptional cases.

*

In some investigations in population genetics the description of an organism is limited to genotype and phenotype for selected loci, in others information on the whole system of chromosomes, called the *organism's karyotype*, is needed. In a normal human karyotype there are 22 pairs of what is called *autosomal chromosomes* (with components diploid in each pair with respect to size and shape) and one pair of sex chromosomes. According to an international convention the latter pair has the number 23; pairs of autosomal chromosomes bear numbers 1 through 22 (according to size and the position of the so-called centromere). The pair of sex chromosomes in females consists of two diploid chromosomes, denoted X. In the pair of sex chromosomes in males there occurs one X chromosome (the same as in females) and a considerably smaller one, denoted Y.

Sometimes karyotypes occur with abnormal chromosome numbers. For instance, instead of a pair of homologous chromosomes a triplet of chromosomes appears.

The description of karyotypes differs considerably in different investigations depending on the information searched for and the type of measurements of values of observable traits. In practical investigations the classification number of a pair of chromosomes is not always observable, and it is not always known which of the chromosomes are homologous. As a rule only measurements of the length of the chromosome arms (that is sections into which the centromere divides the chromosome) are accessible.

The information on partition of the set of chromosomes into homologous pairs and on classification of the pairs according to the accepted convention is necessary when relations between the karyotype and certain diseases are to be investigated.

6.3. Genetic parameters of the human population

The karyotype and the genotype, and also many phenotypic traits do not alter with time. In population genetics we are not interested in description of particular individuals, but in a synthetic description of human groups, be it static or dynamic. In the sequel we begin with a typical static description and with classical research problems connected with it, and next we draw attention to topics relating to dynamic models.

In the static approach we distinguish a certain finite human population (for instance a set of individuals living in a definite moment of time and in a chosen territory, and satisfying certain imposed conditions) and we specify which of the organism's traits are to be considered. For the sake of simplification we will restrict our discussion to genotype G for one selected locus in an autosomal chromosome with a set of alleles $\{\tilde{A}_1, \ldots, \tilde{A}_r\}$ and we will assume that phenotype F is deterministically dependent on G by means of a known function u.

Hence, we consider a finite population with a pair of traits (F, G). Because of the functional relation between F and G the distribution of (F, G) is defined by a bivariate distribution P of genotype G on the set $\{G_{ij} = (\tilde{A}_i, \tilde{A}_j), i, j = 1, \ldots, r\}$. Typical investigation problems which arise in a static situation consist in estimation of the parameters of distribution P and in testing hypotheses concerning this distribution on the basis of the phenotypes observed in a sample of size n, taken from the population. We usually assume that samplings of particular individuals are mutually independent as regards the trait G, so that we obtain a simple random sample (cf. Sec. 1.2) with limited observability depending on function u.

Denote paternal gene and maternal gene by G_1 and G_2, respectively, so that $G = (G_1, G_2)$. In a general case we cannot assume symmetry of distribution P, not even identity of the distributions of paternal and maternal genes (that is, identity of marginal distributions of P). A special role is played by the distribution arising from distributions of genes G_1 and G_2, mixed in equal proportions. For, if the genotype structure ceased to exist, and genes released from genotypes of all individuals would form a set of "free" genes, then the distribution of a gene drawn from this set would become a mixture of distributions of genes G_1 and G_2. In Section 5.5 we spoke about the *primary population of objects* and about the population of clusters or "families" formed from these objects; here we have a population of "free" genes and a population of pairs of genes (i.e., genotypes).

The abstract population of free genes has its physical equivalent in the population of *gametes*, that is of cells originating from division of the paternal cell([1]).

Let t_i be the probability that the free gene (i.e., the gene selected at random from a population of gametes) equals $\tilde{A}_i$. Thus

$$t_i = \tfrac{1}{2}\big(P(G_1 = \tilde{A}_i) + P(G_2 = \tilde{A}_i)\big), \quad i = 1, \ldots, r.$$

([1]) This correspondence results from the course of *meiosis* which consists in random segregation of chromosomes during division of the cell.

If the distribution of the paternal gene G_1 equals the distribution of the maternal gene G_2, then it equals also the distribution of the free gene.

In accordance with Section 5.5 in which the randomly ordered pairs of random variables were discussed, to genotype G corresponds a randomly ordered genotype G^*, i.e., genotype with genes ordered according to a criterion not depending on their origin. The distribution P^* of randomly ordered genotype G is a symmetrization of distribution P. Thus, in accordance with (5.5.1),

$$P^*(G^* = G_{ij}) = \tfrac{1}{2}\big(P(G = G_{ij})+P(G = G_{ji})\big) = \begin{cases} g_{ii} & \text{for} \quad i = j, \\ \tfrac{1}{2}g_{ij} & \text{for} \quad i \neq j, \end{cases} \tag{6.3.1}$$

where

$$g_{ij} = P(G \in \{G_{ij}, G_{ji}\}). \tag{6.3.2}$$

The P^* distribution being symmetrical its marginal distributions are identical and equal to the free gene distribution. Hence,

$$t_i = g_{ii}+\tfrac{1}{2}\sum_{i \neq j} g_{ij}, \quad i = 1, \ldots, r. \tag{6.3.3}$$

The genes G_1 and G_2 in the genotype G are called *independent* if distribution P has independent components, i.e.,

$$P(G = G_{ij}) = P(G_1 = \tilde{A}_i)P(G_2 = \tilde{A}_j), \quad i, j = 1, \ldots, r.$$

Genes G_1 and G_2 are called *coindependent* (cf. Sec. 5.5) if distribution P^* has independent components, i.e.,

$$P^*\big(G^* = (\tilde{A}_i, \tilde{A}_j)\big) = t_i t_j, \quad i, j = 1, \ldots, r. \tag{6.3.4}$$

According to (6.3.1), genes G_1 and G_2 are coindependent if and only if

$$\begin{aligned} g_{ii} &= t_i^2, \\ g_{ij} &= 2t_i t_j, \end{aligned} \quad i, j = 1, \ldots, r. \tag{6.3.5}$$

For a locus with only two alleles $\tilde{A}_1$ and $\tilde{A}_2$ formula (6.3.5) is equivalent to the identity

$$g_{12}^2 = 4g_{11}g_{22}. \tag{6.3.6}$$

The set of distributions of genotypes with coindependent genes can be illustrated together with the remaining distributions of randomly ordered genotypes G^* by means of the de Finetti's diagram (Fig. 6.3.1). This is an equilateral triangle of height equal to 1, in which each of the points represents the distribution of genotype G^* defined by probabilities g_{11}, g_{12} and g_{22} so that g_{11}, g_{12} and g_{22} are respectively equal to the distances of the point from its projections on the sides. Points satisfying (6.3.6) lie on the parabola marked on the diagram by a dashed line.

If the genes G_1 and G_2 are independent and have identical distributions then they are coindependent too. On the other hand, coindependence of genes does not necessarily imply their independence: in case of a locus of three alleles the dis-

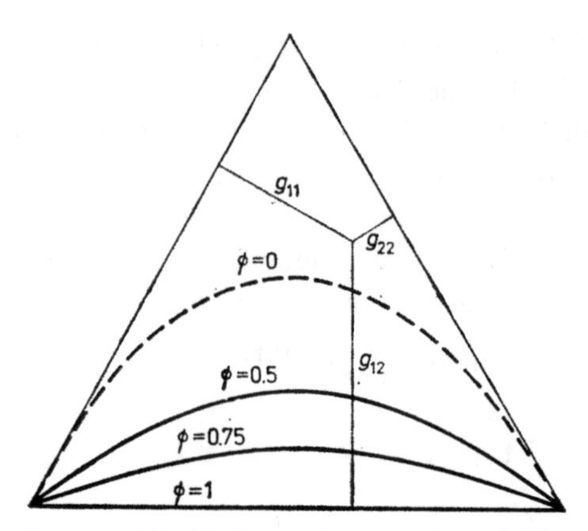

FIG. 6.3.1. De Finetti's diagram. Distributions of genotypes satisfying (6.3.9), for symmetrized correlation coefficient: $\phi = 0$ (coindependent genes), 0.5 and 0.75.

tribution given by matrix (5.5.2) is an example of a distribution of a genotype with genes G_1 and G_2 which have identical distributions and are coindependent but not independent.

In population genetics deviation from coindependence is measured by means of various measures of codependence for discrete distributions, for instance by means of the symmetrized coefficient η^* (formula (5.5.3)). In the case of loci with codominant alleles measures of codependence are observably reducible (hence, their sample analogs are observable), in other cases special measures have to be constructed. On the other hand, measures of deviation from independence are, in general, not observably reducible, even for loci with codominant alleles.

*

In some models of human population development the hypothesis of independence of paternal and maternal genes is a consequence of certain hypotheses on biological and social mechanisms of population development, hypotheses which are of special interest to the investigator. The rejection of the hypothesis of genes independence leads in that case to the rejection of these goal hypotheses. We shall present this in case of a simple dynamic population model, which will be called *primitive generation model.*

As previously, we discuss a single locus on an autosomal chromosome with alleles $\tilde{A}_1, \ldots, \tilde{A}_r$, the phenotype being deterministically dependent on the genotype.

In all the generation models the population history is the history of subsequent generations. These models are a considerable simplification of reality. They are, however, convenient in theoretical considerations, particularly in case when the population development can be described by indicating the state of the initial generation and the law according to which any offspring generation is formed from the parental generation. In the primitive generation model each individual in the offspring generation is treated as though his genotype and sex developed according to a random mechanism which is identical for each child in each generation and acts for the particular children in an independent way. This means a considerable simplification. Randomness is only partly justified by the character of the mechanisms of social and biological mating of pairs and by the formation of gametes, and independent treatment of particular children interferes with the fact that in a finite population the mating of some of the pairs imposes constraints on mating of others. In any case, we have to assume that the size of initial and following generations is sufficiently ("infinitely") large. Moreover, in the primitive generation model we exclude selection, consisting in dependence of the number of offspring in a given parental pair which reaches reproductive age, on the genotypes of this pair. We also omit the migration of individuals, i.e., we do not exclude any person from the process of reproduction and do not include any foreign persons. Further we assume that each of the parents transmits to his progeny the exact copy of one of his genes which eliminates mutation of genes.

Random formation of gametes is based on the following assumptions which are obligatory in the primitive generation model:

1° independent segregation of homologous chromosomes during the division of paternal cells (I Mendel Law); and

2° independence between chromosome segregation in a pair of autosomal chromosomes.

In a primitive generation model defined in this way we distinguish stationary parameters (i.e., parameters assuming the same value in each generation) and we formulate some hypotheses of special interest to investigators. To this aim we consider transition from a given distribution of the pair (G, S), consisting of genotype G and sex S in an arbitrary parental generation, to the distribution of an analogous pair (G', S') in the offspring generation.

Let us treat the parental generation as a set of parental pairs. Let

$$G^{\mathrm{m}} = (G_1^{\mathrm{m}}, G_2^{\mathrm{m}}), \quad G^{\mathrm{f}} = (G_1^{\mathrm{f}}, G_2^{\mathrm{f}})$$

denote the male's (m) and female's (f) genotypes in a randomly selected pair. The distribution of (G, S) specifies the distribution of each of these genotypes separately, because $G^{\mathrm{m}} = (G \mid S = \mathrm{m})$, $G^{\mathrm{f}} = (G \mid S = \mathrm{f})$. Let us introduce indices ε^{m} and ε^{f} where

$$\varepsilon^{\mathrm{m}} = \begin{cases} 1 & \text{if father transmits to child gene } G_1^{\mathrm{m}}, \\ 0 & \text{if father transmits to child gene } G_2^{\mathrm{m}}, \end{cases}$$

and index ε^{f} is defined similarly for the mother. The child's genotype is of the form

$$G' = \left(\varepsilon^{\mathrm{m}} G_1^{\mathrm{m}} + (1-\varepsilon^{\mathrm{m}}) G_2^{\mathrm{m}},\ \varepsilon^{\mathrm{f}} G_1^{\mathrm{f}} + (1-\varepsilon^{\mathrm{f}}) G_2^{\mathrm{f}}\right). \tag{6.3.7}$$

By assumption 1° indices ε^{m} and ε^{f} are independent, and they are also independent of $(G^{\mathrm{m}}, G^{\mathrm{f}})$, and assume values 0 and 1 with probabilities $\frac{1}{2}$. In view of 1° and 2°, the child's sex S' is independent of G', and the probability of giving birth to a male child equals $\frac{1}{2}$. Hence it follows that, irrespective of the distribution of genotype and sex in the initial generation, in each of the following generations the genotype and sex are independent, and thus the distributions of the paternal and maternal genes are identical and invariable for all the generations. Therefore the distribution of the free gene, defined by $t_1, \ldots, t_r$, is a stationary parameter of the primitive generation model. In other words, the population remains in a state of equilibrium with respect to $t_1, \ldots, t_r$.

The distribution of the child's genotype G' depends on the distribution of (G^m, G^f, ε^m, ε^f), and thus, owing to the independence of (G^m, G^f) and (ε^m, ε^f) and to the assumptions on (ε^m, ε^f), it is defined by the distribution of (G^m, G^f) which formalizes the way of mating of parental pairs. In the simplest case we assume independence of parental genotypes G^m and G^f (commonly called *random mating of pairs*) which means that mating does not depend on the partners' genotypes. It follows then from (6.3.7) that genes G_1' and G_2' in genotype G' are independent, and thus also coindependent since the distributions of G_1' and G_2' are identical. The independence of parental genotypes implies therefore stationary coindependence of G_1 and G_2, and thus probabilities t_i and g_{ij} $(i, j = 1, \ldots, r)$ are the same in all generations and satisfy equalities (6.3.4).

The population described by means of the primitive generation model in which genes in the genotype are stationary coindependent, is said to be in *Hardy–Weinberg equilibrium*. This hypothesis of stationary coindependence is referred to as *Hardy–Weinberg hypothesis* (*H–W hypothesis*) for this model.

In the primitive generation model both stationary independence of genes and independence of parental genotypes imply the state of H–W equilibrium (and thus, rejection of the H–W hypothesis causes rejection of both these hypotheses). It should be added that independence of genes does not imply independence of parental genotypes.

Models in which parental genotypes G^m and G^f are not independent have usually a complicated description. This is true for instance when we intend to present formally the influence of kinship on mating of pairs. We will only mention that certain assumptions on the influence of kinship on the distribution of (G^m, G^f) lead to the stationary distribution of the genotype G having the following form:

$$P(G_2 = A_i \mid G_1 = A_j) = \begin{cases} t_i + \phi(1-t_i) & \text{if } \quad i = j, \\ (1-\phi)t_i & \text{if } \quad i \neq j, \end{cases}$$

where ϕ is the parameter assuming values in [0, 1], which characterizes the tendency of mating pairs of a certain degree of kinship. Formally the parameter ϕ is a symmetrized correlation coefficient between G_1 and G_2; in the case of a locus with two alleles it has the properties described in Section 5.2. Probabilities g_{ij} are given by

$$\begin{cases} g_{ii} = \phi t_i + (1+\phi) t_i^2 & \text{if} \quad i = 1, \ldots, r, \\ g_{ij} = 2(1-\phi) t_i t_j & \text{if} \quad i, j = 1, \ldots, r, \quad i \neq j. \end{cases} \tag{6.3.8}$$

Thus, in the case of a locus with two alleles

$$4g_{11}g_{22} = g_{12}^2 + 2\frac{\phi}{1-\phi} g_{12}. \tag{6.3.9}$$

With $\phi = 0$, that is under independence of G_1 and G_2, formula (6.3.8) describes a model in the state of H–W equilibrium.

Graphs of parabolas (6.3.9) with $\phi = 0, 0.5, 0.75$ are presented in de Finetti's diagram (Fig. 6.3.1), cf. p. 144).

The H–W hypothesis is formulated also in generation models other than the primitive one, or even in models in which generations are not distinguished. Here belongs the cyclic model, in which are considered not generations, but a sequence of states of a population in moments separated by the time of duration of the female's reproductive cycle. Jacquard (1974) considers the consequences of removing particular assumptions of the primitive generation model and he states that if independence of parental genotypes holds then no distinct deviations from H–W equilibrium occur. Jacquard also shows that in such generation models and in the cyclic model as well the conclusions concerning the stationarity of particular population parameters are similar to the respective conclusions in the primitive generation model.

The above considerations concerning parameters and hypotheses in static and dynamic models can be generalized to the case of a higher number of loci deriving not necessarily from the same chromosome, and chromosomes corresponding to particular loci can be autosomal or sexual as well. This complicates the notations and statistical methods to be used, but the character of the problems remains the same.

6.4. Estimation of gene probabilities for a single locus

We will consider a single locus with r alleles and a phenotype functionally dependent on the genotype. We assume the population to be in the state of H–W equilibrium. We will deal with estimation of probabilities g_{ij} and t_i (formulas (6.3.2) and (6.3.3)) on the basis of a simple random sample taken from such a population. Since these probabilities satisfy (6.3.5) only estimation of $t_1, \ldots, t_r$ is needed.

Let $(F_1, \ldots, F_k)$ be a sequence of phenotypes dependent on the locus considered (for instance sequence (M, N, MN) for system MN or sequence (A, B, 0, AB) for system AB0), and let p_i $(i = 1, \ldots, k)$ be the probability of appearance of the phenotype F_i in a person selected at random from a population in H–W equilibrium. Owing to (6.3.5) and by applying function u which relates phenotypes to genotypes (cf. Sec. 6.2) we can represent p_i in the form

$$p_i = q_i(t_1, \ldots, t_{r-1}), \quad i = 1, \ldots, k.$$

Let this be illustrated by means of the following examples:

(a) a locus with two codominant alleles $\tilde{\mathrm{M}}$, $\tilde{\mathrm{N}}$,:

$$(F_1, F_2, F_3) = (\mathrm{M}, \mathrm{N}, \mathrm{MN});$$

in view of (6.2.2), (6.2.3) and (6.3.5)

$$\begin{aligned} q_1(t_1) &= t_1^2, \\ q_2(t_1) &= (1-t_1)^2, \qquad (6.4.1) \\ q_3(t_1) &= 2t_1(1-t_1); \end{aligned}$$

(b) a locus with a recessive allele $\tilde{A}_1$ and a dominating allele $\tilde{A}_2$;

$$(F_1, F_2) = (A_1, A_2);$$

in view of (6.2.2), (6.2.3) and (6.3.5)

$$\begin{aligned} q_1(t_1) &= t_1^2, \\ q_2(t_1) &= 1-t_1^2; \end{aligned} \tag{6.4.2}$$

(c) system AB0 with alleles $\tilde{A}$, $\tilde{B}$, $\tilde{0}$;

$$(F_1, F_2, F_3, F_4) = (A, B, 0, AB);$$

in view of (6.2.4) and (6.3.5)

$$\begin{aligned} q_1(t_1, t_2) &= t_1^2+2t_1(1-t_1-t_2), \\ q_2(t_1, t_2) &= t_2^2+2t_2(1-t_1-t_2), \\ q_3(t_1, t_2) &= (1-t_1-t_2)^2, \\ q_4(t_1, t_2) &= 2t_1 t_2; \end{aligned} \tag{6.4.3}$$

(d) system Gm(1, 2) with alleles $Gm^{1,-2}$, $Gm^{-1,-2}$, $Gm^{1,2}$;

$$(F_1, F_2, F_3) = \big(Gm(1, -2), Gm(-1, -2), Gm(1, 2)\big);$$

in view of (6.2.5) and (6.3.5)

$$\begin{aligned} q_1(t_1, t_2) &= t_1^2+2t_1 t_2, \\ q_2(t_1, t_2) &= t_2^2, \\ q_3(t_1, t_2) &= 1-(t_1+t_2)^2. \end{aligned} \tag{6.4.4}$$

Let X_i $(i = 1, \ldots, k)$ denote the number of individuals with the phenotype F_i in a simple random sample of size n. Then

$$(X_1, \ldots, X_k) \sim M_k(n, p).$$

A family of multinominal distributions $\big(M_k(n, p), \ p \in S\big)$, in which S is a certain subset of the set

$$\Sigma = \{p = (p_1, \ldots, p_k)\colon 0 < p_i \leqslant 1, p_i + \ldots + p_k = 1\},$$

is usually called in statistics a *multinominal model* with parameters n and S and is denoted $M_k(n, S)$. In multinominal models the estimation of parameters taking values in a certain set I is considered under an additional assumption that there exists a continuous bijection $q\colon I \to S$. Estimation of the vector $t = (t_1, \ldots, t_{r-1})$ may be treated as a special case of such an estimation in a multinomial model $M_k(n, S)$, if we put

$$\begin{aligned} &I = (0, 1)^{r-1}, \quad q = (q_1, \ldots, q_k), \\ &S = \big\{\big(q_i(t), \ i = 1, \ldots, k\big), \ t \in I\big\}. \end{aligned} \tag{6.4.5}$$

The estimators of the parameter sought for are given by "projecting" the sample frequencies

$$\hat{p} = \left(\frac{1}{n} x_1, \ldots, \frac{1}{n} x_k\right)$$

upon S or by means of what is called the *method of moments* (Bishop *et al.*, 1975). We shall discuss the first approach of which a special case is the ML method.

Let us consider the following types of a real-valued function K defined on $\Sigma \times \Sigma$:

$$K(y, z) = \sum_{i=1}^{k} y_i \ln \frac{y_i}{z_i}, \tag{6.4.6}$$

$$K(y, z) = \sum_{i=1}^{k} \frac{(y_i - z_i)^2}{y_i}. \tag{6.4.7}$$

These two functions are traditionally used for evaluating deviation of the observed data from the model $M_k(n, S)$ in the case of set S given by formula (6.4.5). Namely each vector $\hat{t} \in I$ which satisfies

$$K(\hat{p}, q(t)) = \inf_{t \in I} K(\hat{p}, q(t)) \tag{6.4.8}$$

for a chosen function K is an estimator of t minimizing the deviation of $\hat{p}$ from S. The right-hand side of equality (6.4.8) is called the *deviation of vector* $\hat{p}$ from the multinominal model under consideration. For function K given by formula (6.4.7) the expression $K(\hat{p}, q(t))$ is for each $t \in I$ equal to the chi-square statistic by which the deviation of $\hat{p}$ from $q(t)$ is measured, and thus, estimator $\hat{t}$ minimizes the value of statistic χ^2. Again, for function K given by formula (6.4.6) the estimator sought for maximizes the likelihood function. If certain regularity assumptions concerning q are satisfied then (Birch, 1964) for each $p \in S$ the sequence $\hat{t}^{(n)}$ of maximum likelihood ML estimators of $t = q^{-1}(p)$ has the following property:

$$n^{1/2}(\hat{t}^{(n)} - t) \xrightarrow{\mathscr{L}} N(0, [(A(t))^T A(t)]^{-1}), \tag{6.4.9}$$

where the elements $a_{ij}(t)$ of matrix $A(t)$ are of the form

$$a_{ij}(t) = p_i^{-1/2} \left(\frac{\partial q_i(t')}{\partial t_j} \right) \Bigg|_{t'=t}. \tag{6.4.10}$$

Now let us find the ML estimators of t in examples (a)–(c).

In example (a) the likelihood function is proportional to $\tilde{L}$, where

$$\tilde{L}(t_1; x_1, x_2, x_3) = t_1^{2x_1+x_3}(1-t_1)^{2x_2+x_3}.$$

Let us denote

$$v(t_1; x_1, x_2, x_3) = \frac{\partial \log \tilde{L}(t_1; x_1, x_2, x_3)}{\partial t_1}.$$

Hence

$$v(t_1; x_1, x_2, x_3) = \frac{2x_1}{t_1} + \frac{x_3(1-2t_1)}{t_1(1-t_1)} - \frac{2x_2}{1-t_1} = \frac{x_3 + 2x_1 - 2nt_1}{t_1(1-t_1)},$$

where $n = x_1+x_2+x_3$ and thus, the root of equation $v(t_1; x_1, x_2, x_3) = 0$ is equal to $(x_3+2x_1)/(2n)$ and it is easy to check that $\tilde{L}$ attains its maximum at this point. The estimator $\bar{t}_1$ of probability t_1 having the form

$$\bar{t}_1(x_1, x_2, x_3) = \frac{x_3+2x_1}{2n} \tag{6.4.11}$$

is thus the ML estimator. It is, at the same time, a natural estimator of the probability of gene $\tilde{\text{M}}$ equal to the fraction of genes $\tilde{\text{M}}$ in the sample, because in the total number of $2n$ genes in the sample of size n there are $2x_1$ genes $\tilde{\text{M}}$ in individuals with genotypes $(\tilde{\text{M}}, \tilde{\text{M}})$ and x_3 genes $\tilde{\text{M}}$ in individuals with genotypes $(\tilde{\text{M}}, \tilde{\text{N}})$. Moreover, $\bar{t}_1$ is an unbiased estimator, as in the sample of size n the expected numbers of individuals with genotype $(\tilde{\text{M}}, \tilde{\text{M}})$ and with genotype $(\tilde{\text{M}}, \tilde{\text{N}})$ are equal to nt_1^2 and $2nt_1(1-t_1)$, respectively, and hence $E\bar{t}_1 = t_1$. Therefore, $\bar{t}_1$ is an estimator with minimal variance equal to $t_1(1-t_1)/(2n)$. We have also

$$n^{1/2}(\bar{t}_1-t_1) \xrightarrow{\mathcal{L}} N(0, \tfrac{1}{2}t_1(1-t_1)).$$

Further, the probability t_2 of gene $\tilde{\text{N}}$ is estimated by $\bar{t}_2 = 1-\bar{t}_1$.

For the Polish population the values

$$\bar{t}_1 = 0.5953, \quad \bar{t}_2 = 0.4047 \tag{6.4.12}$$

were found (Szczotka and Schlesinger, 1980) from a sample of size $n = 43\,552$ consisting of 15 417 individuals with phenotype M (35.4%), 7 118 individuals with phenotype N (16.3%) and 21 027 individuals with phenotype MN (48.3%).

In example (b) we have

$$\tilde{L}(t_1; x_1, x_2) = (1-t_1^2)^{x_2} t_1^{2x_1},$$

$$v(t_1; x_1, x_2) = \frac{2(x_1-nt_1^2)}{t_1(1-t_1^2)}$$

and thus the ML estimator is of the form

$$\tilde{t}_1(x_1, x_2) = \left(\frac{1}{n}x_1\right)^{1/2}. \tag{6.4.13}$$

This estimator is asymptotically unbiased and has the following property:

$$n^{1/2}(\tilde{t}_1-t_1) \xrightarrow{\mathcal{L}} N(0, \tfrac{1}{4}(1-t_1^2)).$$

Thus, the asymptotic variance of $\tilde{t}_1$ is equal to $n^{-1}(1-t_1^2)/4$. The quotient of this variance and of the variance of $\bar{t}_1$ in example (a), called *asymptotic efficiency of estimator* $\tilde{t}_1$ *with respect to estimator* $\bar{t}_1$ equals $(1+t_1)/(2t_1)$, and so it is greater than 1. This increase of variance and the appearance of bias is due to the fact that observability is more limited in example (b) than in example (a).

In example (c) the ML estimators of probabilities t_1, t_2 and t_3 cannot be presented as algebraic functions of the numbers of phenotypes, but their values can be found with arbitrary accuracy by the following traditionally used iterative procedure, known as the method of counting genes.

Let F denote the phenotype. In view of (6.3.3) we have

$$\begin{aligned} t_1 = {} & P(G = (\tilde{\mathrm{A}}, \tilde{\mathrm{A}}) \mid F = \mathrm{A})p_1 + \\ & +\tfrac{1}{2}P(G \in \{(\tilde{\mathrm{A}}, \tilde{\mathrm{B}}), (\tilde{\mathrm{B}}, \tilde{\mathrm{A}})\} \mid F = \mathrm{AB})p_4 + \\ & +\tfrac{1}{2}P(G \in \{(\tilde{\mathrm{A}}, \tilde{0}), (\tilde{0}, \tilde{\mathrm{A}})\} \mid F = \mathrm{A})p_1 . \end{aligned}$$

In a similar way we can express t_2 and in view of (6.3.5) we obtain

$$t_1 = \frac{p_1(t_1+t_3)}{t_1+2t_3} + \tfrac{1}{2}p_4,$$

$$t_2 = \frac{p_2(t_2+t_3)}{t_2+2t_3} + \tfrac{1}{2}p_4 .$$

Substituting p_1, p_2 and p_4 in (6.4.14) by the respective sample frequencies we obtain equations equivalent to the equations

$$\frac{\partial \tilde{L}(t_1, t_2, t_3; x_1, x_2, x_3, x_4)}{\partial t_i} = 0, \quad i = 1, 2.$$

This form makes it possible to solve the equations by means of an iterative procedure with an arbitrary starting point within set Σ. The approximate values of the ML estimators for the Polish population, obtained by the method of counting genes, are (Gnot, 1979):

$$\bar{t}_1 = 0.2783, \quad \bar{t}_2 = 0.1530, \quad \bar{t}_3 = 0.5687.$$

Other estimators of probabilities of genes $\tilde{\mathrm{A}}, \tilde{\mathrm{B}}, \tilde{0}$ were obtained by Bernstein. Summing (6.4.3) he found that

$$(t_1+t_3)^2 = p_1+p_3, \quad (t_2+t_3)^2 = p_2+p_3 .$$

Therefore

$$t_1 = 1-(p_1+p_3)^{1/2}, \quad t_2 = 1-(p_1+p_2)^{1/2}, \quad t_3 = p_3{}^{1/2}. \tag{6.4.16}$$

Substituting p_1, p_2, p_3 in (6.4.16) by the respective sample frequencies Bernstein obtained the estimators which will be denoted in the sequel $\hat{t}_1, \hat{t}_2, \hat{t}_3$, and called *Bernestein estimators.*

The Bernstein estimators do not depend explicitly on sample frequency x_3/n of phenotype AB, and their sum does not identically equal 1. They are asymptotically unbiased (Elandt-Johnson, 1971). In calculating the asymptotic variances (as var) of the particular Bernstein estimators we can use the following theorem (Bickel and Doksum, 1977): *if*

$$(X_1^{(n)}, \ldots, X_k^{(n)}) \sim M_k(n, p),$$

then for any function $w\colon \Sigma \to \boldsymbol{R}$ *with continuous derivative and for the respective sequence of random variables*

$$W^{(n)} = w\left(\frac{1}{n}X_1^{(n)}, \ldots, \frac{1}{n}X_k^{(n)}\right)$$

it holds that

$$n^{1/2}\left(W^{(n)}-w(p)\right)\xrightarrow{\mathscr{L}} N(0,\sigma^2), \tag{6.4.17}$$

where

$$\sigma^2 = \left(\sum_{i=1}^{k} p_i\left(\frac{\partial w(s)}{\partial s_i}\right)\bigg|_{s=p}\right)^2-\left(\sum_{i=1}^{k} p_i\left(\frac{\partial w(s)}{\partial s_i}\right)\bigg|_{s=p}\right)^2. \tag{6.4.18}$$

Hence, we obtain as var $\hat{t}_i$ for $i = 1, 2, 3$ equal to $1-(t_2+t_3)^2/(4n)$, $1-(t_1+t_3)^2/(4n)$, $(1-t_3^2)/(4n)$, respectively. Further, in view of (6.4.9) we have for the ML estimator of t_1

$$\text{as var}\,\bar{t}_1 = \frac{t_1}{4n}\left(2-t_1-\frac{t_1^2 t_2}{2(t_3+t_1 t_2)}\right).$$

Thus

$$\frac{1}{2} < \frac{\text{as var}\,\bar{t}_1}{\text{as var}\,\hat{t}_1} < 1. \tag{6.4.19}$$

Identical inequalities hold also for $\hat{t}_2$ and $\hat{t}_3$ so that the asymptotic variance of the ML estimators is lower than the asymptotic variance of Bernstein estimators and higher than one half of this variance.

In example (d) the natural estimators of probabilities t_1 and t_2 are obtained by substituting the respective sample frequencies in formula (6.4.4). Then we have for genes $\text{Gm}^{1,-2}$ and $\text{Gm}^{-1,-2}$

$$\hat{t}_1 = \left(1-\frac{1}{n}x_3\right)^{1/2}-\left(\frac{1}{n}x_2\right)^{1/2}, \quad \hat{t}_2 = \left(\frac{1}{n}x_2\right)^{1/2},$$

and for gene $\text{Gm}^{1,2}$ we put $\hat{t}_3 = 1-\hat{t}_1-\hat{t}_2$. For the Polish population values

$$\hat{t}_1 = 0.1431, \quad \hat{t}_2 = 0.8000, \quad \hat{t}_3 = 0.0569, \tag{6.4.20}$$

were obtained (Szczotka and Schlesinger, 1980, p. 31) from a sample of size $n = 10\,000$ consisting of 2494 individuals with phenotype Gm(1, −2) (24.94%), 6400 individuals with phenotype Gm(−1, −2) (64%) and 1106 individuals with phenotype Gm(1, 2) (11.06%).

6.5. Testing the Hardy–Weinberg hypothesis

Usually acceptance of the assumption that the population is in H–W equilibrium is essential for the geneticist because of the biological implications mentioned previously. This is also the condition for applying the methods of estimation presented in Section 6.4. However, it should be stressed that statisticians can only suggest tests which allow either to reject the H–W hypothesis or to defer the decision (by stating that there are no good reasons to reject the hypothesis). Such tests will be presented in this section.

As previously, we consider a single locus and a phenotype functionally dependent on genotype. We assume to have a simple random sample from a population in which matrix $[g_{ij}]$ and vector $(t_1, \ldots, t_r)$ are stationary. The H–W hypothesis on coindependence of genes in the genotype is thus described by the system of equations (6.3.5).

In case of codomination of all the genes of the locus considered a different phenotype corresponds to each randomly ordered genotype. In such a case the H–W hypothesis is observably reducible (cf. (2.1.10)). We will limit ourselves to the case of a locus with two codominating alleles, as in example (a) in Section 6.4. In examples (b), (c) and (d) the H–W hypothesis is not observably reducible.

In the case of two codominant alleles we test the H–W hypothesis in a multinomial model $M_3(n, \Sigma)$. This task will be reduced to the problem of testing a certain linear hypothesis in an exponential family of bivariate distributions.

The family of bivariate distributions, dominated by the measure ν, will be called *exponential* if the family of its densities can be put in the form $(f_\theta, \theta \in \Theta)$, where

$$\Theta = \left\{\theta = (\theta_1, \theta_2) \in \boldsymbol{R}^2 \colon \int_{\boldsymbol{R}^2} \exp(\theta^T z)\, d\nu(z) < +\infty\right\} \tag{6.5.1}$$

and for some functions $f_1\colon \boldsymbol{R}^2 \to \boldsymbol{R}^+, f_2\colon \Theta \to \boldsymbol{R}^+$

$$f_\theta(z_1, z_2) = f_1(z_1, z_2) f_2(\theta) \exp(\theta_1 z_1 + \theta_2 z_2), \quad (z_1, z_2) \in \boldsymbol{R}^2. \tag{6.5.2}$$

In the case of an exponential family of distributions, uniformly most powerful tests of a linear hypothesis stating that $\theta \in \Theta \cap \Theta_0$ are known for any linear subspace $\Theta_0 \subset \boldsymbol{R}^2$ (Lehman, 1959). In search for such a test to verify the H–W hypothesis we transform vector (X_1, X_2, X_3) into

$$(Z_1, Z_2) = (X_1, 2X_1 + X_3). \tag{6.5.3}$$

Statistic Z_2 is the number of genes $\tilde{\mathrm{M}}$ in a sample of size n consisting of $2n$ genes. The probability of observing the pair (z_1, z_2) can be written down as

$$\frac{n!}{z_1!(n+z_1-z_2)!(z_2-2z_1)!} p_2^n \exp(\theta_1 z_1 + \theta_2 z_2), \tag{6.5.4}$$

where

$$\theta_1 = \ln \frac{4p_1 p_2}{p_3^2}, \qquad \theta_2 = \ln \frac{p_3}{2p_2}. \tag{6.5.5}$$

Formula (6.5.4) was obtained by transforming the probability of observing a triple (x_1, x_2, x_3) drawn according to the distribution $M_3(n, \boldsymbol{p})$. Since from (6.5.5) we can evaluate p_2 as a function of θ_1 and θ_2, and since the set of pairs (θ_1, θ_2) satisfying (6.5.5) for $(p_1, p_2, p_3) \in \Sigma$ is of the form (6.5.1), then considering (6.5.4) and (6.5.2) we can state that the family of distributions of (Z_1, Z_2) is exponential. Moreover, in view of (6.3.6) the H–W hypothesis for the locus under consideration is equivalent to the hypothesis $p_3^2 = 4p_1p_2$, i.e., to the hypothesis $\theta_1 = 0$. Since statistic (Z_1, Z_2) is a one-to-one transformation of (X_1, X_2, X_3) the search for a uniformly most powerful test at level α of the H–W hypothesis in the model

$M_3(n, \Sigma)$ is equivalent to the search for the most powerful test at the same level of the hypothesis $\theta_1 = 0$ in the corresponding exponential family of bivariate distributions.

Denote by P_{θ_1,θ_2} the distribution of (Z_1, Z_2) for a given value of (θ_1, θ_2). Thus, the H–W hypothesis is the set of distributions P_{0,θ_2}. Let E_{0,θ_2} be used for the expected value of any statistic pertaining to distribution P_{0,θ_2}. The test sought for is defined by

$$\Phi(z_1, z_2) = \begin{cases} 1 & \text{if} \quad z_1 < \varkappa_1(z_2) \text{ or } z_1 > \varkappa_2(z_2), \\ \gamma_i & \text{if} \quad z_1 = \varkappa_i(z_2),\ i = 1, 2, \\ 0 & \text{if} \quad \varkappa_1(z_2) < z_1 < \varkappa_2(z_2), \end{cases} \tag{6.5.6}$$

where the thresholds $\varkappa_i\colon \boldsymbol{R} \to \boldsymbol{R}^+$ and the randomization constants $\gamma_i\colon \boldsymbol{R} \to [0, 1]$ for $i = 1, 2$ are found from the following conditions: for any $z_2 \in \boldsymbol{R}^+$

$$\begin{aligned} &E_{0,\theta_2}\big(\Phi(Z_1, Z_2) \mid Z_2 = z_2\big) = \alpha, \\ &E_{0,\theta_2}\big(Z_1 \Phi(Z_1, Z_2) \mid Z_2 = z_2\big) = E_{0,\theta_2}(Z_1 \mid Z_2 = z_2). \end{aligned} \tag{6.5.7}$$

The construction of this test requires the knowledge of the distribution of the random variable $(Z \mid Z_2 = z)$ under the assumption that the H–W hypothesis holds. Denote by C_n a function which assigns numbers $n!/[i_1!(n+i_1-i_2)!(i_2-2i_1)!]$ to the pairs of nonnegative integers (i_1, i_2). Due to (6.5.4)

$$P_{\theta_1,\theta_2}(Z_1 = z_1 \mid Z_2 = z_2) = \frac{4^{-z_1} C_n(z_1, z_2) e^{\theta_1 z_1}}{\sum_{j=j_0}^{j_1} 4^{-j} C_n(j, z_2) e^{\theta_1 j}}$$

where $j_0 = \max(0, z_2 - n)$, $j_1 = [z_2/2]$. Substituting $\theta_1 = 0$ one gets

$$P_{0,\theta_2}(Z_1 = z_1 \mid Z_2 = z_2) = 4^{-z_1} C_n(z_1, z_2) \binom{2n}{z_2}^{-1},$$

and this enables evaluation of the thresholds and of randomization constants of test (6.5.6) (cf. Gnot and Ledwina, 1980).

The same test can be also constructed on an intuitive basis proposed by Haldane (1954). Haldane calculated the conditional probability of observing (x_1, x_2, x_3) in a population in H–W equilibrium, provided that $Z_2 = z_2$. In view of (6.5.4) this probability equals

$$n!\, 2^{x_3} \left(x_1! x_2! x_3! \binom{2n}{z_2} \right)^{-1}, \tag{6.5.8}$$

because the number of genes $\tilde{M}$ in a sample consisting of $2n$ genes is equal to z_2 with probability $\binom{2n}{z_2} t_1^{z_2} (1-t_1)^{2n-z_2}$. Haldane considered the function

$$\begin{aligned} &p(x; x_1, x_2, x_3) \\ &= n!\, 2^x \left(\big(x_1 + \tfrac{1}{2}(x_3 - x)\big)!\, x!\, \big(x_2 + \tfrac{1}{2}(x_3 - x)\big)! \binom{2n}{2x_1 + x_3} \right)^{-1}, \end{aligned}$$

which is monotone at both sides of the only maximum x_0, and he noticed that the value of p at $x = x_3$ is equal to (6.5.8). Intuitively it is clear that a test should be required to reject the H–W hypothesis for particularly low values of expression (6.5.8), and this is equivalent to rejection of H–W for values of x_3 considerably distant from x_0. It is easy to prove that by imposing conditions analogous to (6.5.7) one gets a test identical with the test uniformly most powerful at level α. Critical values of x_3 are tabulated in the paper by Vithayasai (1975).

Some other tests of the H–W hypothesis are based on the estimation of the deviation of the observed triplet (x_1, x_2, x_3) from the H–W equilibrium by means of the function χ^2 of the form $\sum_{i=1}^{3} (X_i - E_{\text{H-W}}X_i)^2 / E_{\text{H-W}}X_i$ where $E_{\text{H-W}}X_i$ denotes the expected value of X_i under the assumption of H–W equilibrium for a given probability t_1 of gene $\tilde{\text{M}}$. Obviously,

$$\begin{aligned} E_{\text{H-W}}X_1 &= 2nt_1^2, \\ E_{\text{H-W}}X_2 &= 2nt_2^2, \\ E_{\text{H-W}}X_3 &= 4nt_1t_2. \end{aligned} \tag{6.5.9}$$

We can estimate $E_{\text{H-W}}X_i$ by substituting for t_1 and t_2 the frequencies of genes $\tilde{\text{M}}$ and $\tilde{\text{N}}$ in the sample, equal to $(2x_1 + x_3)/(2n)$ and $(2x_2 + x_3)/(2n)$, respectively. For a population in H–W equilibrium the function χ^2 adapted in that way has asymptotically the χ^2 distribution with 1 degree of freedom. Thus the H–W hypothesis will be rejected if the observed value of the statistic exceeds the $(1-\alpha)$-quantile of this distribution.

The expected values (6.5.9) can also be replaced by the conditional expected numbers of the particular phenotypes provided that $Z_2 = z_2$ (the expectations referring to a population in H–W equilibrium). They are as follows:

$$\begin{aligned} E_{\text{H-W}}(X_1 \mid Z_2 = z_2) &= \frac{z_2(z_2-1)}{2(n-1)}, \\ E_{\text{H-W}}(X_2 \mid Z_2 = z_2) &= \frac{(2n-z_2+1)(2n-z_2)}{2(2n-1)}, \\ E_{\text{H-W}}(X_3 \mid Z_2 = z_2) &= \frac{z_2(2n-z_2)}{2n-1}. \end{aligned}$$

The properties of test based on this adaptation of function χ^2 have been considered for samples of small size n in the paper by Elston and Forthofer (1977).

For r codominating alleles with $r > 2$ a uniformly most powerful test of the hypothesis H–W does not exist. However, we can construct tests pertaining to function χ^2. We make use of the fact that under certain regularity assumptions on parametrization $\boldsymbol{q} = (q_1, \ldots, q_k)$ discussed in Section 6.4 the ML estimator $\boldsymbol{t} = (t_1, \ldots, t_{r-1})$, denoted by $\bar{\boldsymbol{t}}^{(n)}$, has the property that statistic

$$\hat{T}_n = \sum_{i=1}^{k} \frac{(X_i - nq_i(\bar{\boldsymbol{t}}^{(n)}))^2}{nq_i(\bar{\boldsymbol{t}}^{(n)})} \tag{6.5.10}$$

is asymptotically distributed as χ^2 with $k-r$ degrees of freedom.

*

In real populations certain deviations from the H–W equilibrium must occur and thus, the vector of phenotype probabilities is in practice only approximately equal to vector $\boldsymbol{q}(\boldsymbol{t})$. Hence, if hypothesis H–W is tested by a test having power close to 1 for large n and if the sample size is large then this hypothesis will be practically always rejected, even if the deviation from it is small. Now, in practical investigation, a slight deviation is usually recognized acceptable by investigators. For this reason, instead of formulating and verifying the exact H–W hypothesis we prefer to deal with its approximation, according to which the vector of probabilities of phenotypes is localized "in the proximity" of set $\boldsymbol{S}$ (cf. (6.4.5)).

Bjørnstadt (1975) suggested the following definition of the *epsilon neighbourhood* $\boldsymbol{S}_\varepsilon$ of the set $\boldsymbol{S}$:

$$\boldsymbol{S}_\varepsilon = \{\boldsymbol{p} \in \boldsymbol{\Sigma}\colon d_S(\boldsymbol{p}) \leqslant \varepsilon\} \tag{6.5.11}$$

with respect to a chosen function $d_S\colon \boldsymbol{\Sigma} \to \boldsymbol{R}^+$ which is equal identically to zero on the set $\boldsymbol{S}$ and has a continuous derivative. For instance, d_S can be defined as follows:

$$d_S(\boldsymbol{p}) = \inf_{\boldsymbol{p}' \in \boldsymbol{S}} K(\boldsymbol{p}, \boldsymbol{p}'),$$

where K is a suitably smooth distance in $\boldsymbol{R}^k$ or one of the functions (6.4.6) and (6.4.7). In the case of codominant alleles a natural measure of deviation from the H–W hypothesis (meant as the hypothesis of coindependence of genes) is any symmetrized measure of codependence in the genotype distribution, for instance η^* (formula (5.5.3)). In particular, for two codominating alleles we have

$$\eta^*(p_1, p_2, p_3) = \frac{(4p_1p_2 - p_3^2)^2}{[(2p_1 + p_3)(2p_2 + p_3)]^2}\,. \tag{6.5.12}$$

For the AB0 system function d_S can be defined as an Euclidean distance between $\boldsymbol{p}$ and $\boldsymbol{q}(\boldsymbol{t})$ for vector $\boldsymbol{t}$ given by (6.4.14). It is also possible to make use of the equalities (6.4.16) for $\boldsymbol{p} \in \boldsymbol{S}$ and to define d_S as

$$d(\boldsymbol{p}) = \left|\left((p_1 + p_3)^{1/2} + (p_2 + p_3)^{1/2}\right)^2 - (1 + p_3^{1/2})^2\right|.$$

(Subscript $\boldsymbol{S}$ is here and in the sequel omitted for simplicity sake).

Hypothesis $H_\varepsilon\colon \boldsymbol{p} \in \boldsymbol{S}_\varepsilon$ will be called *approximate H–W hypothesis.* For $\varepsilon = 0$ it is identical with hypothesis H–W.

Let $\sigma_d^2(\boldsymbol{p})$ denote expression (6.4.18) for $w = d$. For each $\boldsymbol{p} \in \boldsymbol{\Sigma}$ the sequence $\sigma_d(\hat{\boldsymbol{p}})$ converges with probability 1 to $\sigma_d(\boldsymbol{p})$ when $n \to \infty$. Therefore it follows from (6.4.17) that for each $\boldsymbol{p} \in \boldsymbol{\Sigma}$

$$\frac{n^{1/2}\left(d(\hat{\boldsymbol{p}}) - d(\boldsymbol{p})\right)}{\sigma_d(\hat{\boldsymbol{p}})} \xrightarrow{\mathscr{L}} N(0, 1). \tag{6.5.13}$$

Thus, if hypothesis H_ε will be rejected when $n^{1/2}\left(d(\hat{\boldsymbol{p}}) - \varepsilon\right)/\sigma_d(\hat{\boldsymbol{p}})$ exceeds the quantile $q_{N(0,1)}(1-\alpha)$ then the asymptotic level of significance of the test will equal α.

A basic drawback of the presented approach is the lack of information on the power of the test and on the rate with which the sequence distribution of statistics dealt with in (6.5.13) converges to the distribution $N(0, 1)$. On the other hand, freedom in choosing function d is of convenience.

Another approach to testing the approximate H–W hypothesis was given by Bednarski (1981). Although very interesting, it will be not presented here because this would require introduction of many new concepts and technical details.

6.6. Final remarks

No monographs on statistical investigation in population genetics are known in the general statistical and genetical literature. However, an important step towards such a monograph is a book by Jacquard (1974) on probabilistic formalization of the evolution of the genetic structure of human populations.

Jacquard considers first an ideal model in which evolution of the genetic structure of a population is due only to the heriditary mechanism, and the population size is infinite. The author investigates the asymptotic state of equilibrium under various assumptions on the dependence of phenotypes and genotypes, on mating of pairs and on demographic parameters of the evolution. The ideal model serves as a reference system for more realistic models. This book deals successively with the influence of migration, mutation, selection, nonrandom mating of pairs and finite size of the population on the equilibrium of the genetic structure, in each case all the remaining assumptions of the ideal model remaining unchanged. Next, the author deals with a model in which all the factors mentioned beforehand act simultaneously. The last part of the book is devoted to the estimation of deviations from the state of equilibrium of the genetic structure on the basis of samples taken from the population.

Jacquard's monograph is a perfect handbook of probabilistic modelling of the genetic structure of human populations. The parameters of this structure are defined in a precise way in regard to a strictly determined population model. The construction of the book is admirable and based on a clear classification of the models considered. However, the statistical subject-matter is restricted to a superficial discussion of a few particular problems of estimation contained in the last part of the book.

Neither do other monographs on probabilistic models in population genetics (for instance Elandt-Johnson, 1971) usually take account of the statistical problems or they discuss them superficially. In statistical papers on population genetics published in statistical journals, only simplified models of the population genetics structure and classical problems related with them are usually considered. Again, statistical aspects of papers published in genetical journals are usually limited to calculations performed on some data with a rather vague interpretation of the results obtained.

Despite these facts, population genetics is a domain in which statistics plays an important role and is a recognized tool of investigation.

The variety of statistical problems in population genetics is immense. In the present chapter we have merely mentioned some simple problems of inference on the distributions of genes and genotypes (in the case of a single locus) on the basis of phenotypes observed in a given sample. More complicated problems are posed in population genetics e.g. for the models considered by Jacquard.

The basic trend in statistical investigation of population genetics deals with inference on parameters of population genetic structure on the basis of phenotypes, but also investigations based on observation of karyotypes are dealt with. The latter will be illustrated here by a typical example of statistical inference on the karyotype of an individual.

Suppose that an unordered set of chromosomes is observed in a single cell and that the aim of the investigation is identification of pairs of homologous chromosomes and ordering these pairs according to the binding classification (mentioned at the end of Sec. 6.2).

We number arbitrarily, from 1 to 46, the observed chromosomes. Let x_i $(i = 1, \ldots, 46)$ denote the results of some observation performed on the i-th chromosome (for instance, x_i can be a pair consisting of the lengths of the longer and shorter arms of the chromosome).

To meet the aim of the investigation we seek a proper permutation of the sequence of chromosomes in the cell, i.e., a permutation $\mu = (\mu(1), \ldots, \mu(46))$, such that for $j = 1, \ldots, 23$ $(x_{\mu(2j-1)}, x_{\mu(2j)})$ is a pair of observations corresponding to the j-th pair of chromosomes according to the binding classification. The number of permutations having this property is 2^{23} (because changing the order in any pair of homologous chromosomes is of course admissible). Such permutations will be called *equivalent*.

The population of cells with initially numbered chromosomes may be divided into disjoint classes, each of them corresponding to one permutation proper for all cells in this class. Thus, each permutation of $\{1, \ldots, 46\}$ designates one class of cells and, clearly, all equivalent permutations designate the same class. The number m of disjoint classes is equal to $46!/2^{23}$. Note that the presented problem may be regarded as a discrimination problem with this number of classes (cf. Chapter 3).

Let I be the index allocating a cell to the proper class. We assume that I has a distribution uniform on the set $\{1, \ldots, m\}$. This assumption concerning the a priori distribution means that the initial numeration of chromosomes is not dependent on the binding classification of pairs of homologous chromosomes. The validity of this assumption depends on the way of numeration accepted by the person performing the observations and on the way of arrangement of the chromosomes in the cell.

Next we assume that the random variables $X_1, \ldots, X_{46}$ representing the results of observations are mutually independent and that for each $j = 1, \ldots, 23$ random variables $X_{\mu(2j-1)}$ and $X_{\mu(2j)}$ have the same density, say γ_j. This means that the observations performed on chromosomes belonging to the same pair are independent and have the same density, and also that the pairs are observed independently. Thus, the density f_μ of $Z = (X_1, \ldots, X_{46})$ in a class of cells in which μ is the proper permutation of the sequence of chromosomes, is determined by densities $\gamma_1, \ldots, \gamma_{23}$ and permutation μ:

$$f_\mu(x_1, \ldots x_{46}) = \prod_{j=1}^{23} \gamma_j(x_{\mu(2j-1)})\gamma_j(x_{\mu(2j)}).$$

If the permutations μ and μ' are equivalent then $f_\mu = f_{\mu'}$. The probability a posteriori of assigning a cell to a class which is determined by permutation μ provided $Z = z$ equals

$$2^{23} f_\mu(z)\Big(\sum_{\mu'} f_{\mu'}(z)\Big)^{-1}. \tag{6.6.1}$$

Let us remind (Sec. 3.2) that the solutions of various probabilistic discrimination problems are based on probabilities a posteriori. In particular in many problems the decision is deferred if the maximal probability a posteriori in classes $1, \ldots, m$ is smaller than a chosen threshold $\varkappa$, and that otherwise one selects any permutation maximizing $f_\mu(z)$ (thus maximizing at the same time the probability a posteriori). With $\varkappa = 0$ we get the Bayes' solution in the case of identical losses caused by incorrect decisions. In such a case it follows from (6.6.1) that the Bayes' solution is the ML estimator of permutation μ.

In practice densities $\gamma_1, \ldots, \gamma_{23}$ are not known and they have to be estimated from auxiliary samples in which (by means of cytologic methods) the error-free classification of pairs of homologous chromosomes is performed. In a parametric model (for, instance under assumption that density γ_j is normal with unknown parameters) the solution of the parametric problem can be adapted by estimating the parameters of this distribution (Bednarski *et al.*, 1978). In a nonparametric model the densities γ_j can be estimated for instance by means of kernel estimators (Habbema, 1976).

However, in practice numerical difficulties preclude the use of the adapted probabilistic solution based on examination of all the possible permutations. It is therefore suggested to begin by dividing the set of chromosomes into several groups, trying to proceed in accordance with the binding classification of pairs, and then to apply the described method to each of the groups separately. Other methods of this kind may be found in the literature. Their properties can be investigated by simulation or experimentally. In the latter instance the classification obtained by a statistical method is compared with the error-free classification carried out by the cytologist.

Chapter 7

Paternity proving[1]

Key words: *paternity proving, putative father, court verdict, paternity exclusion, probability a posteriori, of paternity, critical value, discriminant rule, deferred decision, blood groups, morphological traits, genotype, phenotype.*

7.1. Introduction

Court procedures on paternity proving are a social problem in many countries. For instance, in Poland about a twelve thousand cases of disputed paternity come before the courts each year. The recognition of the true father is of significance both for the child's good, as well as for the good of the defendant male. The recognition of the child's father by the court entitles to set up alimentation and property claims in the name of the child. The rights of the child to set up such claims result from the conviction that the father and the mother should care for the bringing up of the child and ensure suitable material conditions for it. On the other hand a false recognition of paternity causes moral harm to the defendant, and may give rise to serious complications in his private life and also may cause, not only him, but also members of his family to sustain a material loss (for instance in case of legacy division).

The child's good and the right statement on paternity require a penetrating consideration of all possible evidence.

The lego-substantive basis of the writ filed to the court is the conjecture, that the man indicated by the female had an intercourse with her in the period between day 300 and 181 before the birth of the child. So both proof and denial of this fact are extremely difficult.

To the biological proofs belong, above all, determination of blood group traits and anthropological evidence. The most reliable proof in the process is the exclusion of paternity of the defendant on the basis of blood groups. The court averts the complaint if the child has a blood group which, according to the laws of heritability, could not appear if the three persons connected with the writ would form a family.

In cases, in which the paternity of the defendant has not been excluded, the

(1) In the following text terms: establishing paternity, father recognition, paternity recognition, are also used.

probability a posteriori of paternity of the male is estimated on the basis of blood groups and morphological traits.

Statisticians regard the recognition of the father as a discriminant problem. Automatic application of discriminant rules encounters, however, the resistance of the court and the parties concerned, and difficulties in purchasing the sera limit the number of blood group systems determined during the examinations, diminishing the probability of sufficiently reliable verdicts. Therefore, one is usually content with the estimation of the probability a posteriori that the defendant is the father. Such an estimation constitutes additional information for the court.

In Section 7.2 we present schemes of paternity exclusion and of estimation of probability a posteriori of the male's paternity. Against this background we consider the relations of father recognition with the discriminant analysis, referring to Sections 3.2 – 3.4.

In Section 7.3 we discuss the methods of paternity proving on the basis of blood group examination, referring to heritability laws and models of development of human populations (cf. Sec. 6.2 and 6.3) and to estimation of gene probabilities (cf. Sec. 6.4). We analyse in detail the blood group systems AB0 and MN and present the properties of the inference rules on the basis of these systems.

In Section 7.4 we describe the promulgating of verdicts on the basis of anthropological expertises. We present also the results of initial investigations on morphological traits in samples consisting of families, which enable conclusions on paternity also on the basis of those traits.

7.2. Schemes of paternity recognition

The writ concerns three persons: the mother, the child and the male being the "putative father". The writ is usually filed to the court by the mother (on behalf of the child), claiming the recognition of paternity of the defendant male. Cases of paternity exclusion claimed by males (for instance in divorce proceedings) are a rarity. Thus, the male is usually called the defendant, and the writ is determined as just when the male is the father of the child.

Let us consider a population of writs coming from a given area and in a given period of time which enter the courts, adjudging in cases of disputed paternity. We can also speak of a population of triplets consisting of a mother, a child and a man ascribed to writs. The population of systems is divided into two classes: a class of families, that is triplets in which the male is the father of the child, and a class a of false families. The affiliation to a class is described by means of trait I:

$$I = \begin{cases} 1 & \text{when the male is the father of the child,} \\ 2 & \text{in the reverse case.} \end{cases}$$

Let π denote the probability that $I = 1$ (i.e. that a writ randomly selected from the population concerns a family). This is the probability a priori of the male's paternity, ascribed to the said population of writs and characterizing the environment from which they enter the courts. If we are limited to the writs of mothers' complaints, π could be regarded as an index of the females' veracity.

Trait I is not observable. Therefore examination of the mother, the child and the male (or sometimes only the child and the male) are carried out. The joint results of this examination, which are designated further by Z, serve to identify the value of I. These examinations should be selected so that the distributions of Z in both classes be strongly enough differentiated.

If the human genotype were observable, the result Z of the examination carried out should be a trio of genotypes of the mother, the child and the male. The distributions of a trio of genotypes are in both classes different, as in a family the child's genotype depends on the genotypes of the mother and father, whereas in a false family there is no relation between the child's and the male's genotypes. The degree of separability of both distributions depends on which genotypes are taken into consideration, i.e., how many and which of the loci are taken into consideration for defining the genotype.

We assume that the population is in the Hardy–Weinberg equilibrium as regards the selected loci, and that the male and the mother come from the same population, thus their genotypes are chosen at random according to the same distribution. We also assume that in families the matrix of pairs does not depend on the genotypes. Then, in the case of independent loci the distributions of genotype trios in both classes depend only on the gene distributions for the particular loci.

However, the genotypes are at most intermediately observable by means of the phenotypes. The results Z of the investigations thus form trios of appropriately selected phenotypes, composed of heritable traits, that is of traits dependent on the genotype functionally or stochastically (cf. Sec. 6.2).

At paternity testing blood group traits and morphological traits are the first to be taken under consideration. The blood group traits are functionally dependent on the genotype in a way known to the experimenter, while the morphological ones are stochastically dependent and the character of this dependence is usually not known. In the case of some of the morphological traits, difficulties appear already at determining the set of values of the given trait, that is, at classification of its shape. Moreover, the evaluation of affiliation of the shape observed to a definite category is done usually in a subjective way, thus influencing the distribution of the trait. It may also happen that the morphological trait of the child considered depends not only on the environment and on the parents' genotypes, but also on the child's age and sex.

Apart from the trios of values of heritable traits, one frequently includes to the investigation expert's opinions on phenotypic similarities and dissimilarities between the mother, the child and the man involved in the writ.

*

It follows from the laws of inheritance that in a family some triplets of genotypes of the mother, the child and the father are impossible. Such a triplet will be called *incompatible*. On the other hand, in a false family, in view of the independence of the man's and child's genotypes, an incompatible trio may occur with a positive probability. Hence, if the genotypes were observable, then after having observed an incompatible trio the paternity of the man could be excluded.

Let us consider the possibility of paternity exclusion on the basis of phenotype trios observed, under the assumption that the phenotype is a function of the genotype. The set of all the phenotype triplets is then finite. Each observed triplet of mother's-, child's-, and man's phenotypes corresponds to a certain set of genotype triplets. The triplet of phenotypes is called *incompatible* if all corresponding triplets of genotypes are incompatible.

Let A designate a set of incompatible triplets of phenotypes, which in a false family occur with a positive probability, and let $\boldsymbol{Z}$ designate the set of compatible triplets of phenotypes which have a positive probability of occurrence in a family (and the more so in a false family). Beside the sets A and $\boldsymbol{Z}$ there remain such phenotype triplets in which the female cannot be the child's mother.

For any triplet $z \in \boldsymbol{Z} \cup A$, let $f_1(z)$ and $f_2(z)$ be the conditional probabilities of occurrence of this triplet in the class of families and in the class of false families respectively, i.e.

$$f_i(z) = P(Z = z \mid I = i).$$

Then

$$\begin{aligned} z \in \boldsymbol{Z} \quad &\Leftrightarrow \quad f_1(z) > 0 \ \wedge \ f_2(z) > 0, \\ z \in A \quad &\Leftrightarrow \quad f_1(z) = 0 \ \wedge \ f_2(z) > 0, \end{aligned} \tag{7.2.1}$$

i.e., paternity should be excluded on the set A. Trait I is thus conditionally observable in the set A.

Probability W of excluding paternity of a man who is not the father of the child on the basis of the set of phenotypes Z is determined by the formula:

TABLE 7.2.1.
Probabilities $P(I = i, \chi_A = j)$, $i = 1, 2$, $j = 0, 1$.

$\chi_A = j$ / $I = i$	$\chi_A = 0$ (compatible triplet of phenotypes)	$\chi_A = 1$ (exclusion of paternity: incompatible triplet of phenotypes)	Distribution of I
$I = 1$ (family: the man is the father of the child)	$\pi P(Z \in \boldsymbol{Z} \mid I = 1) = \pi$	0	π
$I = 2$ (false family: the man is not the child's father)	$(1-\pi)P(Z \in \boldsymbol{Z} \mid I = 2) = (1-\pi)(1-W)$	$(1-\pi)P(Z \in A \mid I = 2) = (1-\pi)W$	$1-\pi$
Distribution of χ_A	$\pi+(1-\pi)(1-W)$	$(1-\pi)W$	

$$W = P(Z \in A \mid I = 2) = \sum_{Z \in A} f_2(z). \tag{7.2.2}$$

The joint distribution of trait I and the variable χ_A characterizing the affiliation of the phenotype triplet Z to the set A is presented in Table 7.2.1.

This distribution depends on π and on W solely. The closer W to 1, the more similar are: the division of the population into classes of families and of false families and the division based on exclusion of paternity. If $W = 1$, both divisions are identical.

Let us suppose that we can observe the values of k traits each of which is a function of some genotype, the genotypes being independent. Let us denote by $\zeta^{(s)}$ $(s = 1, \dots, k)$ the triplet of phenotypes for the s-th trait. The vector $Z = (\zeta^{(1)}, \dots, \zeta^{(k)})$ has independent components. Let $f_1^{(s)}(z)$ and $f_2^{(s)}(z)$ denote the conditional probabilities of occurrence of the value z of the trait $\zeta^{(s)}$ in the class of families and in the class of false families, respectively, and let $A^{(s)}$ be the set of incompatible triplets of phenotypes for the s-th trait. The events $\{\zeta^{(s)} \in A^{(s)}\}$, $s = 1, \dots, k$, denoting the exclusion of paternity on the basis of particular traits, are independent. Thus, the exclusion of paternity on the basis of Z takes place if any of these events occurs, that is, if there occurs the event

$$\{Z \in A\} = \bigcup_{s=1}^{k} \{\zeta^{(s)} \in A^{(s)}\}. \tag{7.2.3}$$

Therefore

$$\begin{aligned} W = P(Z \in A \mid I = Z) &= 1 - \prod_{s=1}^{k} \left(1 - P(\zeta^{(s)} \in A^{(s)} \mid I = 2)\right) \\ &= 1 - \prod_{s=1}^{k} \Big(1 - \sum_{Z \in A^{(s)}} f_2^{(s)}(z)\Big). \end{aligned} \tag{7.2.4}$$

Probability W increases when the set of the traits analyzed is supplemented by an independent trait which has a positive probability of exclusion of paternity of a man not being the father of the child.

The exclusion of paternity may take place also on the basis of heritable traits which are stochastically dependent on the genotype. In practice this concerns only some traits with a finite set of values but theoretically arbitrary traits can be considered. Therefore, let ν be the suitably chosen measure defined on the set of the results of investigations and let f_1 and f_2 be the probability densities in respect to ν for the results in the class of families and in the class of false families, respectively. Then

$$W = \int_A f_2(z)\, d\nu(z),$$

where A is a set such that its probability equals zero in the class of families, and is as high as possible in the class of false families.

*

An attempt of paternity exclusion is the first indisputable step carried out in father recognition. However, if the paternity is not excluded, there are several possibilities of further procedure ways. Let us first take up the situation in which we want to recognize the exact, or, at least, the estimated value of the probability a posteriori of the man's paternity, i.e., the expression

$$P(I = 1 \mid Z = z) = \frac{\pi f_1(z)}{\pi f_1(z)+(1-\pi)f_2(z)}.$$

If $z \in A$, i.e. if the paternity is excluded, then $f_1(z) = 0$, and the probability a posteriori of a man's paternity equals zero. For $z \in \boldsymbol{Z}$

$$P(I = 1 \mid \boldsymbol{Z} = z) = \left(1+\frac{1-\pi}{\pi}h(z)\right)^{-1}, \tag{7.2.5}$$

where

$$h(z) = \frac{f_2(z)}{f_1(z)}, \quad z \in \boldsymbol{Z}. \tag{7.2.6}$$

In paternity problems the density ratio $h(z)$ is usually called the *critical value of* z, and its reciprocal the *odds of paternity*. The probability a posteriori of a man's paternity decreases when the critical value increases or when the probability a priori of paternity decreases.

Estimation of the probability a posteriori reduces to the estimation of probability a priori π and of the critical value $h(z)$.

In practice π is frequently not estimated, instead it is assumed that $\pi = 0.5$. Since π is usually higher than 0.5 (in studies carried out in Poland in the fifties it has been found that π approximately equals 0.7), such an assumption results in underestimating the probability a posteriori. For instance, with $h(z) = 0.1$ the probability a posteriori equals 0.909 with $\pi = 0.5$, and 0.959 with $\pi = 0.7$.

The estimation of π is carried out on the basis of the available material of previously considered writs. If for those writs it were known whether the man is the father of the child, the writs would constitute an auxiliary sample fully observable, and it would be possible to estimate π as a fraction of fathers in the available writs. However, since paternity is observable only if $z \in \boldsymbol{A}$, the material of the writs considered is used less directly. We use the fact that according to Table 7.2.1

$$P(Z \in A) = W(1-\pi), \tag{7.2.7}$$

hence

$$\pi = 1-\frac{P(Z \in A)}{W}. \tag{7.2.8}$$

The estimator $\hat{\pi}$ of parameter π is obtained from (7.2.8) by replacing probabilities W and $P(Z \in A)$ by their estimators. A natural estimator of $P(Z \in A)$ is the fraction of paternity exclusions in the set of the previously examined writs. On the other hand, probability W can be estimated by replacing probabilities $f_2(z)$ in (7.2.2) by the estimators $\hat{f}_2(z)$. Thus, if we estimate f_1 and f_2 on the basis of suitably collected data, and $P(Z \in A)$ on the basis of the examined writs, then we can calculate $\hat{\pi}$ and $\hat{h} = \hat{f}_2/\hat{f}_1$ as well, and this in turn makes it possible to estimate the probability a posteriori for any z.

In the most simple cases the estimation of f_1 and f_2 is not difficult. If, for instance, z is a triplet of phenotypes of mother, child and man for a single trait dependent functionally on the genotype in the case of a single locus with r alleles of probabilities $t_1, \ldots, t_r$, then f_1 and f_2 will be known functions of the arguments $z, t_1, \ldots, t_r$. This is presented in detail in the next section for the blood group systems AB0 and MN. Usually the parameters $t_1, \ldots, t_r$ are not known. However, for a population in the state of Hardy–Weinberg equilibrium, and under the assumption of independent random mating, they can be estimated from auxiliary data (Sec. 6.4). In the case of a single trait functionally dependent on several loci the procedure is analogous.

Now, let us consider a single phenotypic trait with a continuous distribution, which depends stochastically on the genotype. In such a case we sometimes build a parametric model assuming that the densities f_1 and f_2 are known functions of z and of certain real-valued parameters. For instance, in what is called the normal model, f_1 is the density of a three-dimensional normal distribution with identical marginal distributions, in which the phenotypes of mother and father are independent, and the correlation coefficient between the phenotypes of child and mother is the same as the correlation coefficient between the phenotypes of child and father. On the other hand, f_2 in this model is the product of the density of a two-dimensional normal distribution of the phenotypes of mother and child, and a one-dimensional normal distribution of the man's phenotype. The unknown parameters: the mean and variance of the phenotypic trait considered and the correlation coefficient of the mother's and child's phenotypes can be estimated on the basis of auxiliary data.

For some traits it is difficult to build a proper parametric model; therefore a nonparametric model is accepted, in which f_1, f_2 and π are unknown. In this case for estimating f_1, f_2 and π use is made of auxiliary data obtained by examining a set of families. Density f_1 is estimated by means of the appropriate frequencies, and for estimating f_2 we multiply the density estimator of the mother's and child's phenotypes by the density estimator of the phenotype of a man randomly selected from the set of fathers.

The estimation in a nonparametric model can be performed for any set of heritable traits, but it is usually more expensive and more time-consuming than the estimation in a parametric model. The non-parametric estimation can be applied for single traits or for a couple of traits, but here the number of traits is limited by the cost of estimation and organizational aspects (the size of auxiliary samples consisting of families should rise rapidly with the increase of the number of traits).

If we consider r independent heritable traits, then also the triplets of phenotypes are independent. It follows that f_i $(i = 1, 2)$ for $Z = (\zeta^{(1)}, \ldots, \zeta^{(s)})$ is the product of densities $f_i^{(s)}$ $(s = 1, \ldots, r)$. Then, the critical value h for Z is the product of critical values for $\zeta^{(1)}, \ldots, \zeta^{(s)}$.

The blood group traits for independent loci can be taken as an example of independent heritable traits. Morphological traits are usually divided into independent groups of traits, whereas the traits within one group are usually dependent. Thus, the set of considered traits consists of independent subsets (sometimes single-element ones) of dependent traits. Some difficulties appear in the case of a subset of dependent traits, for which f_1 and f_2 cannot be jointly estimated in the way described above on the basis of an auxiliary set of families (for example because of the cost of the study). Sometimes an initial reduction of this group of traits is carried out. Namely, a transformation of the initial data is sought for, for which the distributions of transformed data in the class of families and in the class of false families are separated as much as possible.

The classic method of data reduction consists in estimating on the basis of the auxiliary material the critical value for each trait separately; next, their product (or any other symmetrical function increasing in respect to each argument, e.g. the sum of logarithms) is taken into consideration. This transformation is usually called *discriminant function.* Of course, the estimation of distributions of transformed data requires additional empirical investigations on subsidiary data. In many instances repeated empirical investigation (concerning estimation of critical values of particular traits and estimation of distributions of transformed data) can be performed.

Apart from the classic method various natural methods of reduction are used, based on the assumption that in a family the similarity between child and his father is usually higher than the similarity between child and man in a false family. In estimating this similarity, the similarity of the child to the mother is usually taken into account, as well as the similarity between monozygotic twins, and the influence of the child's age and sex on the trait under examination. Some examples are discussed in Section 7.4.

It results from the above considerations that owing to the initial studies carried out on the auxiliary material the probability a posteriori of a certain part of serological and morphological data can be estimated. The estimated values of the probabilities a posteriori, and an analysis of the remaining transformed data are the basis for the anthropologist to prepare his report to be presented to the court. Example is given in Section 7.4.

*

Usually courts do not know how to interpret the supplied estimate of the probability a posteriori in cases of writs in which it is positive (that is, the paternity is not excluded). In our opinion the court should be advised on the relation between various decision rules based on the posterior probability and the credibility of these decisions. Therefore, father recognition on the basis of biological evidence should be treated as a discriminant problem, and the decision rule which solves the problem, as well as the properties of this rule, should be presented to the court.

The court, of course, would not be obliged to proceed in accordance with any suggested rule, treating individually the whole of the evidence material for each of the writs. However, the knowledge of the rule and of its consequences might serve as reference point for the court, and could lead to a higher uniformity and objectivism of the court's verdicts. Among the rules considered also rules with suspended decision should be provided.

In Section 3.2 various probabilistic discriminant problems have been presented. At the beginning of Section 3.2 it was pointed out that if the supports of densities f_1 and f_2 are not the same, then a cut down of both densities should be performed together with a proper modification of the probability a priori. Then the symmetric difference of both supports serves for taking faultless decisions. Such a situation is typical for problems of father recognition. The set A, i.e., the symmetric difference of the supports of densities in both classes, serves for paternity exclusion in a faultless way. However, if paternity is not excluded we have to deal with a discriminant problem in which the density in the class of families still equals f_1, the modified density in the class of false families equals $f_2' = f_2/(1-W)$, and the modified probability a priori equals

$$\pi' = P(I = 1 \mid Z \in \boldsymbol{Z}) = \frac{\pi}{\pi+(1-\pi)(1-W)} . \tag{7.2.9}$$

As seen, π' increases from π to 1 with W increasing from 0 to 1.

Basing on (7.2.5) and (7.2.9) the change over of f_2 and π into f_2' and π' does not change the probability a posteriori of the man's paternity in the set $\boldsymbol{Z}$, because

$$\frac{1-\pi'}{\pi'} f_2' = \frac{1-\pi}{\pi} f_2 .$$

The solution of a discrimination problem is expressed by means of thresholds for the selected discriminant function, for instance for the function $1-P(I = 1 \mid Z = z)$ or for the critical value f_2/f_1. The properties of this solution are described by means of expressions $a_{12}, a_{21}, a_0, b_1, b_2$ (defined in Sec. 3.2) which denote the probability of unjustified paternity exclusion, the probability of unjustified paternity recognition, the probability of suspended decision, the quotient of probabilities of wrong and correct decisions in the class of families, and the quotient of probabilities of wrong and correct decisions in the class of false families, respectively.

If we do not allow for a suspended decision, then the probability of a wrong decision when paternity is not excluded, equals $\pi' a_{12}+(1-\pi')a_{21}$. If W is large then the rule minimizing the probability of a wrong decision does not differ much from the rule recognizing as the father each defendant whose paternity has not been excluded. The latter rule will be called *feministic*. The probability of an error for this rule equals, of course, $1-\pi'$.

In practice instead of f_1, f_2' and π' we must use their estimators, based on $\hat{f}_1$, $\hat{f}_2$, $\hat{\pi}$ and $\hat{W}$. In solving a particular discriminant problem on the basis of all available blood group systems and morphological traits difficulties arise which are analogous to those related to the estimation of probability a posteriori.

For illustration purposes, in the following section solutions of various discriminant problems are presented, based on two blood group systems AB0 and MN only. It was found that, with these scarce data, reliable solutions are reached only in few extremal cases.

7.3. A STUDY OF BLOOD GROUP TRAITS

One of the more complicated blood group systems employed in paternity testing is the AB0 system (cf. pp. 139–140), that is a system of four alleles $\tilde{A}_1, \tilde{A}_2, \tilde{B}, \tilde{0}$, from among which $\tilde{0}$ is recessive against each of the remaining alleles, $\tilde{A}_1$ dominates over $\tilde{A}_2$ and codominates with $\tilde{B}$, and $\tilde{A}_2$ codominates with $\tilde{B}$. The probabilities of genes $\tilde{A}_1, \tilde{A}_2, \tilde{B}, \tilde{0}$ in the considered population will be denoted by p_1, p_2, q, r, respectively.

We will call *concordant* any two blood systems possessing the same scheme of domination and codomination over a set of alleles. Concordant systems differ at most by the values of probabilities of alleles occurrence; some of them may equal zero, and thus, concordant blood systems may have a different number of alleles.

Among the systems used in Poland for paternity recognition concordant with AB0 are systems with two codominant alleles:

$$\text{MN}, \quad \text{Hp}, \quad \text{Gc}, \quad \text{PGM}_1, \quad \text{AK}, \quad \text{ADA}, \quad \text{EsD}, \tag{7.3.1}$$

system with two alleles, from which the second dominates over the first:

$$\text{Kell}, \quad \text{Sese}, \quad \text{Gm(1)}, \quad \text{Km(1)}, \tag{7.3.2}$$

and system Gm(1, 2), with three alleles traditionally denoted by $\text{Gm}^{-1,-2}$, $\text{Gm}^{1,2}$, $\text{Gm}^{1,-2}$. They correspond with genes $\tilde{0}, \tilde{A}_1, \tilde{A}_2$ but there is no equivalence for gene B (i.e. $q = 0$). On the other hand, AB0 is concordant neither with system Rh which has a very complicated scheme of domination and codomination, nor with systems AP and GPT with three codominant alleles and system C3 with four codominant alleles.

First of all we will present the inference on paternity based on system AB0. The presented scheme of inference may be applied for any blood systems concordant with AB0 by substituting respective alleles and values of gene probabilities.

TABLE 7.3.1.
Relation between genotype and phenotype, and phenotype distribution in the AB0 system.

Genotypes	Phenotype	Phenotype probability
$\tilde{0}\tilde{0}$	0	r^2
$\tilde{A}_1\tilde{A}_1, \tilde{A}_1\tilde{A}_2, \tilde{A}_1\tilde{0}$	A_1	$p_1(p_1+2p_2+2r)$
$\tilde{A}_2\tilde{A}_2, \tilde{A}_2\tilde{0}$	A_2	$p_2(p_2+2r)$
$\tilde{A}_1\tilde{B}$	A_1B	$2p_1q$
$\tilde{A}_2\tilde{B}$	A_2B	$2p_2q$
$\tilde{B}\tilde{B}, \tilde{B}\tilde{0}$	B	$q(q+2r)$

In the AB0 system six phenotypes occur: 0, A_1, A_2, B, A_1B, A_2B. Thus, there are as many as 6^3 possible triplets for three persons connected with a single writ. For given probabilities p_1, p_2, q, r of genes $\tilde{A}_1, \tilde{A}_2, \tilde{B}, \tilde{0}$ it is possible to calculate probabilities $f_1(z)$ and $f_2(z)$ for each triplet, using the laws of inheritance and relations between the genotype and the phenotype shown in Table 7.3.1. We assume that the man and the mother come from the same population characterized by the values of p_1, p_2, q, r. We also assume that we are dealing with a primitive generation model (cf. Sec. 6.3) so that mating of parental pairs does not depend on the parental genotypes (such a population is called *panmictic*).

For instance let $z = (A_2, A_1, A_1)$, that is the mother has phenotype A_2 (thus genotype $\tilde{A}_2\tilde{A}_2$ or $\tilde{A}_2\tilde{0}$), and both child and father have phenotype A_1 (thus each of them has one of three possible genotypes: $\tilde{A}_1\tilde{A}_1$, $\tilde{A}_1\tilde{A}_2$, $\tilde{A}_1\tilde{0}$). Columns 2, 3 and 4 of Table 7.3.2 contain genotypes of a child whose parents possess phenotypes as shown above. The genotypes of a child possessing phenotype A_1 are distinguished in the table by means of a frame. On the basis of Tables 7.3.1 and 7.3.2

$$\begin{aligned} f_1(A_2, A_1, A_1) &= p_1^2p_2^2 + \tfrac{1}{2}2p_1p_2^3 + \tfrac{1}{2}2p_1rp_2^2 + 2p_1^2p_2r + \\ &\quad + \tfrac{1}{2}4p_2^2p_1r + \tfrac{1}{4}4p_1p_2r^2 \\ &= p_1p_2(p_2+2r)(p_1+p_2+r). \end{aligned}$$

In a false family the blood group of the true father of the child is not known, therefore all possible genotypes of the father should be considered. For that reason in Table 7.3.2 genotypes of a child are presented, mother's phenotype of which is A_2,

TABLE 7.3.2.

Set of genotypes of child whose mother possesses phenotype A_2; in frames genotypes of child with phenotype A_1.

Father's phenotype	0	A_1			A_2		A_1B	A_2B	B	
Father's genotype / Mother's genotype	$\tilde{0}\tilde{0}$	$\tilde{A}_1\tilde{A}_1$	$\tilde{A}_1\tilde{A}_2$	$\tilde{A}_1\tilde{0}$	$\tilde{A}_2\tilde{A}_2$	$\tilde{A}_2\tilde{0}$	$\tilde{A}_1\tilde{B}$	$\tilde{A}_2\tilde{B}$	$\tilde{B}\tilde{B}$	$\tilde{B}\tilde{0}$
	(1)	(2)	(3)	(4)	(5)	(6)	(7)	(8)	(9)	(10)
$\tilde{A}_2\tilde{A}_2$	$\tilde{A}_2\tilde{0}$	$\boxed{\tilde{A}_2\tilde{A}_1}$	$\boxed{\tilde{A}_2\tilde{A}_1}$	$\boxed{\tilde{A}_2\tilde{A}_1}$	$\tilde{A}_2\tilde{A}_2$	$\tilde{A}_2\tilde{A}_2$	$\boxed{\tilde{A}_2\tilde{A}_1}$	$\tilde{A}_2\tilde{A}_2$	$\tilde{A}_2\tilde{B}$	$\tilde{A}_2\tilde{B}$
	$\tilde{A}_2\tilde{0}$	$\boxed{\tilde{A}_2\tilde{A}_1}$	$\boxed{\tilde{A}_2\tilde{A}_1}$	$\boxed{\tilde{A}_2\tilde{A}_1}$	$\tilde{A}_2\tilde{A}_2$	$\tilde{A}_2\tilde{A}_2$	$\boxed{\tilde{A}_2\tilde{A}_1}$	$\tilde{A}_2\tilde{A}_2$	$\tilde{A}_2\tilde{B}$	$\tilde{A}_2\tilde{B}$
	$\tilde{A}_2\tilde{0}$	$\boxed{\tilde{A}_2\tilde{A}_1}$	$\tilde{A}_2\tilde{A}_2$	$\tilde{A}_2\tilde{0}$	$\tilde{A}_2\tilde{A}_2$	$\tilde{A}_2\tilde{0}$	$\tilde{A}_2\tilde{B}$	$\tilde{A}_2\tilde{B}$	$\tilde{A}_2\tilde{B}$	$\tilde{A}_2\tilde{0}$
	$\tilde{A}_2\tilde{0}$	$\boxed{\tilde{A}_2\tilde{A}_1}$	$\tilde{A}_2\tilde{A}_2$	$\tilde{A}_2\tilde{0}$	$\tilde{A}_2\tilde{A}_2$	$\tilde{A}_2\tilde{0}$	$\tilde{A}_2\tilde{B}$	$\tilde{A}_2\tilde{B}$	$\tilde{A}_2\tilde{B}$	$\tilde{A}_2\tilde{0}$
$\tilde{A}_2\tilde{0}$	$\tilde{A}_2\tilde{0}$	$\boxed{\tilde{A}_2\tilde{A}_1}$	$\boxed{\tilde{A}_2\tilde{A}_1}$	$\boxed{\tilde{A}_2\tilde{A}_1}$	$\tilde{A}_2\tilde{A}_2$	$\tilde{A}_2\tilde{A}_2$	$\boxed{\tilde{A}_2\tilde{A}_1}$	$\tilde{A}_2\tilde{A}_2$	$\tilde{A}_2\tilde{B}$	$\tilde{A}_2\tilde{B}$
	$\tilde{A}_2\tilde{0}$	$\boxed{\tilde{A}_2\tilde{A}_1}$	$\tilde{A}_2\tilde{A}_2$	$\tilde{A}_2\tilde{0}$	$\tilde{A}_2\tilde{A}_2$	$\tilde{A}_2\tilde{0}$	$\tilde{A}_2\tilde{B}$	$\tilde{A}_2\tilde{B}$	$\tilde{A}_2\tilde{B}$	$\tilde{A}_2\tilde{0}$
	$\tilde{0}\tilde{0}$	$\boxed{\tilde{A}_1\tilde{0}}$	$\boxed{\tilde{A}_1\tilde{0}}$	$\boxed{\tilde{A}_1\tilde{0}}$	$\tilde{A}_2\tilde{0}$	$\tilde{A}_2\tilde{0}$	$\boxed{\tilde{A}_1\tilde{0}}$	$\tilde{A}_2\tilde{0}$	$\tilde{B}\tilde{0}$	$\tilde{B}\tilde{0}$
	$\tilde{0}\tilde{0}$	$\boxed{\tilde{A}_1\tilde{0}}$	$\boxed{\tilde{A}_1\tilde{0}}$	$\tilde{0}\tilde{0}$	$\tilde{A}_2\tilde{0}$	$\tilde{0}\tilde{0}$	$\tilde{B}\tilde{0}$	$\tilde{B}\tilde{0}$	$\tilde{B}\tilde{0}$	$\tilde{0}\tilde{0}$

and framed are these genotypes which a child of phenotype A_1 can have. Since in a false family man's genotype is independent by assumption from the pair of mother's and child's genotypes, the following holds:

$$\begin{aligned} f_1(A_2, A_1, A_1) &= (p_1^2 p_2^2 + 2p_1^2 p_2 r + \tfrac{1}{2} 2p_1 p_2^3 + \tfrac{1}{2} 4p_1 p_2^2 r + \\ &\quad + \tfrac{1}{2} 2p_1 p_2^2 r + \tfrac{1}{2} 4p_1 p_2 r^2 + \tfrac{1}{2} 2p_1 p_2^2 q + \\ &\quad + \tfrac{1}{2} 4p_1 p_2 qr) p_1 (p_1 + 2p_2 + 2r) \\ &= p_1^2 p_2 (p_2 + 2r)(p_1 + 2p_2 + 2r). \end{aligned}$$

On the other hand, for a child with blood group B it follows from Table 7.3.2 that it cannot be the child of a mother with phenotype A_2 and of a father with phenotype A_1; therefore $f_1(A_2, B, A_1) = 0$. Then for $z = (A_2, B, A_1)$ paternity of the man is excluded; on the basis of Table 7.3.3 the probability of such a triplet z in a false family equals:

$$f_2(A_2, B, A_1) = p_1 p_2 r q (p_1 + 2p_2 + 2r).$$

TABLE 7.3.3.

Set A for the blood system AB0: triplets of mother's, child's and man's phenotypes which result in exclusion of paternity.

Phenotype of		
mother	child	man
0, A_1, A_2, B	0	A_1B, A_2B
0, A_2, A_2B, B	A_1	0, A_2, A_2B, B
0, B	A_2	0, A_1B, B
A_1, A_2, A_2B	A_2	A_1B
A_1	A_1B	0, A_1, A_2
A_2B, B	A_1B	0, A_2, A_2B, B
A_1B	A_1B	0, A_2
A_1, A_2	A_2B	0, A_1, A_2
A_1B, B	A_2B	0, A_1B, B
A_2B	A_2B	0
0, A_1, A_2	B	0, A_1, A_2

For some pairs of woman's and child's blood groups the maternity of the woman is also impossible, regardless the phenotype of the father (for example, mother with phenotype 0 cannot have a child with phenotype A_1B or A_2B). There are six such pairs of mother's and child's phenotypes:

$$\begin{aligned} &(A_1B, 0), \quad (A_2B, 0), \quad (A_1B, A_2), \\ &(0, A_1B), \quad (A_2, A_1B), \quad (0, A_2B). \end{aligned} \tag{7.3.4}$$

Since we assume that the woman relevant to the writ on father recognition is the mother of the child, and since we do not allow gene mutation, then each of the 36 possible triplets with such a pair of woman's and child's phenotypes and with any man's phenotype has zero probability, in a family and in a false family as well.

The triplets z for which $f_1(z) = 0$ and $f_2(z) > 0$, form set A. It has 68 elements, presented in Table 7.3.3. The remaining 112 triplets of phenotypes for which both $f_1(z)$ and $f_2(z)$ are positive, form the set $\boldsymbol{Z}$. The elements of this set are presented in column 2 of Table 7.3.4.

Probabilities $f_1(z)$ and $f_2(z)$ are the functions of arguments p_1, p_2, q, r. For Poland the following estimates of p_1, p_2, q, r (called *gene frequencies*) were found:

$$\hat{p}_1 = 0.2230, \quad \hat{p}_2 = 0.0560, \quad \hat{q} = 0.1538, \quad \hat{r} = 0.5672. \tag{7.3.5}$$

They were obtained by the Bernstein method described in Chapter 6 for three alleles $\tilde{A}$, $\tilde{B}$, $\tilde{0}$ (cf. (6.4.16)). By substituting frequencies (7.3.5) for p_1, p_2, q, r the estimated values $\hat{f}_1(z)$ and $\hat{f}_2(z)$ of probabilities $f_1(z)$ and $f_2(z)$ were obtained, and estimated critical values $\hat{h}(z) = \hat{f}_2(z)/\hat{f}_1(z)$ for $z \in \boldsymbol{Z}$ were computed. The values $\hat{h}(z)$ can be read from column 3 in Table 7.3.4 (as critical values $h(z)$ in system AB0 with p_1, p_2 q, r equal to gene frequencies (7.3.5)). Since it was found that $\sum_{z \in \boldsymbol{Z}} f_2(z) = 0.78537$, then by formula (7.2.2)

$$\hat{W}_{AB0} = 0.21463, \tag{7.3.6}$$

where W_{AB0} is the probability of paternity exclusion of a man not being the father of the child, obtained on the basis of system AB0. Therefore, if π, p_1, p_2, q and r were equal to their estimated values (7.3.5) then the probability of paternity exclusion of a man on the basis of system AB0 would equal $0.3 \cdot 0.2146 = 0.0644$, the whole matrix of probabilites in Table 7.2.1 would have the form

$$\begin{bmatrix} 0.7 & 0 \\ 0.2356 & 0.0644 \end{bmatrix},$$

and the conditional paternity probability of a man under the condition that paternity has not been excluded would equal

$$\pi' = \frac{0.7}{0.7 + 0.3 \cdot 0.7854} = 0.7482.$$

The elements of set $\boldsymbol{Z}$ are arranged in column 2 of Table 7.3.4 in nondecreasing order of critical values $\hat{h}(z)$ (column 3). From columns 4 and 5 estimated values of the probabilities a posteriori of the man's paternity may be read. They are computed according to formula (7.2.5) for $\hat{\pi} = 0.5$ and $\hat{\pi} = 0.7$, respectively.

The probabilities with which the particular critical values appear in the

TABLE 7.3.4.

Critical values $h(z)$ for $z \in Z$ in the AB0 system, with $p_1 = 0.2230$, $p_2 = 0.0560$, $q = 0.1538$, $r = 0.5672$, and probabilities of critical values in the class of families and in the class of false families.

No.	Examined triplets: blood groups of mother, child and man	Critical value	Probability a posteriori of man's paternity		Probability of critical value $h(z)$		
			for $\pi = 0.5$	for $\pi = 0.7$	in class of families	in class of false families	in class of false families in distribution truncated to set Z
	$z \in Z$	$h(z)$	$P(I = 1 \mid Z = z)$		$g_1(h(z))$	$g_2(h(z))$	$g_2'(h(z))$
(1)	(2)	(3)	(4)	(5)	(6)	(7)	(8)
1	A_1B A_2B A_2 B A_2B A_2 0 A_2 A_2 B A_2 A_2	0.10697	0.90337	0.95617	0.01934	0.00207	0.00263
2	A_1B A_2B A_2B B A_2B A_2B 0 A_2 A_2B B A_2 A_2B	0.11200	0.89928	0.95420	0.00477	0.00053	0.00068
3	A_1 A_2 A_2	0.18888	0.84113	0.92511	0.00525	0.00099	0.00126
4	A_2B A_2B A_2B	0.20980	0.82658	0.91750	0.00015	0.00003	0.00004
5	A_1 A_2 A_2B	0.21394	0.82377	0.91601	0.00120	0.00026	0.00033
6	A_1 A_1B B A_1 A_2B B A_2 A_2B B A_1 B B A_2 B B 0 B B	0.27479	0.78444	0.89464	0.07940	0.02182	0.02778
7	A_1 A_1B A_1B A_1 A_1B A_2B A_1 A_2B A_1B A_2 A_2B A_1B A_1 A_2B A_2B A_2 A_2B A_2B 0 B A_1B A_1 B A_1B A_2 B A_1B 0 B A_2B A_1 B A_2B A_2 B A_2B	0.30760	0.76476	0.88353	0.03073	0.00945	0.01203
8	A_2B A_2B B	0.37485	0.72735	0.86159	0.00096	0.00036	0.00046

(1)	(2)	(3)	(4)	(5)	(6)	(7)	(8)
9	A_1B A_1B A_1B	0.37680	0.72632	0.86097	0.00235	0.00089	0.00113
10	0 A_1 A_1 A_2 A_1 A_1 A_2B A_1 A_1 A_2B A_1B A_1 B A_1 A_1 B A_1B A_1	0.38723	0.72086	0.85766	0.11393	0.04412	0.05617
11	A_2B A_2B A_2	0.40075	0.71390	0.85343	0.00030	0.00012	0.00015
12	A_2B A_2B A_1B	0.41960	0.70442	0.84758	0.00030	0.00012	0.00016
13	A_2B A_1B A_1B B A_1B A_1B 0 A_1 A_1B A_2 A_1 A_1B A_2B A_1 A_1B B A_1 A_1B	0.44600	0.69156	0.83953	0.02071	0.00924	0.01176
14	A_2 A_2 A_2	0.55661	0.68653	0.83634	0.00343	0.00157	0.00200
15	0 0 0 A_1 0 0 A_2 0 0 B 0 0	0.56720	0.63808	0.80445	0.18248	0.10350	0.13178
16	B B B	0.48459	0.63108	0.79965	0.03164	0.01850	0.02355
17	A_2B A_2 0 A_2B A_2 A_2	0.62320	0.61607	0.78921	0.00335	0.00208	0.00265
18	A_1B A_1B A_1	0.65430	0.60448	0.78100	0.00647	0.00423	0.00539
19	A_1B A_1B B	0.67322	0.59765	0.77608	0.00380	0.00256	0.00326
20	A_2 A_2 0	0.67417	0.59731	0.77584	0.01123	0.00757	0.00964
21	A_2 A_2 A_2B	0.70588	0.58621	0.76774	0.00057	0.00041	0.00052
22	A_1 A_1 A_1	0.70951	0.58496	0.76683	0.08806	0.06248	0.07955
23	A_1B B 0 A_2B B 0 A_1B B B A_2B B B	0.72100	0.58106	0.76394	0.02231	0.01608	0.02048
24	A_1B A_1B A_2B	0.75360	0.57026	0.75587	0.00030	0.00022	0.00028
25	B B 0	0.84199	0.54289	0.73483	0.03568	0.03004	0.03825
26	A_1B A_1 0 A_1B A_1 A_1 A_1B A_1 A_2	0.84620	0.54165	0.73386	0.02456	0.02078	0.02646
27	B B A_1B B B A_2B	0.94252	0.51480	0.71228	0.00850	0.00801	0.01020
28	A_1 A_1 0 A_1 A_1 A_2	1.01043	0.49741	0.69782	0.07329	0.07405	0.09429

(1)	(2)			(3)	(4)	(5)	(6)	(7)	(8)
29	A_1	A_1	A_1B	1.16378	0.46215	0.66722	0.01124	0.01308	0.01665
30	0	0	A_2	1.19040	0.45654	0.66218	0.02203	0.02623	0.03340
	A_1	0	A_2						
	A_2	0	A_2						
	B	0	A_2						
	A_1	A_2	0						
31	A_2B	A_2	A_2B	1.24640	0.44516	0.65182	0.00007	0.00009	0.00012
32	0	0	B	1.28820	0.43702	0.64430	0.04948	0.06374	0.08116
	A_1	0	B						
	A_2	0	B						
	B	0	B						
33	A_2B	A_2	B	1.41538	0.41401	0.62244	0.00075	0.00106	0.00135
34	A_1B	B	A_1B	1.44200	0.40950	0.61805	0.00184	0.00266	0.00338
	A_1B	B	A_2B						
	A_2B	B	A_1B						
	A_2B	B	A_2B						
35	0	0	A_1	1.46940	0.40496	0.61359	0.08842	0.12993	0.16543
	A_1	0	A_1						
	A_2	0	A_1						
	B	0	A_1						
	0	A_2	A_1						
	B	A_2	A_1						
	A_1	A_2	A_1						
	A_2	A_2	A_1						
	A_2B	A_2	A_1						
	A_1B	A_2B	A_1						
	B	A_2B	A_1						
36	A_1B	B	A_2	1.51318	0.39790	0.60661	0.00136	0.00206	0.00266
	A_2B	B	A_2						
37	A_2	A_2	B	1.53114	0.39508	0.60379	0.00304	0.00466	0.00594
38	A_1B	A_1	A_1B	1.69240	0.37142	0.57960	0.00147	0.00249	0.00317
	A_1B	A_1	A_2B						
39	B	B	A_2	1.76711	0.36139	0.56904	0.00352	0.00622	0.00792
40	A_1B	B	A_1	1.86784	0.34869	0.55540	0.00543	0.01014	0.01291
	A_2B	B	A_1						
41	A_1B	A_1	B	1.92185	0.34225	0.54835	0.00299	0.00575	0.00732
42	A_1	A_1	A_2B	2.02087	0.33103	0.53588	0.00163	0.00328	0.00418
43	B	B	A_1	2.18128	0.31434	0.51684	0.01403	0.03059	0.03895
44	A_1	A_1	B	2.29485	0.30350	0.50416	0.01646	0.03778	0.04810
45	A_1	A_2	B	2.70358	0.27000	0.46325	0.00109	0.00295	0.00375
46	A_2B	A_2B	A_1	5.50500	0.15373	0.29768	0.00011	0.00059	0.00075

class of families and in the class of false families (in the AB0 system with probabilities p_1, p_2, q, r equal to the estimated values (7.3.5)), are denoted by g_1 and g_2 and given in columns 6 and 7 of Table 7.3.4. Therefore, $g_i(a)$ is equal to the sum of $\hat{f}_i(z)$ for z such that $\hat{h}(z) = a$. Of course,

$$\sum_{z \in Z} g_2(\hat{h}(z)) = \sum_{z \in Z} \hat{f}_2(z) = 1 - \hat{W}_{AB0} = 0.78537.$$

In column 8 the values of function $g_2' = g_2/0.78537$ are tabulated, which correspond to the density of the critical value $\hat{h}(z)$ in the class of false families truncated to the set Z. The distribution functions corresponding to densities g_1, g_2, g_2', denoted G_1, G_2, G_2', are presented in Fig. 7.3.1. As seen, the differentiation of the dis-

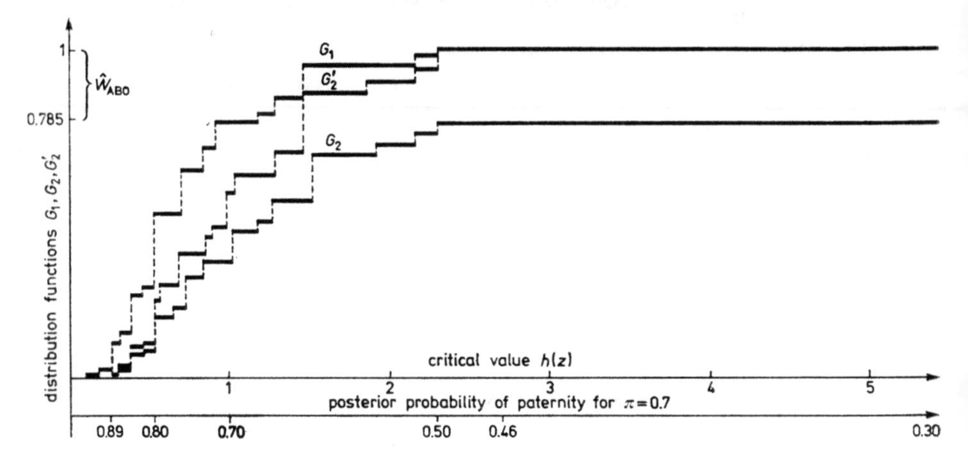

FIG. 7.3.1. Distribution functions G_1, G_2, G_2' corresponding to densities g_1, g_2, g_2' from Table 7.3.4.

tribution functions G_1 and G_2', representing the classes of families and of false families respectively, is very small. Moreover, the probability of obtaining a critical value larger than 2.29 is very small. Therefore, it follows from the relation between the critical value and the probability a posteriori of a man's paternity for $\pi = 0.7$ (presented in Fig. 7.3.1 by means of double scaling of the abscissa axis) that the probability a posteriori found on basis of system AB0 is, in practice, always higher than 0.3.

On the basis of Fig. 7.3.1 one can also see that if paternity is not excluded, the AB0 system provides hardly any information on the conjectured paternity. This is clearly confirmed by the properties of discriminant rules, constructed on the basis of $\hat{g}_1, \hat{g}_2$ and $\hat{\pi}'$. For illustration, let us discuss some of the rules from Section 3.2 for g_1, g_2 and π' equal respectively to their estimated values (under assumption that $\pi = 0.7$).

If one minimizes the probability of a wrong decision (Example 3.2.1) when the deferred decision is not allowed, then on the basis of Table 7.3.4 the solution δ^{FB} consists in rejecting paternity for the critical value $h = f_2/f_1$ exceeding 2.33, i.e, with the probability a posteriori of the man's paternity lower than 0.5. The probability α_{12} of rejecting paternity in the class of families equals under this rule 0.0012, and probability α_{21} of erroneous paternity recognition equals as much as 0.9955. Further, the conditional probability of wrong decision under the condition that paternity is not excluded equals 0.2516; this figure is almost identical with the analogous probability for the feministic rule, equal to $1-\pi' = 0.2518$ (see remark made at the end of Section 7.2).

On the other hand, if α_{12} and α_{21} are required to be identical and minimal (Example 3.2.4B), then paternity should be rejected for critical values exceeding 0.6066, i.e., for probability a posteriori lower than 0.7639. Then we would obtain $\alpha_{12} = \alpha_{21} = 0.3731$.

Now, let us come back to the problems from Example 3.2.3, in which the deferred decision is allowed and where the quotients b_1 and b_2 of probabilities of wrong and correct decisions in the class of families and in the class of false families are bound by β_1 and β_2, respectively. In these problems β_1 and β_2 must satisfy condition (3.2.22). From Table 7.3.4 it follows that this condition is of the form

$$\beta_1\beta_2 > \inf_{z\in Z} h(z)\cdot\inf_{z\in Z}\frac{1}{h(z)} = 0.136197\cdot 0.142668 = 0.019431.$$

For instance, if we fix $\beta_1 = 0.04$ then β_2 should be higher than 0.486; this means that we cannot secure a sufficiently reliable decision simultaneously in both classes even if the probability of deferred decision is very high. With $\beta_1 = \beta_2 = 0.2$ the condition (3.2.22) is met. The solution of such a problem consists in rejecting paternity when $h > 2.2949$ and on recognizing paternity when $h < 0.1070$. This means that the probability a posteriori must be very high (higher than 0.9562) in order that paternity could be recognized, and may be relatively high (but lower than 0.5042) if paternity should be rejected. The probability of deferred decision is as high as 0.9887.

Thus, the properties of discriminant rules based solely on the **AB0** system are highly unsatisfactory.

We shall now show to what extent the situation will be improved if the AB0 system is supplemented by the **MN** system with two codominating alleles **M** and N with probabilities p and q, estimated in Poland by $\hat{p} = 0.5953$ and $\hat{q} = 0.4047$ (cf. (6.4.12)).

TABLE 7.3.5.

Relation between genotypes and blood groups, and the distribution of blood groups in the MN system

Genotype	Blood group	Blood group probability
$\tilde{M}\tilde{M}$	M	p^2
$\tilde{M}\tilde{N}$	MN	$2pq$
$\tilde{N}\tilde{N}$	N	q^2

In accordance with Table 7.3.5 in the MN system there are three blood groups: M, N, MN. Therefore there are 27 triplets of blood groups for mother, child and man.

The assumption of the woman's maternity excludes from consideration such triplets z in which the mother's and child's blood groups form pairs (M, N) or (N, M). Paternity is excluded if $z \in \boldsymbol{A}$, where

$$\boldsymbol{A} = \{(\text{M}, \text{M}, \text{N}),\ (\text{M}, \text{MN}, \text{M}),\ (\text{N}, \text{N}, \text{M}),\ (\text{N}, \text{MN}, \text{N}),\ (\text{MN}, \text{N}, \text{M}),\ (\text{MN}, \text{M}, \text{N})\}.$$

The estimated probability of paternity exclusion of a man not being the father of the child equals in the MN system

$$\hat{W}_{\text{MN}} = 0.1829$$

TABLE 7.3.6.

Critical values $h(z)$ for $z \in \boldsymbol{Z}$ in the blood group system MN with $p = 0.5953$ and $q = 0.4047$ and probabilities of critical values in the class of families and in the class of false families.

No.	Examined triplets: blood groups of mother, child and man	Critical value	Probability a posteriori of man's paternity		Probability of critical value $h(z)$		
			for $\pi = 0.5$	for $\pi = 0.7$	in class of families	in class of false families	in class of false families in distribution truncated to set $\boldsymbol{Z}$
	$z \in \boldsymbol{Z}$	$h(z)$	$P(I = 1 \mid \boldsymbol{Z} = z)$		$g_1(h(z))$	$g_2(h(z))$	$g_2'(h(z))$
(1)	(2)	(3)	(4)	(5)	(6)	(7)	(8)
1	M, MN, N MN, N, N N, N, N	0.4047	0.7119	0.8522	0.1243	0.0503	0.0616
2	M, M, M MN, M, M N, MN, M	0.5953	0.6268	0.7967	0.2690	0.1601	0.1960
3	M, MN, MN MN, N, MN N, N, MN	0.8094	0.5527	0.7425	0.1829	0.1480	0.1811
4	MN, MN, M MN, MN, MN MN, MN, N	1.0000	0.5000	0.7000	0.2409	0.2409	0.2948
5	M, M, MN MN, M, MN N, MN, MN	1.1906	0.4566	0.6621	0.1829	0.2177	0.2665

Therefore by (7.2.4) and (7.3.6) the estimated probabilities of paternity exclusion on the basis of both systems AB0 and MN jointly equals

$$\hat{W}_{\mathrm{AB0+MN}} = 0.3583.$$

If this value is taken for probability W then the matrix of probabilities in Table 7.2.1 assumes the form

$$\begin{bmatrix} 0.7 & 0 \\ 0.1925 & 0.1075 \end{bmatrix}$$

and probability π' (cf. (7.2.9)) equals 0.7843. The estimated critical values for the MN system are tabulated in column 2 of Table 7.3.6, which is analogous to Table 7.3.4 for the AB0 system.

The product of critical values in both systems AB0 and MN truncated to the set of triplets for which paternity is not excluded, has a joint distribution which

TABLE 7.3.7.

Probabilities concerning the exclusion of paternity of a man not being the father of the child, for 16 selected blood group systems.

i	Blood group system	Probability of paternity exclusion of a man not being the father of the child when the i-th blood group system has been examined W_i	Probability of paternity exclusion of a man not being the father of the child when i initial blood group systems have been examined $W_{1+\cdots+i}$	Probability of making an error with the feministic rule, based on i initial blood group systems (with $\pi = 0.7$), $0.3(1-W_{1+\cdots+i})$	Conditional probability of paternity under the condition that paternity was not excluded, computed for $\pi = 0.7$ on the basis of i initial blood group systems according to (7.2.9)
(1)	(2)	(3)	(4)	(5)	(6)
1	AB0	0.2146	0.2146	0.2356	0.7482
2	MN	0.1829	0.3583	0.1925	0.7843
3	Rh	0.3028	0.5526	0.1342	0.8391
4	Kell	0.0355	0.5684	0.1295	0.8439
5	Sese	0.0142	0.5745	0.1277	0.8458
6	Hp	0.1793	0.6508	0.1048	0.8698
7	Gc	0.1667	0.7090	0.0973	0.8891
8	Gm(1,2)	0.1036	0.7392	0.0782	0.8995
9	AP	0.2725	0.8103	0.0569	0.9248
10	PGM_1	0.1539	0.8395	0.0482	0.9356
11	GPT	0.1887	0.8698	0.0391	0.9472
12	AK	0.0326	0.8760	0.0372	0.9495
13	ADA	0.0622	0.8892	0.0332	0.9547
14	EsD	0.0801	0.8913	0.0326	0.9555
15	Km(1)	0.0524	0.8970	0.0309	0.9577
16	C3	0.1240	0.9098	0.0271	0.9628

can be retrieved from Tables 7.3.4 and 7.3.6. Then we find that the properties of discriminant rules become improved to a very small degree only. For instance for the rule δ^{FB} (Example 3.2.1B) we obtain $a_{12} = 0.0176$, $a_{21} = 0.9347$, and the probability of a wrong decision equals 0.2154 (hence it is almost identical with $1-\pi' = 0.2157$). Instead, in problems from Example 3.2.3 the upper bounds β_1 and β_2 must meet the condition $\beta_1\beta_2 > 0.0066$, and with $\beta_1 = \beta_2 = 0.2$ the probability of a deferred decision equals 0.9490. Thus, inclusion of the MN system results virtually only in increasing chances of paternity exclusion.

Table 7.3.7 shows the influence of inclusion of the remaining independent blood group systems, mentioned at the beginning of this section. Let $W_{1+\dots+i}$ denote the probability of paternity exclusion of a man not being the father of the child when i initial systems from column 2 in Table 7.3.7 have been examined. This probability increases from 0.215 with $i = 1$ to 0.910 with $i = 16$ (column 4). The probability of a wrong decision under the feministic rule based on i initial systems, equaling $(1-\pi)(1-W_{1+\dots+i})$ (cf. Table 7.2.1) decreases for $\pi = 0.7$ from 0.236 to 0.027 (column 5). The conditional probability of paternity under condition that paternity was not excluded, computed for $\pi = 0.7$ according to (7.2.9), increases from 0.748 to 0.963 (column 6). Complementing to unity the numbers in column 6 one gets the conditional probabilities of making an error with the feministic rule, under the condition that paternity is not excluded (the unconditional probability is given in column 5).

All the rules of inference based on 16 blood group systems mentioned in Table 7.3.7 result only in several percent of erroneous decisions or in a high percent of deferred decisions. To avoid this drawback a larger set of blood group systems should be examined, or morphological traits should be included into consideration.

7.4. Anthropological evidence

In a panmictic population the mechanism of inheritance of blood group systems discussed in the previous section as well as the gene frequencies conditioning the particular blood groups are known. The patterns of their inheritance are not complicated and are typical for monogenic traits. The situation is simplified further by the independence of the particular blood group systems and their invariance with age of the individual.

Probabilities a posteriori of a man's paternity can be also established on the basis of many other observable heritable traits the model of inheritance of which is not yet known. Usually these are polygenic traits (depending on many loci), in which the phenotype depends stochastically on the genotype, and the environment and variability in individual development exert a considerable influence on the formation of the given trait (for instance body height). There are also polygenic traits which change only slightly under the influence of environmental factors. For example the traits of iris structure or of the auricle, and the traits of the dermal ridges do not change in the postnatal period. Therefore, questions arise, which of the morphological traits are strongly enough genetically determined to be used for inference on the similarity of genotypes, and for indicating diagnostic values of these traits. The answers are yielded by anthropological studies. The following studies are carried out:

1° population studies on variability of traits considered in the given population, and, if necessary, their variability with age—this enables comparison of these traits in children and adults;

2° familial studies for assessment of the joint distribution of the considered traits of the child and its parents;

3° studies of monozygotic twins, yielding information on the range of phenotypic differences between individuals with identical genotypes.

In very small children the morphological traits are not yet enough developed, and sometimes it is difficult to infer on their final shape. Therefore the lowest age for examination of a child has been fixed to three years.

*

The basic material for an anthropological expert evidence consists of results of direct comparison of the forms of numerous morphological traits of the child with analogous traits of its mother and putative father. About 200 traits from 10 morphological regions are considered, namely:

1) the general features of the brain case (e.g. the relief of the cranial vault, the arching of the occiput, the general horizontal outline of the head);

2) the general features of the face (e.g. convexity of the zygomatic arches, height and breadth of the face);

3) the eye region (e.g. shape of the palpebral folds, height of the upper and lower lid);

4) the iris (e.g. colour, crypts, contraction furrows);

5) the mouth region (e.g. height and profile of the upper and lower lips, the shape of the red of the lips, the outline of the chin);

6) the nose (e.g. height and breadth, convexity of nasal alae, the shape of the basis of the nasal septum);

7) the auricles (e.g. shape of the helix and concha, the shape of the antihelix, the relief of the ear lap);

8) hair (e.g. the outline and colour of the eyebrows, the course of neck hair, the shape and colour of head hair);

9) build of hands and feet (e.g. breadth and length, shape of fingers, toes and finger nails);

10) dermal ridges or dermatoglyphs (e.g. patterns on fingers and toes, the course of main lines on palms, dermal ridge counts forming the particular patterns (cf. Fig. 7.4.1)).

Traditional, verified in many years' practice and commonly used in anthropological evidence, procedure is based on analysis of similarities and dissimilarities of particular traits between child and putative father or mother, on the basis of direct estimation by the expert. After detailed and simultaneous examination of the

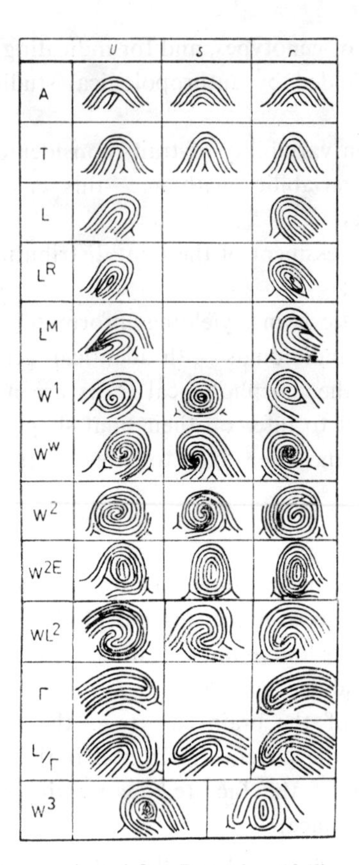

FIG. 7.4.1. Classification of finger prints (after Rogucka, 1968).

triplet of persons connected by the writ, several groups of traits are distinguished, beginning from traits of strongest support for paternity up to traits of strongest evidence against it:

group 1: traits testifying for resemblance of child to putative father, differently developed in mother;

group 2: traits formed in a similar way in all the three persons examined;

group 3: traits measured on at least the ordinal scale, where the child's trait is half way between the traits of the mother and the putative father;

group 4: traits formed in a similar way in child and mother, and differently in the putative father;

group 5: traits distinctly different in child than in mother and putative father.

The ordering of groups seems to be obvious. For instance traits in group 4 testify weaker against the man's paternity than traits in group 5, as similarity of the child only to its mother is frequently met in true families, and dissimilarity of child to both parents is relatively rare.

The anthropological expert evidence is based, above all, on what is called evidence value of group 1 and group 5.

The evidence value of group 1 is determined on the basis of the number of traits in this group and on the dependence between the traits, on the degree of dissimilarity of child to mother for particular traits and on frequencies with which a given trait value observed in the child occur in the population. Hence, the expert uses the results of population studies, which yield information on the variablity range and distribution of the trait examined, on sexual dimorphism, on age changes, on relations between traits and on results of familial studies. If a given category of the child's trait appears in the population so rarely that its coexistence in the child and the man not being its father can be regarded as practically impossible, then the coexistence of this trait in the child and the putative father speaks strongly for his paternity. For instance, the frequency of a high number of crypts in the iris is equal to 5% population. Therefore, a high number of crypts observed in the child and its putative father connected with no crypts in the mother is a highly significant evidence indicating the man's paternity.

Complex similarities of the child to the putative father in a given morphological region are evidence of high value. In Fig. 7.4.2 the complex similarity of child and man in a set of iris-structure traits is presented.

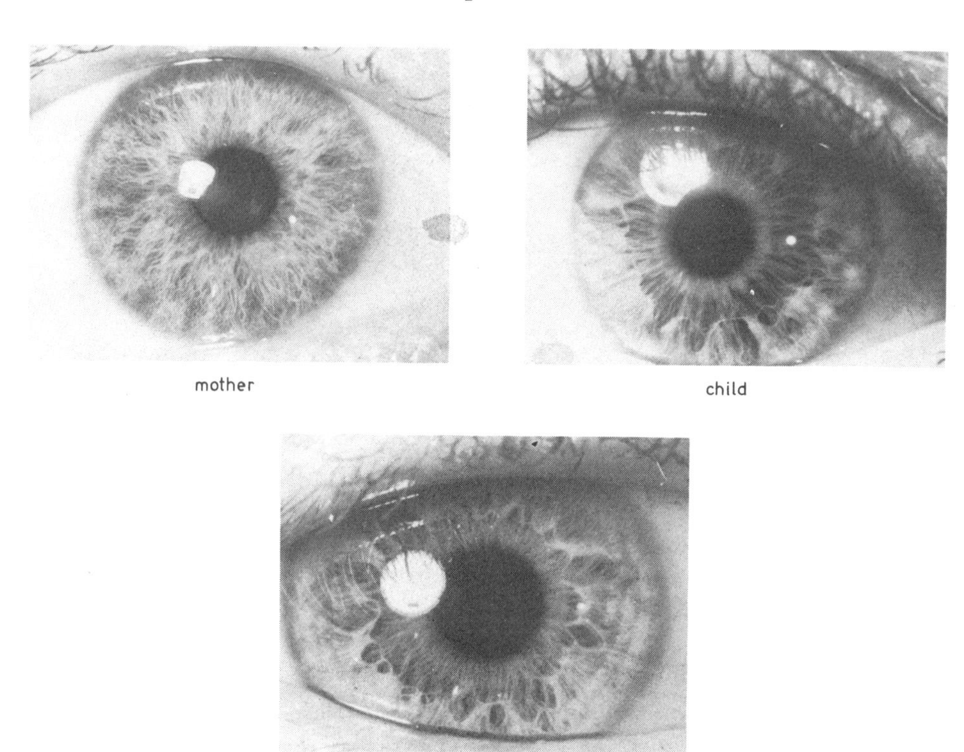

FIG. 7.4.2. Similarity of child to putative father with respect to iris-structure traits.

The evidence value of group 5 consisting of traits in which the child differs from its mother and from its putative father is also analysed. In particular if familial studies show that a certain category of child's trait appears very rarely under a certain combination of other categories of father's and mother's trait, then such a dissimilarity is of high evidence value, speaking against the man's paternity. An example of such a significant dissimilarity is the convex relief of the ear lobe in the child together with a concave one in the mother and putative father. Figure 7.4.3 presents the course of palmar dermal ridges distinctly different in the child as compared with the course in the mother and putative father.

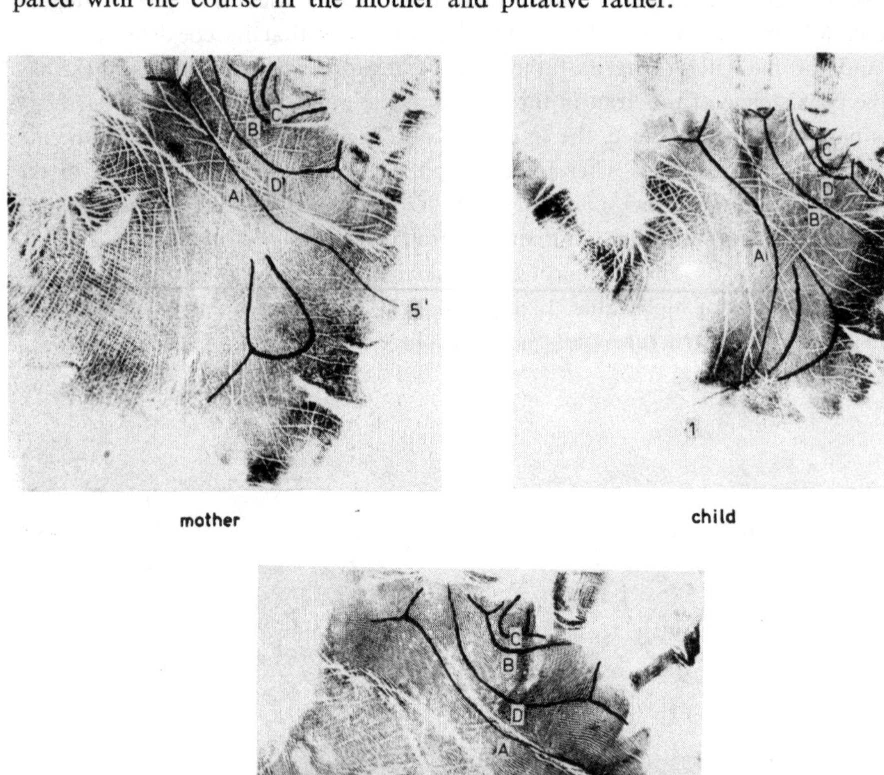

FIG. 7.4.3. Traits in the course of palmar dermal ridges in which child differs from mother and from putative father: 1) The terminations of the main lines D, C, B, A and the loop in 7 (IV) palmar area. 2) Radial arc on the *hypothenar*.

In the process of classification of traits to particular groups the expected developmental changes in the child should be taken into consideration. For example, a concave nose profile in the child may alter with age into a straight profile, but not vice versa.

The relations between traits influence the evidence value of a group of traits since highly codependent traits carry together the same information as each of them separately.

On the basis of a detailed analysis of group 1 and group 5 the anthropologist estimates the chance of the man's paternity, usually by means of the following seven point scale:

"paternity practically certain" (+3);
"paternity very probable" (+2);
"paternity probable" (+1);
"paternity undecided" (0);
"paternity improbable" (−1);
"paternity very improbable" (−2);
"paternity practically impossible" (−3);

The terms in quotation-marks correspond to the interpretation given by the anthropologist to the particular evaluations. Numerous efforts have been undertaken to unify the way of classifying the examined triplets of individuals, and to eliminate the expert's subjectivism. Schwidetzky (1956) has formulated postulates regarding the way of classification as follows:

+3: the child shows many significant similarities to the putative father, these similarities being at the same time considerably higher to the putative father then to the mother;

+2: the similarities between the child and the putative father are numerous, and among them there appear similarities in traits rarely occurring in the population, or complex similarities of morphological regions;

+1: among the traits in which the child differs from the mother there are many traits showing similarity of the child to the putative father, but there are no similarities with a significant evidence value;

0: among the traits in which the child differs from the mother there is, more or less, the same number of traits similar in the child and the putative father as traits dissimilar in the child and the man; however, these similarities or dissimilarities have no significant evidence value;

−1: among the traits in which the child differs from the mother, there are many traits differentiating the child from the putative father, but there are no dissimilarities of highly significant evidence value;

−2: among the traits differentiating the child from the mother there are, in general, traits distinctly differentiating the child from the putative father, but the similarities between the child and the man are of low evidence value;

−3: in the child no similarities of high evidence value were found, the general similarity to the mother being low, but there are numerous dissimilarities of a high evidence value between the child and the putative father.

Schwidetzky studied a group of 366 families and the same number of triplets composed of a mother, her child and a randomly chosen man. She analysed in every triplet 70 morphological traits and estimated the chance of paternity (categories +3,

TABLE 7.4.1.
Empirical distributions (in percents) of the anthropological evaluation of the chance of paternity in a population of families and in a population of false families from a sample of size 366 (Schwidetzky, 1956).

Mark	−3	−2	−1	0	+1	+2	+3
Families	0.0	0.0	0.0	3.8	47.5	39.6	9.0
False families	3.3	21.6	54.9	19.4	0.8	0.0	0.0

+2, ..., −3) according to the postulates presented above. The results of this study published by Schwidetzky (1956) are given in Table 7.4.1.

The obtained results are astoundingly good. Suppose that the distributions in the populations of true and false writs are identical with those observed by Schwidetzky, and the man is automatically recognized as the father whenever he gets a positive mark and as a stranger to the child whenever his mark is negative. Then the man not being the father of the child would be recognized as the father with probability 0.008 the father of the child would never be unjustly recognized as a stranger, and the probability of a deferred decision would not exceed 0.194. This seems to prove that very good results could be achieved if every triplet of individuals connected by a writ would be examined by means of a sufficiently high number of morphological traits (the same for each of the writs) and the results would be classified by properly trained experts.

*

In the sequel we present an example of anthropological expert evidence elaborated by means of the traditional descriptive method. In the text particularly important traits are underlined.

Expert evidence No ...

Group of traits appearing similarly in the child and the putative father, and differently in the mother (group 1).

1. Head length greater than in mother.
2. Greater face height.
3. Poorly developed palpebral folds.
4. Greater height of orbital part of upper eyelids.
5. Lack of mesodermal pigment in the irises.
6. Large number of crypts in the irises, and therefore lack of radial furrows, and small fragments of contraction furrows.
7. Numerous spots of connective tissue in circumference of the peripheral zone of the irises.
8. Nasal shaft narrow.
9. Nasal wings thinner.
10. Weaker convexity of nasal alae.

11. Longer rear part of basis of nasal septum.
13. Shallower alo-nasal fissures.
14. Greater height of upper lip.
15. Straight profile of upper lip.
16. Lower height of the upper red of the lip.
17. Lesser fleshiness of the lower red of the lip.
18. Poorly marked chin-lip fissure.
19. Weaker bend of upper part of helix and therefore a shallower and broader concha in the upper part of the auricles.
20. Stronger bend of lower part of helix, and deeper and narrower concha in the lower part of the auricles.
21. Broad incisura intertragica of auricles.
22. Relief of ear lobes.
23. Stronger accreted ear lobes.
24. Divergent direction of hair stream on neck.
25. Dextrorotary hair whorl on head top.
26. Higher total number of dermal ridges forming patterns on digital pulps.
27. Course of main lines D, C and B, dermal ridges and appearance of loop in the seventh area in both palms.
28. Appearance of loop with additional triradius in eleventh area in both palms.
29. Dermal-ridge whorls in third area of soles.
30. Appearance of tibial loop in fifth area of right sole.

Group of traits formed differently in child in relation to mother and putative father (group 5).

1. Lower than in mother and putative father, auricular height of head.
2. Straighter outline of free margin of nasal alae.
3. Fluent outline of mouth fissure.
4. Appearance of dermal ridge whorl on pulp of third right-hand finger.
5. More horizontal course of dermal-ridge main line A on both palms.
6. Appearance of dermal-ridge arches on pulps of five toes.
7. Dermal-ridge fibular arch in distant part of second area on right sole.
8. Dermal-ridge distal loop in fourth area of right sole.

The above listing shows that the number of traits of resemblance of the child to the putative father (group 1) exceeds by far the number of traits differentiating the child from the mother and from the putative father as well (group 5).

The similarities of the child to the defendant appear in nine out of ten of the considered morphological regions. A particularly high degree of similarity was found in the build of the orbital region, in colour and structure of the iris, in build of the nose and auricles and in the system of dermal ridges. In these regions the child resembles more the defendant than the mother. The appearance of a high number of traits similar in the child and the defendant in a morphological region increases the evidence value, because such an aggregation decreases the possibility of accidental appearance of this set of traits in two nonrelated individuals. Form of 13 underlined traits show a high degree of similarity. Let it be stressed that the appearance of dermal-ridge loops with an additional triradius in the eleventh area of the palm is a trait rarely found in the Polish population (below 10%).

Only single traits from the three morphological regions, and a set of traits from the system of dermal ridges (appearing singly in 4 subsets) are included into the group of traits formed differently in the child as compared with analogous traits in the mother and the putative father. However, these traits show considerable differences in the child in relation to the mother and the defendant (underlined traits). As shown by familial investigations, differences of this kind do appear between children and their parents as nonheritable modifications or as traits phenotypically not manifested in the parents. Therefore, the evidence value of this group of traits is in this case not very high indeed.

A considerable prevalence of the number of traits of group 1 over that of group 5, the appearance in group 1 of a considerable number of collective similarities of the child to the putative father, the large number of traits of a high degree of similarity, and the appearance of a trait rarely occurring in the Polish population, enable to formulate the following conclusion:

Final conclusion:

The paternity of ....in relation to the juvenile.............. is very probable.

Apart from the above conclusion the probability a posteriori of paternity of the defendant was calculated on the basis of 12 blood group systems and a result of 95.6% was obtained. This result increases the verisimility of paternity evaluated on the basis of morphological traits.

Signatures of two experts.

In the above presented example of a paternity proving case the anthropological evidence was supplied together with the probability a posteriori based on 12 blood group systems. However, it is usually advisable to supplement the anthropological evidence by the critical value or the probability a posteriori based on the whole complex of blood group traits and morphological traits examined in the child, mother and putative father. Clearly, this is possible only in the case of such a complex of traits for which previously relevant studies were carried out on an additional material of families and false families. In the next section we describe studies of this kind performed in Poland in 1967 – 1984.

7.5. Final remarks

The application of blood group traits and morphological traits in cases of father recognition has already its over fifty years long history.

In the initial period examinations of blood groups tended only towards the exclusion of paternity of unjustly summoned men. Since in the interwar period only a small number of blood group traits could be subjected to examination, the odds for paternity exclusion were rather small. However, due to the rapid development of serology (above all after the World War II) many new blood group systems were discovered. This increased the availability of blood-group-systems' examinations not only for exclusion of paternity of the unjustly summoned men but also for estimation of the a posteriori probability of paternity. This estimation was advocated, above all, by such Polish scientists as Ludwig Hirszfeld, Hugo Steinhaus and Józef Łukaszewicz (Hirszfeld, 1948; Steinhaus, 1954; Hirszfeld and Łukaszewicz, 1958; Łukaszewicz, 1956), who indicated the way of calculation of what is called "index of females' veracity", i.e., the a priori probability of paternity. This probability, calculated for Poland by Steinhaus (1954), amounted to 70%, what means that in 30% of cases the women indicated false fathers of their children. On the other hand, many authors, as for instance Essen-Möller (1938), assumed arbitrarily for the probability a priori the value 0.5, and do not try to estimate it.

Studies on gene frequencies conditioning the various blood groups in a given population made it possible to make tables of critical value or of relative chances of paternity, necessary for calculation of the probability a posteriori. For preparation of an expert evidence only tables for the native population of the examined persons should be used. Tables of critical values or relative chances of paternity for selected sets of blood groups in the German population were prepared by Hummel (1961), and Hummel, Ihm and Schmidt (1971), and in the Polish population by Orczykowska-Świątkowska and Świątkowski (1970), and Szczotka and Schlesinger (1980). The latter investigators compiled also tables of relative chances of paternity for pairs child–man. These tables are helpful when child's mother has died before the verdict concerning paternity, and her blood group is not known.

In connection with studies on transplantation the system of tissue concordance, called the HLA system (*human lymphocyte system* A) has been discovered in man. The three loci of this system denoted by HLA–A, HLA–B and HLA–C are well enough elaborated to be used in cases of father recognition. The number of factors discovered in each of the loci is very high and amounts nowadays to 15 for HLA–A, 27 for HLA–B and 8 for HLA–C. These factors are inherited codominantly with the exception of only one in each of the loci, which represents factors not discovered so far (designated by the symbol "blank"). The inheritance model of the alleles of each of the loci separately is identical with the model of inheritance in the AB0 system without subgroups, the recessive allele 0 corresponding at the same time to the allele "blank". The number of phenotypes in the HLA system is very high (the particular loci have 85, 352 and 29 phenotypes respectively). The loci of the HLA system are, however, strongly linked, therefore the calculation of a posteriori probabilities of paternity is not commonly performed. This is caused by difficulties in estimating the distributions of pairs and triplets of alleles, called *haplotypes*, and by computation difficulties. Instead, the HLA system is used for paternity exclusion because of its high probability of exclusion of a man not being the father (over 95%). Objective difficulties together with realization of this time-consuming and expensive investigations hinder the introduction of this system into common use. Moreover, the examination of HLA can be carried out in elder children only, as the amount of blood collected for investigation is larger than in classic systems.

The first anthropological evidence in father recognition was delivered by Otto Reche in Vienna in 1926 (Reche, 1926). Poljakoff (1929) gave the first theoretical basis regarding paternity recognition by means of morphological traits. Since that time numerous papers have been published on the methodology of this research. A specially large contribution in this field is due to German and Austrian resear-

TABLE 7.5.1.
Estimated odds of paternity for the combination of three categories of breadth of the *incisura intertragica*.

Category combinations of traits of man mother and child	Number of families	Percent of families	Estimated probability in class of false families %	Estimated odds of paternity
(1)	(2)	(3)	(4)	(5) = (3):(4)
1 1 1	18	2.10	0.83	2.530
1 1 2	10	1.13	1.09	1.037
1 1 3	1	0.11	0.58	0.190
1 2 1	13	1.47	0.70	2.067
1 2 2	15	1.69	2.24	0.754
1 2 3	6	0.68	1.46	0.466
1 3 1	6	0.68	0.20	3.400
1 3 2	10	1.13	1.04	1.086
1 3 3	3	0.34	1.75	0.194
2 1 1	36	4.07	3.47	1.172
2 1 2	40	4.52	4.56	0.991
2 1 3	14	1.58	2.40	0.658
2 2 1	31	3.50	2.91	1.203
2 2 2	87	9.83	9.35	1.051
2 2 3	30	3.38	6.09	0.555
2 3 1	10	1.13	0.85	1.329
2 3 2	49	5.54	4.36	1.271
2 3 3	62	7.01	7.31	0.959
3 1 1	21	2.37	4.11	0.577
3 1 2	52	5.88	5.39	1.091
3 1 3	38	4.29	2.83	1.516
3 2 1	19	2.15	3.44	0.625
3 2 2	95	10.73	11.05	0.974
3 2 3	103	11.64	7.20	1.617
3 3 1	2	0.22	1.00	0.220
3 3 2	34	3.84	5.15	0.746
3 3 3	80	9.05	8.64	1.047
	885			

chers, for instance Weninger (1935), Geyer (1938), Schade (1954), Keiter (1956, 1960), Schwidetzky (1956). Among the recent papers worth mentioning are Czechoslovakian (Ferák and Spevárová, 1972) and Polish (Orczykowska–Świątkowska, 1972, 1977; Hulanicka, 1973, 1981; Szczotkowa and Szczotka, 1976, 1981; Szczotkowa, 1977, 1979, 1985; Lebioda, 1983; Orczykowska–Świątkowska, Lebioda, Szczotka and Szczotkowa, 1985).

In 1967 – 72 the Institute of Anthropology of the Polish Academy of Sciences carried out familial investigations including many morphological traits and blood group traits. Among others 9 traits of the auricle have been elaborated: the physiognomic length and breadth, the bend of the upper and lower helix, the development of the transverse arch of the *anthelix*, the *crus superior anthelicis*, the upper tubercle of the *antitragus*, the breadth of the *incisura intertragica*, the relief of the ear lobe (Szczotkowa, 1977, 1979, 1985). For this study traits showing small age variability were chosen, and those strongly age-dependent were omitted. For each chosen trait 3 categories were distinguished, and the distribution of combinations of these categories in families has been estimated (column 3 in Table 7.5.1). Next, the estimated probabilities of particular combinations of categories in false families have been calculated as product of the frequency of the given combination of mother–child categories and the frequency of the man's category (column 4 in Table 7.5.1). Eventually, the estimated odds of paternity for the given combination of categories have been calculated (column 5).

The results obtained were used to find empirical distributions of the sum of natural logarithms of relative chances of paternity for the whole set of auricle traits in the class of families and in the class of false families (Fig. 7.5.1). This served

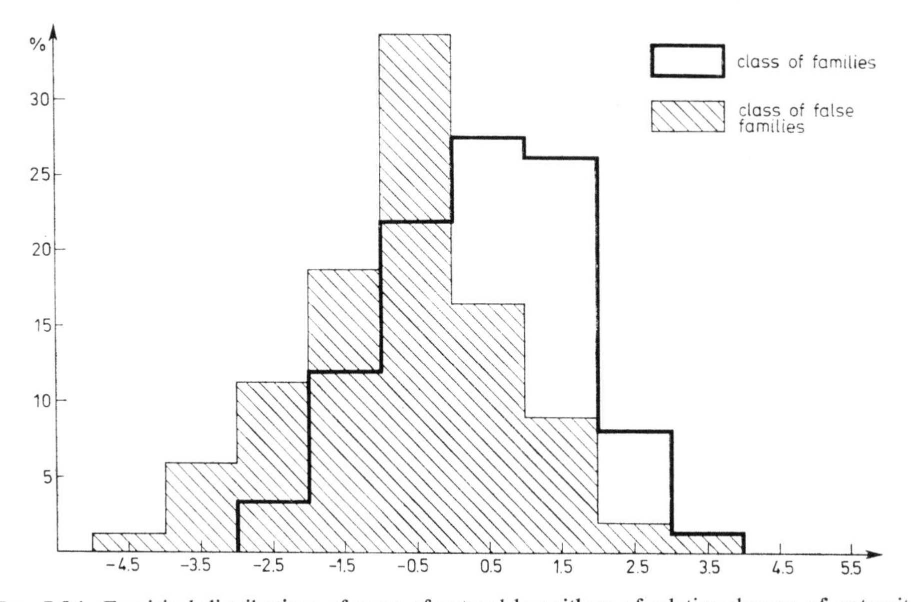

FIG. 7.5.1. Empirical distributions of sums of natural logarithms of relative chances of paternity for 9 traits of the auricle, for families and for false families.

further for preparing a table which assigns values of a posteriori paternity probabilities to the sums of logarithms of relative chances of paternity. Considering the large number of families (885) these estimates can be regarded as close to the true values.

In a similar way tables of estimated odds of paternity were set up for the set of ten nose traits: the breadth of the nasal shaft, the shape of the wing fissures, the thickness and convexity of the alae, the transition of the nasal septum into the upper lip, the breadth of the nose top, the development of the naso-alae fissure, the length of the rear part of the septum base and the breadth of the frontal part of the nasal septum (Lebioda, 1983), and also for the set of five iris traits: intensity of crypts, number of radial fissures and contraction furrows, localization of the iris frill, and the amount of pigment (Lebioda and Orczykowska-Świątkowska, 1985).

Further thirteen measurable traits of the head have been investigated: t–v, g–op, eu–eu, ft–ft, zy–zy, go–go, n–gn, n–sto, n–sn, al–al, ch–ch, en–en, ex–ex. The investigation of this set of traits was carried out in a different way than in the case of the systems discussed previously, because it concerned continuous traits of considerable age variability and sexual dimorphism. Therefore a proper standardization of the traits was needed. For each of the traits the mean and standard deviation were calculated in various categories of age and sex (from the 3rd to the 70th year of life for males, and from the 3rd to the 60th year of life for females), and in each case the set of values has been divided into 13 intervals of length equal to one half of the standard deviation. Moreover, values of the correlation coefficients between father's and child's traits, and between mother's and child's traits have been calculated; the observed values belonged to [0.3, 0.5]. For each trait the odds of paternity were approximated in the following hypothetical situation: the correlation coefficients between mother's and child's trait, and between father's and child's trait are assumed to have the same value, say r; the correlation coefficient between mother's and father's traits equals zero; the joint distribution of standardized traits of father, mother and child (i.e., in the class of families) equals $N_3(0, \Sigma)$, where

$$\Sigma = \begin{bmatrix} 1 & 0 & r \\ 0 & 1 & r \\ r & r & 1 \end{bmatrix},$$

and the joint distribution of these traits in the class of false families is the product of distributions $N(0, 1)$ and $N_2(0, 0, 1, 1, r)$. Under such an assumption the odds of paternity for any r and for any combination of three standardized trait values can be calculated numerically. The next step was calculation of tables of odds of paternity. They involved $r = 0.3, 0.4, 0.5$, and all possible triplets and values formed from the set $\{0, \pm 0.5, \ldots, \pm 2.5, \pm 3\}$, the elements of which represent the standardized midpoints of the intervals splitting the set of values of each trait in each age–sex category. These tables served for estimation of distributions of the sum of natural logarithms of odds of paternity for the whole set of traits characterizing the head. Next, a posteriori probabilities of paternity were assigned to these sums of values (Szczotka and Szczotkowa, 1981, Tables 41 and 42).

The next studied morphological region was the set of 70 traits of dermal ridges including patterns and ridge counts on fingers, patterns on toes, and patterns and ridge counts on palms and soles (Orczykowska-Świątkowska, and Krajewska, 1984). Because of the high number of traits and the variety of dermatoglyphic patterns in this case a special integer-valued index called ISD (*index of similarity of dermatoglyphs*) was introduced. This index expresses in a complex way the degree of similarity between child and father. Note that child to mother similarity is considered only in cases when child's trait differs from father's trait. The ISD takes into account the frequency of the given trait in the population as well. The distributions of ISD in families and in false families were investigated, thus giving a basis for assigning probabilities a posteriori to particular ISD values.

Besides, Orczykowska-Świątkowska investigated 10 blood group traits: AB0, MN, Rh(CcC^W Dd Ee), Kell, Se, Gm(1, 2), Hp, Gc, AP, PGM_1 (Orczykowska-Świątkowska, Lebioda, Szczotka and Szczotkowa, 1985). Earlier tables of estimated odds of paternity for each of these traits separately have been prepared by Szczotka and Schlesinger (1980). For this purpose the gene frequencies conditioning various blood groups in the Polish population (calculated from materials collected by the Institute of Immunology and Experimental Therapy of the Polish Academy of Sciences in Wrocław) were used. The tables by Szczotka and Schlesinger served for estimating odds of paternity for the product of relative chances of paternity in the class of families and in the class of false families for the whole set of 10 blood group traits.

The tables of estimated odds of paternity, compiled for each of five morphological regions (auricle, nose, iris, head, dermal ridges) and for the set of then blood group traits, were used by the group of authors: Orczykowska-Świątkowska, Lebioda, Szczotka and Szczotkowa (1985) for determining empirical distributions of the sum of natural logarithms of these odds, jointly for all the six systems. For this investigation use has been made of 152 triplets child–mother–father (formed from 75 families ([1]), examined in the years 1978/79, and for 148 triplets child–mother–randomly selected man). In this study advantage was taken of former experiments which show that within the particular sets some of the traits are dependent but no dependences were found between morphological sets (Hulanicka, 1973; Orczykowska-Świątkowska, 1977) or between morphological sets and the set of blood group traits (Schaeuble, 1961; Orczykowska-Świątkowska, 1977).

It was found that after having examined the six systems one is able to pronounce on paternity with rather small errors. However, anthropologists are of the opinion that anthropological evidences delivered to the court should not be limited to a value of probability a posteriori calculated on the basis of the said six systems. This probability should serve only as auxiliary evidence, and the basis for evidence, in the case of nonexclusion of paternity by blood groups, must be composed of the ana-

([1]) The sample size is rather small, but complex investigation of whole families is very difficult

lysis of similarities and dissimiliarities of a large number of traits in child, mother and putative father described in Section 7.4. However, this analysis ought not to be limited to the traits included in the said six systems, as this would cause a considerable loss of information. For instance, there may occur a very characteristic similarity of child to putative father (and only to him) in the form of a flatten rim of auricle (*helix taeniata*) or in the build of hands or feet, i.e., in traits which were not included into these systems.

In court jurisdiction only exclusion of paternity based on blood group traits is abided by. In other cases the biological evidence has only an auxiliary character for the court, but an extremely high or low probability a posteriori has, however, a decisive influence on the court's verdict. In highly doubtful cases the court orders sometimes, following a suggestion of experts, an additional examination of the HLA system.

*

The assumptions considered in this chapter seemed to be appropriate in Poland in the 80's. However it should be noted that the present techniques allow us to „read" very long DNA strands so that paternity of a father can be proved with an extremely high degree of certainty (although, at present at least, at a cost that might be prohibitive). Moreover, in some populations it is appropriate to assume that a woman (or her legal advisor) has general knowledge of blood groups inheritance and that it is relatively easy to know the AB0 blood type of any person. As a consequence, a woman of blood type 0 whose child has blood type A (say) is not likely to bring suit against a man with blood type 0 or B, knowing, that he will immediately be excluded.

Additional information of this sort—if available—should be incorporated into the model.

Chapter 8

Studies on sister cells

Key words: *cell cycle, generation time, control points, switch mechanism, sister cells, macronuclear* DNA (Ma–DNA), *heterochromatin, euchromatin, stochastic dependence, randomization, monotone dependence function, correlation coefficient, transition probability model, survival function.*

8.1. Introduction

In cell biology particularly interesting to us is the answer to the question: how does growth of the cell occur and its reproduction leading to the formation of two daughter cells of similar nature. Interest in cell proliferation is the result of the fact that, in spite of accumulation of the results of multidirectional investigations, a univocal concept is lacking which would describe the mechanism controlling a normal cell cycle. Neither is a uniform interpretation available which would explain uncontrolled cell growth leading to pathological changes in the organism.

For understanding the cell proliferation mechanisms appropriate "markers" of the cell cycle are necessary, that is easily observable facts which would allow to establish every moment of the interdivision stage. To such markers belong phenomena connected with division of the cell nucleus and the cell contents.

Mitosis is a process leading to precise division of genetic information from the maternal cell to the newly forming daughter cells. Its course is concomitant with other prominent and well noticeable morphological changes in the nuclear chromatin (differentiation of chromosomes). The process of cytokinesis, on the other hand, involves the separation of the cytoplasmic material and ends in physical separation of the daughter cells. Both these processes are the result of ending of numerous important, yet directly unobservable events occurring in the course of the cycle.

Howard and Pelc (1953), using a labelled DNA synthesis precursor, demonstrated that doubling of the genetic material (nuclear chromatin) takes place only in a definite phase of the interdivision stage. This period in which DNA replication occurs has been termed *phase* S (from *synthesis*) and has become the next after mitosis distinguishable period of the cycle. Thus, the whole cell cycle has been divided into the following phases: G_1—the presynthetic phase preceding the period of DNA synthesis (from *gap one*), S—the period of DNA synthesis, G_2—the post-synthesis phase (*gap two*) and M and D that is mitosis and cytokinesis, respectively. The division of the cycle is shown diagramatically in Fig. 8.5.2 (cf. p. 215).

In view of the division of the cell cycle into phases and the possibility of distinguishing some of them, recently many models have been devised for trails of elucidation of the mechanisms controlling the passage of the cell through the successive phases of the cycle. Most of these models concern as already mentioned the phenomena of DNA synthesis and mitosis, treating one of them or both as "control points" in the cell cycle, related either with initiation of DNA replication or initiation of cell division.

Analysis of the duration time of the successive phases is very difficult. First of all, establishment of the phase in which the cell is at the given moment is not impossible, but usually it requires fixation of the cell. Moreover, the cells exhibit a high variability of the time of occurrence of the particular phases as well as of the time of passage through the whole cycle, and their resistance to changes in the environment is very low.

Since in most cells the joint duration of phases S, G_2, M and D is relatively constant as compared with the duration time of phase G_1, particular attention of investigators is focussed on the latter phase. Some believe that it is here that the main point of cycle control lies, connected with the initiation of DNA synthesis. Transition through this point makes the cell enter into a sequence of events ending in cell division. The nature of this transition remains, however, obscure.

A basic feature on which investigations of the cell cycle are based is its duration named usually generation time and denoted GT. This time describes simply the lifetime of the cell and is determined by the moments of separation of the daughter cells in two subsequent divisions. This is one of the traits of the cycle which can be directly observed and measured.

Since the duration of the cell cycle is not strictly genetically determined, it would seem that it reflects the differences in the rate of the course of intracellular events (not observable by the investigator), conditioning a faster or slower passage of the cell through the successive steps of the cycle. Hence the tendency to concluding on what is going on in the cell during the cycle on the basis of changes in GT duration. It is, namely, known that the duration time of the cycle is the basic and observable response of cells to changes in the environmental conditions.

Changes in the duration of the cycle of experimental cells in comparison to the GT of control ones allow conclusions as to the site of action of the factors applied and the cycle disturbances evoked by them. Such concluding, however, should be based on correct evaluation of the cycle length both in control and experimental cells. Since the investigations are mostly carried out on cell populations, a knowledge of the differences between them in GT is essential for the searcher and so is correct determination of its "average" value serving for comparisons.

The main problems connected with the cycle duration concern the choice of an appropriate method of its estimation, adapted to the material used and of parameters describing GT distribution. The choice of these parameters is essential since on their basis conclusions are reached concerning the mechanisms governing the cycle.

The variability of cycle duration in cells derived from the same population is very wide. The GT distributions obtained for cells from the same population and the same conditions at various times are also characterized by such variability which excludes in general investigations on randomly chosen cells.

Trials of finding a convenient experimental model run along two lines. The first one deals with synchronized cell populations. Various chemical substances or drastic changes of physical conditions are applied as synchronizers (e.g. thermic shock). These methods, however, are not without influencing the course of the cycle, modifying it as compared with that of normally growing cells. The degree of synchrony is rather low and may soon be lost. What is more, synchronization concerns only certain definite processes in the cells (e.g. passing through division or DNA synthesis) and not the entire cellular metabolism.

The second research line concerns investigation of groups of cells characterized by a possibly low variability in respect to the features of interest. Here belong *sister cells*, that is those arising in each division of the mother cell. Since the siblings arise from a common mother, their similarity in many traits is doubtlessly greater than of randomly taken cells. Thus, it is only important whether this similarity is sufficient to make possible comparison of their behaviour under the influence of various stimuli.

This chapter presents the possibility of utilisation of daughter cells in some types of experiments on the biology of the cell cycle.

Section 8.2 describes briefly statistical problems in evaluation of the influence of a selected factor on given objects in the studied population. Various experimental models have been considered, especially those based on pairs of objects, with various assumptions concerning their observability. Care was taken to choose an adequate parameter by means of which the influence of the given factor may be evaluated. This section contains general statistical considerations exploited in the further part of this chapter. In Section 8.3 the choice of the parameters to be measured and corresponding experimental schemes is especially considered for populations of sister cells.

Section 8.4 describes investigations concerning the influence of starvation of the cells on the distribution of the feature GT in the population in experiments with sibling cells of *Chilodonella steini* (*Ciliata*). The relationships between several other sister cell traits are also discussed in this species.

Section 8.5 gives tentative conclusions concerning the phenomena of regulation of the cell cycle phases, on the basis of GT study, especially in sibling systems.

In Section 8.6 some chosen papers from the pertinent literature are critically discussed, above all those concerning some statistical models of the cell cycle.

8.2. Study of a population of pairs of objects

We will now deal with statistical evaluation of the influence of a chosen factor on objects in the studied population in various simplified models with the adoption of various experimental schemes.

For each object in the population we measure a certain feature. The result of the measurement may depend on whether the object is submitted or not to the action of a given factor. The feature assigning the result of measurement to the object is denoted by V when the factor is not active, and by V^+ when it acts on the objects of the population. Thus, the given population of objects is formally described by the joint distribution of features V and V^+.

The measured feature should be adjusted to the studied factor so that the distribution of the pair (V, V^+) would describe the effect of the factor on the objects in the population in a way supplying the searcher with information essential for him.

For instance let traits V and V^+ assume real values and the following equation is fulfilled

$$V^+ = aV+b+\varepsilon, \tag{8.2.1}$$

where a and b are constants, $a \neq 0$, and ε is a random variable independent of V, normally distributed with zero mean and constant variance σ_ε^2. The influence of the factor is then expressed by means of the constants a, b and σ_ε^2 to which the investigator assigns an appropriate interpretation. In another case we can for instance assume that there exists a positive dependence between V and V^+ (distribution of the pair (V, V^+) belongs to the family QD^+ defined in Section 5.3 or to the chosen subset of this family). The influence of the studied factor may then be expressed by means of a chosen monotone dependence measure, for instance Kendall's τ, the correlation coefficient or the monotone dependence function.

Thus, the aim of the study is determined as a certain parameter of the distribution of the pair of traits (V, V^+).

If there exist no limitations of the observability of the values of the traits (V, V^+) for the particular objects and if a simple random sample can be taken from the population of objects in respect to this pair of features, we are dealing with a typical situation of parameter estimation on the basis of a simple fully observable random sample. In general, however, various limitations occur which make conclusions impossible in such an experimental scheme.

Above all it may happen that the given object can be studied only once, thus, we can only once observe the value of trait V or V^+. In such a case we deal with two populations of objects: the one primarily considered and the one subjected to the influence of the factor. In the former we observe the values of V and in the latter those of V^+. The aim in view must then solely be the function of distributions of traits V and V^+ since other parameters of the distribution of (V, V^+) are in this situation not observably reducible (cf. Sec. 2.1). For instance in model (8.2.1) parameters a, b and σ_ε^2 may be the object of investigation, but a measure of monotone dependence between V and V^+ cannot be the aim. If possible, we then take for estimation from the former and latter population two simple random samples, not necessarily equal in size. Such experimental schemes referred to as *models of two independent samples* are, however, useful only, when the influence of the given factor is sufficiently pronounced in differentiating the distributions V and V^+ and if we are able to select an appropriate parameter of these two distributions, on the basis of which we evaluate the influence of the factor. Practically, the difference in the values of the expectations of V and V^+ is most frequently estimated. This, however, is only rational under appropriate assumptions concerning the distribution of V and the influence of the factor expressed by the distribution of V^+.

*

Let us assume that formalization of the problem as based on two independent samples does not suit our purpose, and the values of V and V^+ cannot be simultaneously observed on the same object. There is a solution to such a situation when, instead of the population of objects, we consider the population of clusters of objects (cf. Sec. 5.5) and when the objects belonging to the same cluster differ but little. We will further limit our considerations to a population of pairs of objects, calling objects from the same pair *sisters*.

Let us assume that in the sample taken from the population of pairs we can observe each pair of sisters separately. We shall now consider the variants of investigation of each single pair:

(i) Both sisters are examined without the action of the factor; we record the results of measurement of the given feature (V) for the first and second sister. Such assignment of the results of measurements to the pairs of sisters will be denoted as (U_1, U_2).

(ii) Both sisters are examined when subjected to the influence of the factor (thus, in respect to V^+); we record the results of measurements for the first and second sister. Assignment of these results to pairs of sisters will be denoted as (Z_1, Z_2).

(iii) Only the second sister is subjected to the action of the factor. Assignment of the results of measurements for the pairs of sisters is denoted as (U_1, Z_2).

(iv) Only the first sister is subjected to the action of the factor. Analogously as in (iii) we deal here with the assignment (Z_1, U_2).

In the variants the ordering of the sisters in the pair plays an essential role, and the distributions of the pairs (U_1, U_2) and (Z_1, Z_2) connected with the ordering adopted need not of course be symmetrical. The distribution is symmetric only when the ordering of the sisters in the pair is independent of the measured feature of the objects, that is of V for the pair (U_1, U_2) and of V^+ for the pair (Z_1, Z_2). We endeavour in general to substitute by a pair of sisters a single object, when the latter cannot be studied more than once. It would, therefore, be desirable that one sister be an ideal copy of the second one as regards the measured feature and the tested factor, so that $U_1 = U_2$ and $Z_1 = Z_2$. Thus, in practice such features are useful for which there exists a dependence between U_1 and U_2 and between Z_1 and Z_2 and the distributions of the pairs (U_1, U_2) and (Z_1, Z_2) are symmetrical.

If there is no such symmetry randomization of the objects may first be done, that is they may be ordered according to an arbitrarily chosen feature independent of V and of V^+, or ordered randomly by way of auxiliary drawing. In connection with the symbols used in Section 5.5 assignment of results of measurements to pairs of siblings ordered independently of V and V^+ will be denoted by:

(v) (U_1^*, U_2^*) in the situation described in variant (i) (when neither of the sisters is subjected to the action of the factor);

(vi) (Z_1^*, Z_2^*) in the situation described in variant (ii) (when the factor acts on both sisters),

(vii) (U_1^*, Z_2^*) in the situation described in variant (iii) (when the factor acts upon the second sister).

The variant with the pair (Z_1^*, U_2^*) is not mentioned separately because it is of course equivalent to variant (vii): (U_1^*, Z_2^*) has the same distribution as (U_2^*, Z_1^*).

If in the sample taken from the sibling pairs population a possibility exists of observing each pair separately, the applicability of the particular variants (i) – (vii) is dependent on whether:

1° there exists a possibility of observing the ordering of the sisters in the pair, relevant to the searcher;

2° there exists a possibility of separate studying of each of the sisters in the pair (i.e., of isolation of objects from the pair); this means that the factor can act only on one (chosen) sister in the pair.

When neither of these conditions is fulfilled, the experiment may only be performed according to variants (v) and (vi). When only condition 1° is not fulfilled, variant (vii) is additionally applicable. When only condition 2° is not fulfilled, all variants are applicable with the exception of (iii), (iv) and (vii).

In specific studies we want to compare the behaviour of objects subjected to the action of the factor with that in the case when this factor is not active. We can, thus, consider plans of an experiment taking into account only variant (iii) in which the first sister serves as control for the second one subjected to the influence of the tested factor or, analogously, only variant (iv) or only (vii) as well as schemes of an experiment taking into account two or more variants chosen from among (i) – (vii). For instance if conditions 1° and 2° are not fulfilled, we can get some information on the action of the factor by studying two groups of objects: one according to variant (v) and the other one according to variant (vi). If, however, only condition 2° is fulfilled, we can plan an experiment taking into account either only variant (vii) or supplementing it by variant (v) and eventually (vi).

When planning experiments, it should not only be taken into account which of the conditions 1° and 2° is fulfilled, but the parameter of distribution of observable data by means of which we wish to estimate the influence of the given factor should also be considered. If we observe only the realization of the pair of random variables (U_1^*, Z_2^*) (variant (vii)), we evaluate deviations of the distribution of (U_1^*, Z_2^*) from a symmetrical distribution (we particularly check whether the chosen parameter of the distribution of the difference $Z_2^* - U_1^*$, e.g. the expected value, is different from zero). If we confront variants (vii) and (v), we compare the chosen parameters of (U_1^*, Z_2^*) with the corresponding parameters of (U_1^*, U_2^*) (e.g. one can study the influence of the factor and the kind and strength of dependence between the variables in these distibutions).

Thus, if there exists a possibility of taking simple random samples for variants chosen from among (i) – (vii) (according to 1° and 2°), we are then dealing with classical problems of estimation of the selected parameters.

It is difficult usually in practice to ensure fulfillment of this condition. It should be especially considered that, beside the factor under study, there may also exist the influence of many other disturbing factors which are only partly under control, while their influence is difficult to eliminate. Thus, measurements are performed on samples connected with the given uncontrolled system of experimental conditions. If this cannot be changed by suitable planning the experiment, such parameters must be found for the distribution of observable data which are not susceptible to the influence of uncontrolled factors.

8.3. Sister cells as an experimental system

The influence of various chemical substances or changes in physical conditions on the cell cycle is usually evaluated on the basis of changes in the duration of the cell GT (trait V^+) in reference to the control population (trait V) not subjected to the action of the tested factor. In experiments based on the model of two independent random samples a possibly precise control of the conditions is important, so that the estimation should concern exclusively the influence of the tested factor. Therefore, control of the following conditions is necessary:

1. To ensure identical physical conditions such as temperature, light, volume and composition of culture medium, oxygen conditions etc., so that the tested population would differ from the control one only by the influence of the tested factor and that the latter effect could be evaluated repeatedly without change of external conditions (control of this condition is essential not only in investigations concerning the biology of the cell cycle).

2. Experiments should be performed with cells from the same developmental phase of the population. It is, namely, known that, in dependence on the phase, the rate of cell division changes. Development of the cell culture starts from the period of gradual increase of this rate, continues over a period when it reaches the maximal and constant level up to the period when it gradually begins to diminish. Its rate is the reflection of the duration time of the cell cycle, and this in turn is dependent on the different rates of syntheses occurring in the cells. Thus, if the scientist is interested in changes in the duration of the cycle phases or retardation of divisions under the influence of the tested factor, it is indispensable that the control and the experimental sample be derived from a population which is in the same developmental phase. Fulfillment of this condition allows correct estimation of the influence of the tested factor on the cell cycle and ensures the possibility of comparison of the results in repeated replications of the experiment.

In investigations of the cell cycle most frequently cells are used derived from the middle phase of culture growth known as the logarithmic growth phase. It is characterized for the given conditions and cells by a constant and maximal rate of cell division and exponential increment of the number of cells per time unit. Then the duration of the cell cycles is the shortest and differences between them in the population smallest.

Owing to the constant division rate and exponentially changing number of cells, this phase, especially in microorganisms, is easy to standardize. In tissue cultures it is much more complicated, especially when the proliferation ability of the cells is not identical or when we deal with what is called a *resting stage* when cells, under the influence of definite external stimuli start to proliferate. In spite of major technical difficulties in these experiments and an in general greater nonhomogeneity of the studied material, investigations on tissue cultures are carried on very intensively, because it is in these cells that uncontrolled growth occurs.

3. Experiments should be performed with cells from even-aged strains, since it is known that the age of the strain measured by the number of cell divisions it has undergone, has an essential influence on the rate of division, the duration of the cell cycle and cell mortality. In many microorganisms changes in the strain age are connected with observable changes in the ability of undertaking the sexual process. In tissue culture there are of course no such changes. The age of the strain is essential, especially in long term experiments or those performed at considerable time intervals.

4. An appropriate choice of the control and experimental samples, above all if the population from which the cells originate is characterized by a high variability of the studied feature. Most important here is the knowledge of the range of variability of this feature in cells not subjected to the action of the factor. When the variability of the feature is wide, as it usually is, and the experimental samples are small, the influence of the factor may be completely masked by the natural variability of the material. If there are interactions between the particular elements of the sample and if they depend on the size of the sample (this being in general the case in a cell population) this situation must be taken into account, both in planning the experiment (similar sizes of control and experimental samples) as well as in the description of the effect of the studied factor on the cell population.

The above mentioned requirements complicate greatly investigations on the cell cycle. Control of all conditions is usually difficult and obliges the searcher to perform many control experiments in order to rule out the influence of uncontrollable factors. Moreover, even if this influence is eliminated (directly or indirectly) the situation is still far from ideal. The natural variablity of biological material itself and its changes in time make the choice of stable and standardized parameters which could serve as instruments for comparison impossible.

All this in general excludes the possibility of conducting investigations according to the scheme of two independent samples and inclines to the use of experimental systems consisting of sibling cells. Such systems are characterized by a lower variability of many characters. Moreover, in many cases the strength and character of the dependence existing between the values of the chosen traits of both sisters allow to select appropriate measures of dependence or of codependence as the aim in view. Of course the above mentioned conditions for the experiments should also apply here.

We take for investigations cells from a strain of definite age and developmental phase of culture, treating this strain as a set of sibling cells. We disregard at the same time other relations resulting from the kinship of the cells. It would be preferable to treat the chosen samples of cell pairs as simple random samples from the given strain at the given moment of its development and under established experimental conditions. This, however, is not always possible on account of the conditions which are beyond our control. This is illustrated by the studies presented in the following paragraph.

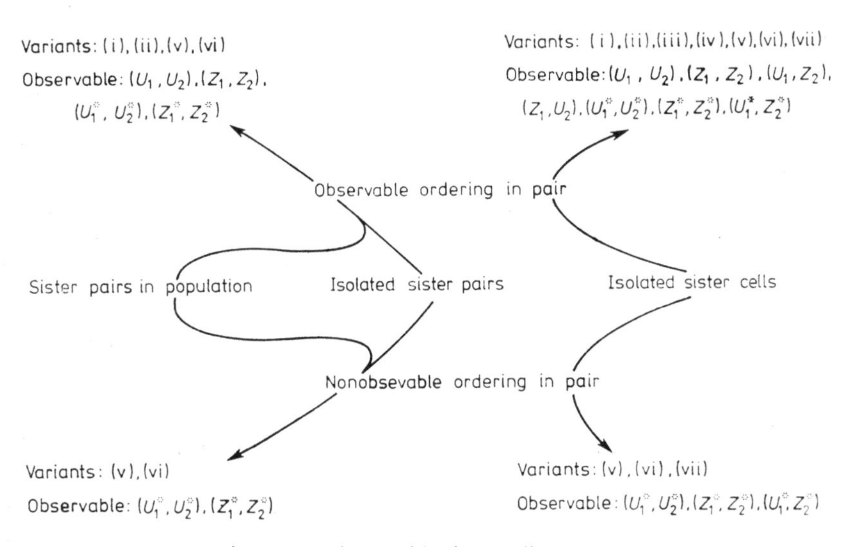

FIG. 8.3.1. Scheme of experimental variants with sister cells.

In Fig. 8.3.1 the scheme is shown of experimental variants for sister cells. According to Section 8.2 the choice of the variant depends on the possibility of observing the ordering in the pair (condition 1°) and on the possibility of isolating a single sister from the pair (condition 2°). Figure 8.3.1 thus presents the division following condition 1° (upper and lower parts) and according to condition 2° (right and left sides). Division into "isolated sister pairs" and "sister pairs in the population" is also distinguished in the figure; this referring to the interaction between the cells. The experiments performed in isolated sister pairs supply information only on the hypothetic behaviour of the cell population consisting of isolated pairs. The applicability of the particular variants remains of course the same, but the interpretation of the parameters treated as the main object is different.

8.4. Investigation of *Chilodonella steini* (*Ciliata*, *Kinetophragminophora*) sister cells

Many studies on the biology of the cell cycle are performed on microorganisms from among which *Ciliata* are frequently used. It is so because the cells constituting independent unicellular organisms are somewhat easier to culture than cells derived from tissues. They, namely, do not require in general a definite constant temperature nor strictly controlled oxygen conditions and the presence in the medium of growth factors. As evolutionally primary, and at the same time not susceptible to dependencies of various type, characteristic for cells from multicellular organisms, they may exhibit some simpler regulation mechanisms. Nevertheless, in spite of a high similarity in the course of many phenomena in these different types of cells, generalization of conclusions deduced in investigations on microorganisms should be advanced with circumspection.

The way of material preparation and the choice of conditions for the experiments will be described for the ciliate *Chilodonella steini.*

Physical conditions. The ciliates were cultured in a thermostat at constant temperature of $20 \pm 1°C$ under conditions of continuous illumination with white light of 500 lux intensity in medium of constant volume (5 ml). They received as food diatoms of the genus *Navicula* given in excess in a suspension of constant volume.

Development phase of the culture. The ciliates were chosen randomly in batches of 100 cells and placed on new medium at 24-h intervals for 5 – 6 days. After this period the culture entered the logarithmic phase of growth which continued for a further 24-h period. The rate of division in such a culture was about two divisions per 24 h. These were basic cultures from which cells were isolated for the experiments proper. All the here reported results concern cells derived from this phase of culture development.

Age of strain. In the developing strain three periods can be distinguished. At first the cells are in the phase of sexual immaturity (young strain). Then no events are observed connected with the conjugation process (the sexual process in ciliates consisting in pairing of two cells and parallel going through the nuclear transformations leading in general to the formation of new genetic combinations). The appearance in the ciliate population of the first conjugating pair denotes the reaching by the strain of the phase of sexual maturity (that is they enter their mean age). This phase is characterized by the ability of the cells to intensive conjugation. Subsequently, the strain passes into the period of ageing, the beginning of which is much more difficult to determine. At this moment, namely, the frequency of conjugation changes and the genetic combinations arising as its result are less viable. The duration of the particular periods differs for various strains. In the case of the relevant strains the period of sexual immaturity lasts over about 400 – 500 cell generations, this corresponding more or less to a culture period of one year in the laboratory. The results presented in this section concern the majority of cells originating from young strains in which the sexual process had not been observed or strains in the early maturity phase.

Mode of conducting the experiments. They were performed on isolated, singly growing cells. The GT of the generations of such cells is significantly longer as compared with that of cells growing in a population. This mode was chosen because, in the case of sister cells, it allows to use the experimental variant shown on the right side of the diagram 8.3.1. Cells were cultured in a constant 0.5 ml volume of medium, receiving 0.3 ml of liquid with diatoms (control groups) or without them (starved groups) so that all the groups would undergo the same manipulations in the course of the experiments.

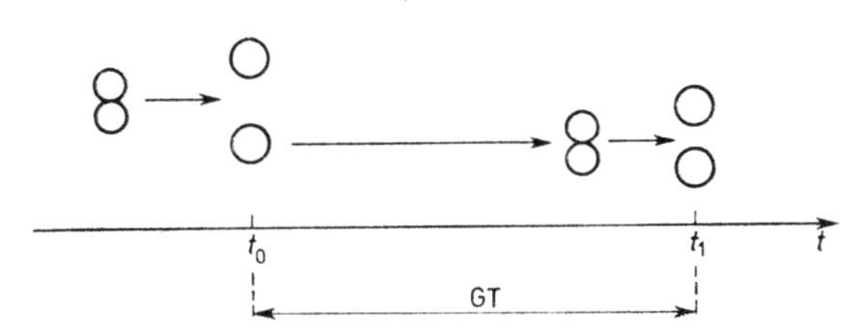

FIG. 8.4.1. Scheme of GT readings.

The duration of the cell cycle (GT) was measured directly as the period elapsing between two successive divisions that is separation of the daughter cells in the first division (moment t_0) and the moment of separation of daughter cells in the next division (moment t_1) as shown schematically in Fig. 8.4.1. Duration of GT was measured with an accuracy up to 5 min (the relative error did not exceed 3%).

The GT was measured on an interval scale (the same in all experiments).

The main aim of the here presented investigations was estimation of the influence of various periods of cell starvation on the duration of cell cycles. Starvation was chosen as the tested factor because there is no unequivocal opinion as to its influence on cell cycles. It is generally believed that starvation causes uniform prolongation of GT duration. *Chilodonella steini* is a convenient species for experiments on starvation, because, in contrast to the objects used so far in these investigations, it is highly selective in its nutrition, this warranting actual starvation.

The experiments proper were preceded by studies on unstarved control cells. The aim was to determine the variability of GT and the range of variability and stationariness of the parameters describing the GT distribution of unstarved cells. Three random samples were taken from the same strain in the logarithmic phase of growth at various time intervals, containing 180, 105 and 104 cells. The GT distributions obtained are shown in Fig. 8.4.2; the medians and quartile deviations (equal to one half of the difference between the third and first quartile) were calculated. Such parameters of location and dispersion (instead of the routinely used arithmetic mean and standard deviation) were chosen because of the distinct skewness of the distributions.

As seen, the distributions in particular samples differ widely. It was, therefore, first of all checked, whether they can be treated as simple random samples from cell populations with identical GT distributions. For this purpose simulation studies were performed consisting in that the results for the samples were pooled and the

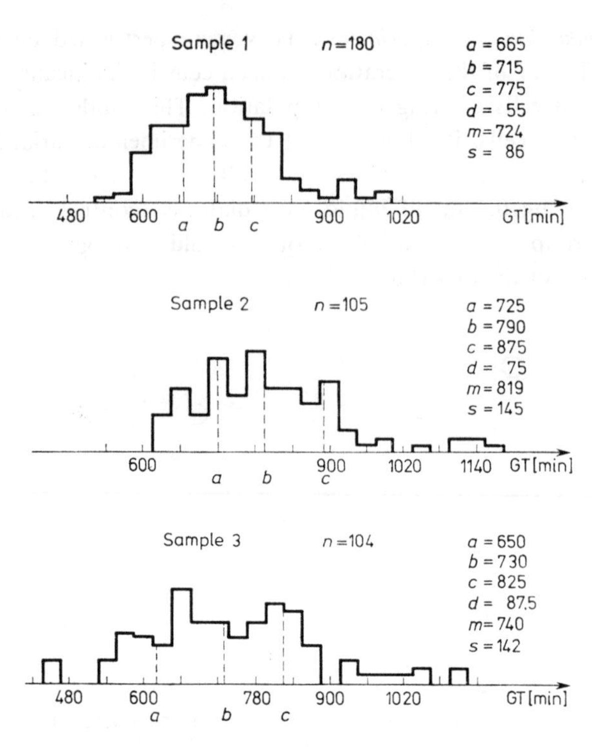

FIG. 8.4.2. Empirical distribution of GT (in minutes) for three independent samples from the same strain and distribution parameters: a—first quartile, b—median, c—third quartile, $d = (c-a)/2$—quartile deviation, m—arithmetic mean, s—standard deviation; sample size is dnoted by n.

obtained set of results was repeatedly randomly divided into three parts of the same size as the primary samples. Each time the empirical distributions were calculated and so were their parameters. When comparing the parameters of simulative divisions with those of the first samples it was found that the latter samples cannot be treated as random. For instance quantile deviations in the sample of size 180 are larger in all simulative divisions than 60 and lie around 72.5, whereas in the primary sample this deviation is equal to as little as 55. In the second sample (size 105) the distribution is distinctly shifted to the right as compared with those in the simulative divisions (for instance the medians differ from the arithmetic mean of the medians of simulative distributions by as much as 50).

The nonstationariness of the GT distribution in the studied strains is due to the influence of unknown and uncontrolled factors in the experiment (to which belongs probably the varying rate of nutrition uptake). Nonstationary are also the location and dispersion parameters which markedly differ from sample to sample. Particularly wide is the variability of the arithmetic means and standard deviations the values of which are also shown in Fig. 8.4.2. The variability of GT expressed by the ratio of the quartile deviation to the median is very high in all distributions (e.g. in the third sample it is about 12%).

It was, therefore, decided to use sister cells for the experiments since in those of *Chilodonella steini* a tendency is observed to undergoing cell division after cycles of similar duration.

Division in *Chilodonella steini* as in most ciliates is transversal. The furrow arising at mid length of the cell separates it into two more or less similar daughter individuals: the anterior one called *proter* and the posterior one—*opisthe.* The sisters differ in a natural way in certain details of their surface and the age of certain cortical structures. These details are, however, well visible only in preparations made by special techniques and cannot be noted in vivo. Accurate observations of freshly separated cells allowed to distinguish minute differences in shape between the newly formed sisters. The proter has a distinctly truncated posterior pole, whereas the opisthe is more regular and almost spherical. Although these differences disappear soon (about 3 min after separation), in this short time the proter and opisthe can be distinguished in any sister pair.

The properties of *Chilodonella steini* cells thus make possible the choice of any experimental variant in the 8.3.1 scheme as they can grow singly and their ordering in the pair is observable. However, since distinction among cells complicates the already difficult technique of the experiments, it is more convenient and natural to conduct them without observing the ordering. Then the variants (v), (vi), and (vii) are applicable.

In view of the nonstationariness of location and dispersion parameters of GT and their high variability in the samples, it was decided to investigate the dependence or codependence of GT of sister cells.

*

We shall now follow the successive steps of the investigations.

1. First of all the codependence of sister cell GTs was examined according to variant (v), that is without the influence of the tested factor. The experiments were run without distinguishing the proter and opisthe; instead of the values of (U_1, U_2) (that is the GT of the proter and opisthe) the values of $(U^{\min}, U^{\max})$ were recorded, where $U^{\min} = \min(U_1, U_2)$, and $U^{\max} = \max(U_1, U_2)$. This way of recording was most convenient for the experimenter. The symmetrized distribution of (U_1, U_2), that is the distribution of (U_1^*, U_2^*), is of course the same as the symmetrized distribution of $(U^{\min}, U^{\max})$.

Samples from four different strains a, b, c, d were subjected to analysis. They comprised 73, 36, 29 and 21 pairs. Thus, in each strain the sequence

$$(u_i^{\min}, u_i^{\max}), \quad i = 1, \dots, n,$$

was obtained, where n denotes the size of the sample for the given strain. Then the symmetrized correlation coefficient $\widehat{\text{cor}}^*(U_1, U_2)$ and the symmetrized monotone dependence function $\hat{\mu}^*_{U_1 U_2}$ (cf. Sec. 5.5) were determined for the samples. Figure 8.4.3 shows diagrams of the $\hat{\mu}^*_{U_1 U_2}$ function and the values of the correlation coefficient $\widehat{\text{cor}}^*(U_1, U_2)$ for the studied strains.

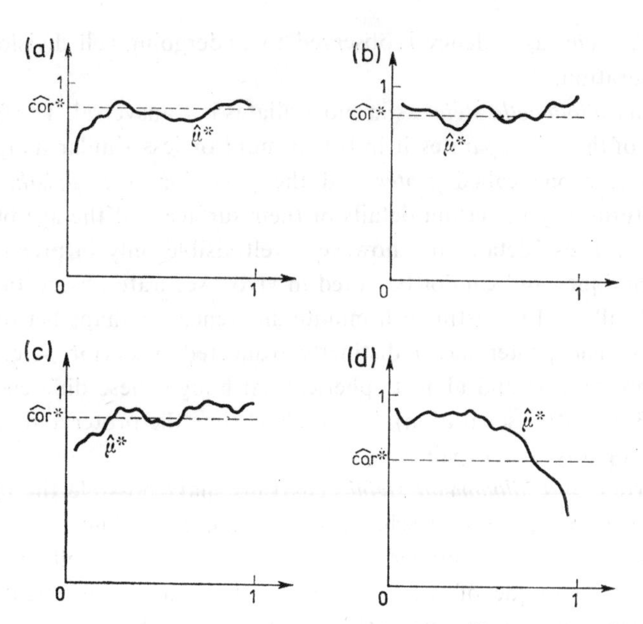

FIG. 8.4.3. Plots of symmetrized monotone dependence functions $\hat{\mu}^*_{U_1, U_2}$ and values of symmetrized correlation coefficients $\widehat{cor}^*(U_1, U_2)$ for sister pairs from four different strains, (a), (b), (c), (d)

The considerations in Section 5.5 concerning characterization of the regular linear codependence between two random variables by means of the symmetrized monotone dependence function and symmetrized correlation coefficient allow appropriate interpretation of the obtained diagrams. The diagrams for the first three strains indicate the existence of a strong positive linear codependence of the GTs of both sisters:

$$(U_1^*, U_2^*) \in \text{LRLF}^*,$$

where LRLF* is the family of symmetrical distributions with linear regression, defined in Section 5.5. For these strains the symmetrized correlation coefficient cor* is the proper parameter for evaluation of the codependence of U_1 and U_2. For the fourth strain (d), however, the function $\hat{\mu}^*$ is positive and decreasing. It may, thus, be expected that the codependence of GTs is positive and strong, but becomes weaker with the prolongation of the generation time. The biological causes of the different character of the codependence between the GTs of both sisters in this strain remain unclear. Such a character of their relation, however, excludes the possibility of assuming the experimental system of interest to us. Further experiments were, therefore, run with strains with a regular linear codependence between the GTs of both sisters.

2. The influence of starvation on the dependence between the GTs of both sisters was investigated according to variant (vii), that is the dependence between U_1^* and Z_2^*. The results shall be presented for one of the strains with a regular linear codependence between U_1^* and U_2^* (strain a). Immediately after cell division one

randomly chosen cell of the pair was separated and subjected to starvation for a definite time period. The GT of the "control" sister (not starved) was recorded as the value of U_1^* and the time of the starved sister as Z_2^*. In this way three groups of sister pairs were studied, consisting of 28, 16 and 19 pairs, respectively (to begin with the sizes of the groups were equal (30)). The starvation periods were 2, 6 and 10 h, respectively. For each group the correlation coefficient $\widehat{\text{cor}}(U_1^*, Z_2^*)$ and the monotone dependence function $\hat{\mu}_{U_1^*, Z_2^*}$ were determined as shown in Fig. 8.4.4.

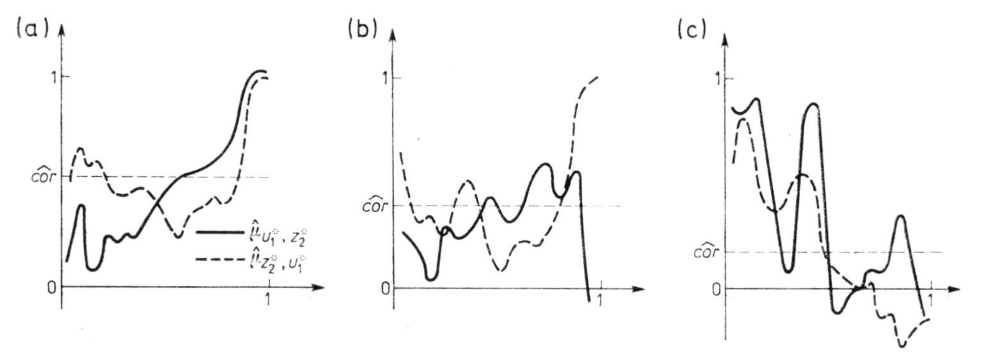

FIG. 8.4.4. Plots of monotone dependence functions $\hat{\mu}_{U_1^*, Z_2^*}$, $\hat{\mu}_{Z_2^*, U_1^*}$ and values of correlation coefficients $\widehat{\text{cor}}(U_1^*, Z_2^*)$ for GTs with starvation periods of: (a) 2 h, (b) 6 h, (c) 10 h.

Comparison of the obtained results (Fig. 8.4.4) with those from the variant (v) (Fig. 8.4.3(a)) indicates that starvation greatly disturbs the similarity of GTs, influencing both the character and the strength of their codependence. The longer the period of starvation the weaker is the dependence between U_1^* and Z_2^* and the wider the deviation from the regular linear dependence model. In the third group this dependence is very weak and not monotone.

The sizes of the examined groups are, it is true, small, but still close to those of the groups previously studied according to variant (v). Thus, the regular linear dependence in the population of sister pairs not subjected to starvation is so strong that it is noticeable even in small samples. Therefore, the differences observed in such samples between the codependence of U_1 and U_2 and the dependence of U_1^* and Z_2^* may be considered as resulting from starvation. Moreover, if the consequences of starvation shown in Fig. 8.4.4 would be side effects due to the small size of the samples, it would be difficult to explain the distinct relation between the duration of the starvation period and the degree of differentiation of the GTs of both sisters.

3. It had further to be checked whether, in the population of cells not subjected to the tested factor, the ordering of sister pairs in the sequence proter–opisthe can be considered as irrelevant to GT. If this hypothesis would be true, the distribution of the pair (U_1, U_2) (that is of the pair of GTs for the proter and opisthe) would be symmetric and equal to the dependence between U_1 and U_2.

Therefore, 34 pairs of cells were examined with distinction in each pair of the proter and opisthe and recording of $(u_{1,i}, u_{2,i})$, $i = 1 \ldots, 34$, which are values of traits U_1 and U_2 in the particular pairs. The correlation coefficient $\widehat{\text{cor}}(U_1, U_2)$ and the monotone dependence functions $\hat{\mu}_{U_1,U_2}$ and $\hat{\mu}_{U_2,U_1}$ determined for this sample are shown in Fig. 8.4.5(a), and $\widehat{\text{cor}}^*(U_1, U_2)$ and $\hat{\mu}^*_{U_1,U_2}$ in Fig. 8.4.5(b).

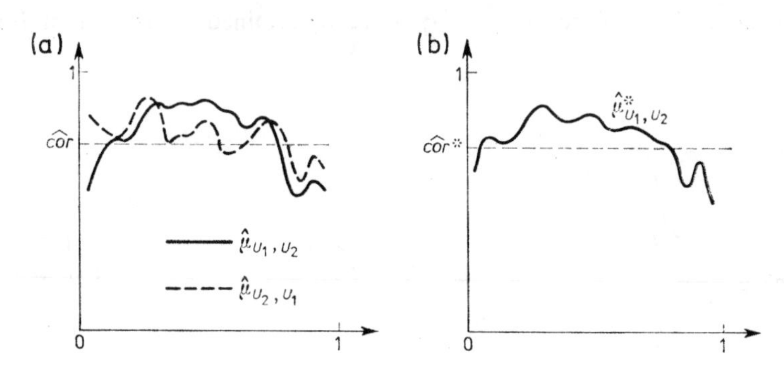

FIG. 8.4.5. Plots of monotone dependence function and correlation coefficients values between GTs for 34 pairs of *Chilodonella steini* sisters: (a) for proter–opisthe ordering, (b) for random ordering.

The functions in Fig. 8.4.5(a) differ but little from one another and are close to the function identically equal to the correlation coefficient. The results as a whole bring evidence that the character and strength of the dependence between U_1 and U_2 and between U_1^* and U_2^* are very similar. It would seem, therefore, that GT can be preliminarily considered as independent or almost independent of the ordering proter–opisthe.

*

The relation existing between the GTs of sister cells is probably caused by their endogeneous similarity in view of their arising from a common "mother cell". This similarity is also the cause of the strong dependence observed between the values of other features connected with cell division.

In most eukaryotic cells the division of the cell nucleus is a very precise process leading to a situation in which both the newly forming cells receive exactly the same genetic material in the same quantity. In ciliates the situation is different. One of their characteristic features is the occurrence in one cell of two kinds of nuclei—what is called a *micronucleus* or generative nucleus and a *macronucleus* or a vegetative one. The micronucleus does not interest us in the present consideration. The macronucleus because of its vegetative function in most ciliates contains a much larger amount of chromatin. Its division described as amitosis gives the impression of random separation of chromatin granules. Hence, situations are frequent when the chromatin levels in the newly formed sister cells are unequal. Thus, it is necessary to check the relation between the sister cells in respect to this feature.

The structure of the macronucleus in *Chilodonella steini* supplies, moreover, information on the separation of the functional chromatin (hetero- and euchromatin) fractions for the nuclei of the sister cells.

The cytoplasm and the remaining cell components are assigned to the daughter cells more or less equally. The newly forming sister cells thus differ slightly in the content of the particular components. In view of the strong dependence of sister cell GTs we are interested in the dependence between them in total protein content.

Investigations were, therefore, performed on the dependence between the macronuclear DNA (Ma–DNA) level determined as trait C and of its fractions—heterochromatin (trait H) and euchromatin (trait E) and total protein level (P) in sister cells.

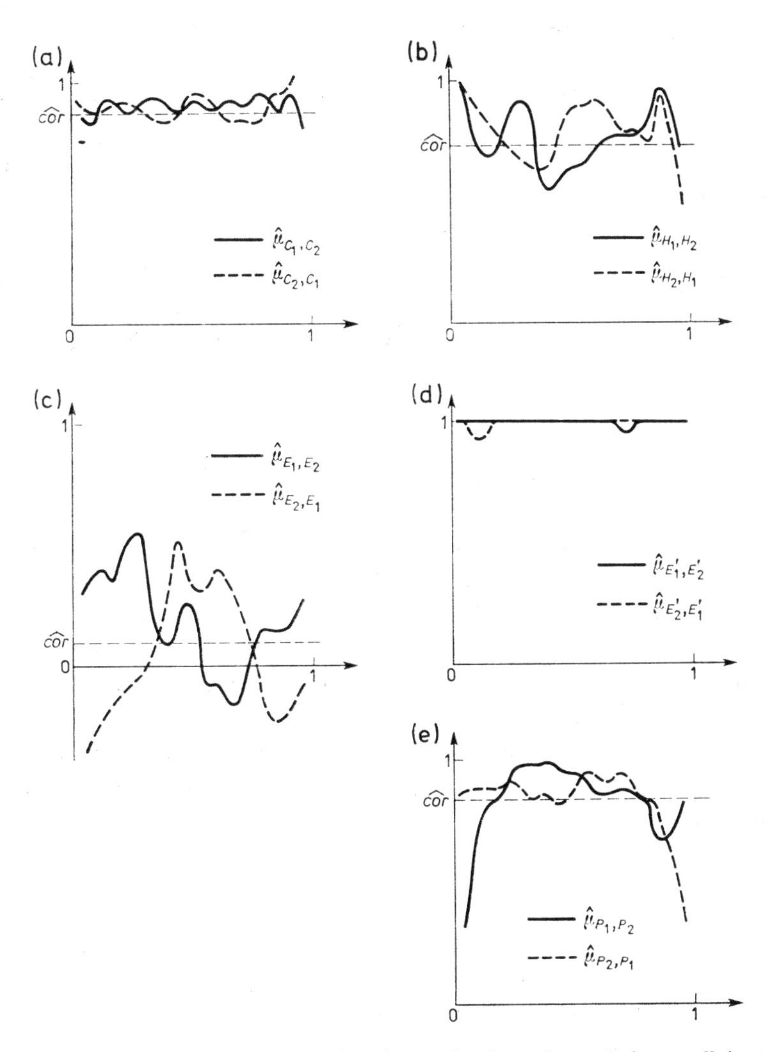

FIG. 8.4.6. Plots of monotone dependence function and values of correlation coefficients for pairs of features: (a) (C_1, C_2), (b) (H_1, H_2), (c) (E_1, E_2), (d) (E'_1, E'_2), (e) (P_1, P_2).

The here described investigations were performed by Radzikowski. The Ma–DNA, hetero- and euchromatin levels and total protein were determined by cytophotometric measurements in suitably stained cells. As regards DNA the Feulgen reaction was run, and for proteins DNFB staining was applied.

All experiments were performed according to variant (i) with observance of proter–opisthe ordering.

Samples of the following sizes were analyzed: 35 cells for Ma–DNA (pair (C_1, C_2)), 18 for heterochromatin (pair (H_1, H_2)), 18 for euchromatin in cells fixed for 3 h after division (pair (E_1, E_2)), 18 for euchromatin in cells fixed 4 h after division (pair (E_1', E_2')) and 35 for total protein (pair (P_1, P_2)). In Fig. 8.4.6 plots of both functions are presented for each sample (e.g. $\hat{\mu}_{C_1,C_2}$ and $\hat{\mu}_{C_2,C_1}$) and the correlation coefficient values are given. These diagrams are interpreted on the basis of characterization of regular linear dependence between two random variables presented in Section 5.4. Plots for (C_1, C_2), (H_1, H_2), (P_1, P_2) indicate the existence of a strong regular linear dependence between the studied characters. Comparison of the monotone dependence curves for both orderings indicates the existence of a symmetry in the joint distribution of the pairs (C_1, C_2), (H_1, H_2) and (P_1, P_2), this proving that the ordering of the cells in the pair may be considered as irrelevant or almost so to the features C, H and P.

As regards euchromatin, the situation is different. For cells fixed 3 h after division, the curves of monotone dependence functions indicate a lack of monotone dependence between E_1 and E_2 and lack of symmetry of the distribution of (E_1, E_2). On the other hand, the curves obtained for cells fixed 4 h after division indicate the existence of a very strong regular linear dependence between E_1', E_2'.

The different picture of dependence between the euchromatin levels is probably connected with the migration of the nuclear material during reconstruction of the macronucleus after cell division. After three hours the process of reconstruction of the nuclear structure is not yet ended and the whole euchromatin material did not pass to its proper place. It seems as if the process of migration in sibling nuclei were nonsynchronous and this affects at once the strength and character of the studied dependence. After four hours, when the nucleus is already reconstructed and the whole chromatin occupies the right place, the dependence becomes very strong and regularly linear.

Since euchromatin is only a small fraction of nuclear DNA, the strength and character of the dependence between C_1 and C_2 will be determined by the behaviour of heterochromatin.

*

The above presented results indicate that the monotone dependence function and the correlation coefficient jointly constitute a stable parameter of the sibling cell population, which may be used in investigation of the influence of exogenous factors on the cell cycle (as shown on the example of the influence of starvation on GT) or in other types of experiments concerning regulation of the cell cycle.

8.5. Sister systems in investigations of the cell cycle

It is generally accepted that regulation and modification of the rate of cell reproduction occurs during their passage from mitosis to the synthesis of nuclear DNA. Slowing down of this rate is caused by a longer persistence of the cell in the G_1 phase, whereas the sum of durations of S, G_2, M and D phases remains more or less constant, changing only very slightly even when the changes in growth rate are wide.

Investigations on many cell types demonstrated that the mechanism(s) controlling the entrance of the cell into the phase of DNA synthesis acts as a "switch" within phase G_1 which, in dependence on its position, allows (position "on") or does not allow (position "off") the initiation of DNA synthesis in the cell. It is believed that the "permission" to start synthesis automatically leads the cell into the further stages of the cell cycle.

In experimental situations with lack of important nutrient substances or growth factors the switch takes the position "off" and the cell "stops" in phase G_1, remaining in it until the necessary substances are supplied and the switch takes the position "on". There is a controversy concerning the problem whether there exists one switch for all the blocking factors in phase G_1 or whether there are different ones for various factors. In the light of many experimental results the hypothesis of the existence of one switch for all blocking substances can hardly be plausible. It is also very difficult to prove the existence of numerous switches for different factors. These difficulties are connected in the first place with the impossibility of ascertaining the connection between the direct site of action of the blocking factor and the observable behaviour of the cell, thus, the effect of its action. In other words, the site(s) sensitive to the action of various factors in the course of phase G_1, arresting the cells in the latter phase are not accessible to direct observation. It is, therefore, difficult to establish whether the behaviour of the cells exposed to the given factor is certainly caused by the operation of the switch mechanism. The impossibility of observing the susceptible sites and the uncertain relations between the effect of the factor and the site of its action make difficult the verification of the hypotheses postulating the existence within the cycle of one or more switches.

Notwithstanding whether the "switch" hypothesis is accepted or not the opinion generally prevails that regulation of the cell cycle occurs in phase G_1. Hence in investigations on cell reproduction particular attention is devoted to this phase in which specific and unique phenomena are searched for which would disclose the mechanism of cycle blocking. In the case of cell cycles deprived of a distinguishable phase G_1, it is believed that the control mechanisms appear in other phases of the cycle.

Although it is generally considered that the period between division and DNA synthesis in the cell is most variable and important of all phases of the cycle, yet interpretation of phase G_1 is controversial. Some authors consider that it is not a part of the cycle, only a manifestation of the time gap in the course of passing of the cell from division to DNA synthesis. According to these authors, this time is necessary for the cell to attain for instance its critical size or weight or critical plasma–

nucleus ratio indispensable for setting the switch in position "on". They also believe that in cells which are components of tissues the period G_1 is not the result of restriction of the division rate, but it is rather a physiological break induced by regulatory signals coming from the cell microenvironment and acting on the switch mechanism. Other investigators consider that prolonged phase G_1 causes the cell to enter a specific phase of rest denoted as G_0. The cell may remain in this phase as long as necessary and either return to the cycle under favourable conditions or die. The problem here is, however, the difficulty of determining the limit between the prolonged G_1 period and the reversible or irreversible period G_0, and also the fact that in most cases there are no observable features distinguishing these two stages.

It should be mentioned that some tissues of the organism are composed of cells characterized by a nuclear DNA level similar to that in phase G_1. It is considered, however, that this fact is due to their functional differentiation and not to arrest of their development in phase G_0. Such differentiated cells (e.g. neurones) "do not return" to the cell cycle, and such tissue belongs to nonproliferating tissues.

Phase G_1 thus seems to be the period in the cell cycle in which, in dependence on various specific conditions, the further fate of the cells is decided. They may pass through further steps of the cycle leading to division and entering of their progeny into the next G_1 phase; they may differentiate for performing definite functions while remaining outside the cycle; finally they may enter phase G_0 remaining in it for any length of time and then die or return to the normal cell cycle. These possibilities are shown schematically in Fig. 8.5.1.

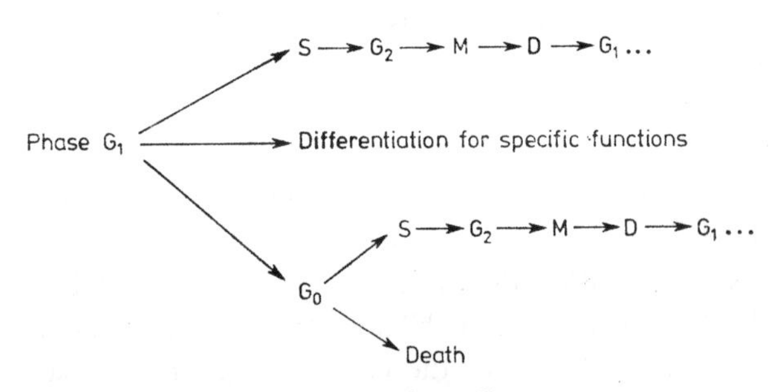

FIG. 8.5.1. Different fates of cells after leaving phase G_1.

The differences in the GTs of cells with identical genotypes directs the attention of investigators to the stochastic nature of the process determining the duration of various functional periods in the cell cycle. From these observations a third group of views evolved concerning G_1 phase. A model was suggested (Smith and Martin, 1973) in which the variability of the GTs of cells is supposed to be the consequence of occurrence in the cycle of a single phenomenon of probabilistic nature. This concerns populations in which the majority of cells is capable of proliferation. If we assume the existence of a critical phenomenon in phase G_1 the whole cycle may be conventionally divided into two periods:

(i) from the beginning of the cycle to the occurrence of the critical event (i.e., to what is called the *critical point*) denoted as *compartment A*,

(ii) from the critical point to the end of the cycle denoted as *compartment B*.

GT is, therefore, the sum of the times of the cell remaining in compartments A and B denoted by T_A and T_B:

$$\mathrm{GT} = T_A + T_B. \tag{8.5.1}$$

The relations between the conventional division of the cell cycle and division on the basis of the critical point are presented in Fig. 8.5.2. Compartment A covers

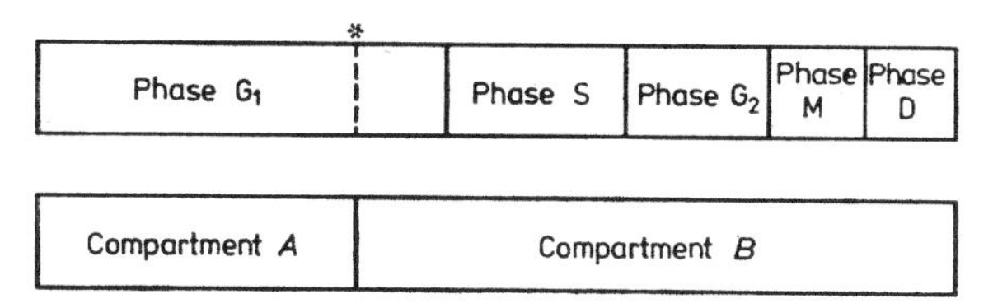

FIG. 8.5.2. Relations between two ways of dividing the cell cycle (* denotes critical point).

a considerable part of the conventional phase G_1 up to the critical point (denoted $*$), compartment B comprises the remaining part of phase G_1 and phases S, G_2, M and D. Each cell in a proliferating population thus passes through the critical point in every cell cycle.

It is assumed in the Smith–Martin model that the chance of the cell passing through the critical point at time $(t, t+\Delta t)$ from the moment of the beginning of the cycle is the same for each t, provided the cell has survived to moment t:

$$P\big(T_A \in [t, t+\Delta t) \mid T_A \geqslant t\big) = \lambda\Delta t + o(\Delta t),$$

where $\lim_{\Delta t \to 0} \big(o(\Delta t)/\Delta t\big) = 0$ and λ is the intensity of transition describing the given cell population. The constant intensity λ characterizes the exponential distribution with parameter λ, thus,

$$T_A \sim E(\lambda). \tag{8.5.2}$$

In a vague formulation the condition (8.5.2) is "an assumption of constant probability of the cell passing from compartment A to compartment B". A model of the cell cycle in which two compartments, A and B, are distinguished, and it is assumed that the probability of passing from A to B is constant, is termed a *model of constant probability of transition* (abbreviated to "model TP").

*

The mechanisms governing the cell cycle can be studied by way of evaluation of the influence of the chosen factor on the cell population. Under assumption that the cycle consists of compartments A and B and that the GTs of compartments A and B are observable, changes can be investigated in the distribution of (T_A, T_B) caused by the factor in question. The problem of investigation of these changes by observing the sum T_A+T_B is analogous to the problems discussed in the first part of Section 8.2, but much more complex. Let T_A^+ and T_B^+ denote the time during which the cell remains in compartments A and B, respectively, in the population of cells subjected to the influence of the given factor. Hence, the pairs (T_A, T_B) and (T_A^+, T_B^+) correspond to V and V^+ considered in Section 8.2. Thus, suitable assumptions have to be adopted on the distribution of (T_A, T_B) and (T_A^+, T_B^+) as well. Further, we have to evaluate the influence of the factor by comparing the chosen parameters of both distributions and giving a suitable biological interpretation to the said comparison. Then experiments have to be planned aiming at estimation of the chosen parameters, under the existing limitations of observability.

This will be illustrated with the specially simplified assumptions concerning distributions. Let us, therefore, assume that condition (8.5.2) holds and that

$$T_B = t_0 \quad \text{for a certain unknown constant } t_0. \tag{8.5.3}$$

Condition (8.5.3) refers to the experimentally ascertained lower variability of the time during which the cell dwelt jointly in phases S, G_2, M and D as compared with the variability of the time spent in phase G_1. Let us assume, moreover, that the factor acts in the same way on all cells of the population, causing a prolongation or shortening of the time T_B or a change in the intensity of transition from compartment A to compartment B. This means that the constant t_0 and the intensity λ may change, but the model of the pair (T_A^+, T_B^+) continues to be a constant transition probability model.

Let (t_0, λ) and (t_0^+, λ^+) be the parameters of distribution of (T_A, T_B) and (T_A^+, T_B^+). If the assumptions concerning distributions would be correct, and the values of the parameters known, their comparison could lead to interesting biological conclusions. Especially when $\lambda \neq \lambda^+$, it may be considered that the given factor influences the appearance of the critical phenomenon, and the case $\lambda^+ = \lambda$ and $t_0^+ \neq t_0$ can be interpreted as follows: the factor does not influence the mechanism of transition through the critical point, but it affects the cell in compartment B. Estimation of t_0 and λ with observance of $T_A + T_B$ and of parameters t_0^+ and λ with observance of $T_A^+ + T_B^+$ is possible since in the considered model these parameters are observably reducible.

The above mentioned assumptions on the distributions are, however, in distinct disagreement with the results of the investigation. If the distribution were actually determined by (t_0, λ), the "survival function" of GT, defined as 1 minus the distribution function of GT, is equal to

$$\alpha_{t_0,\lambda}(t) = P(\mathrm{GT} \geqslant t) = \begin{cases} 1 & \text{if} \quad t \leqslant t_0, \\ e^{-\lambda(t-t_0)} & \text{if} \quad t \geqslant t_0. \end{cases} \tag{8.5.4}$$

Thus, $\alpha_{t_0,\lambda}(t)$ denotes the probability that the cell will not divide ("survive") up to moment t_0. The plot of the logarithm of $\alpha_{t_0,\lambda}(t)$ is shown in Fig. 8.5.3 by the hatched line. The results of experiments are usually presented graphically by the plot of function $\hat{\alpha}(t)$, $t > 0$, where $\hat{\alpha}(t)$ denotes the percentage of cells in the sample which survived at least time t. In the plot the logarithmic scale is used on the ordi-

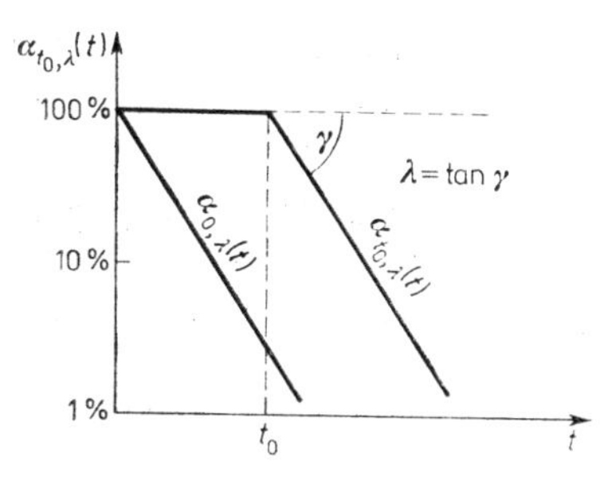

FIG. 8.5.3. Plot of survival function $\alpha_{t_0,\lambda}(t)$ (i.e., percentage of cells with GT $\geqslant t$, which is expected in model TP), for $t_0 = 0$ and some $t_0 > 0$.

nate axis. In practice plots of $\hat{\alpha}(t)$ usually deviate distinctly from the hatched line in Fig. 8.5.3; a typical plot of survival function $\hat{\alpha}(t)$ prepared for 104 *Chilodonella steini* cells is shown in Fig. 8.5.4.

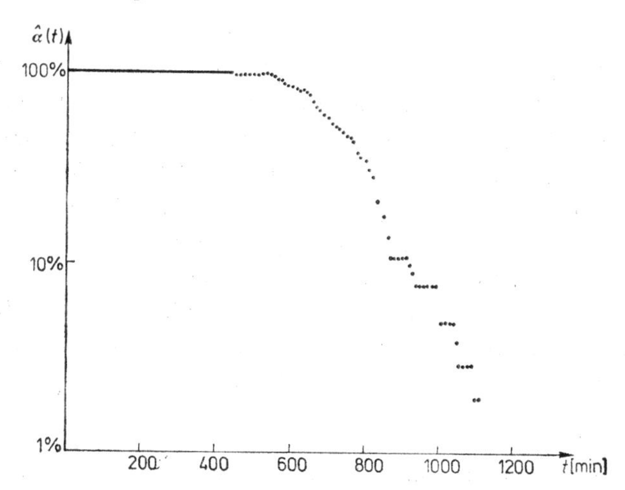

FIG. 8.5.4. Plot of survival function $\hat{\alpha}(t)$ for sample consisting of 104 *Chilodonella steini* cells.

Therefore, the question arises whether, on the basis of the obtained results, the assumption of exponential distribution of time T_A can be valid, provided that condition (8.5.3) be substituted by more sophisticated assumptions as for instance

$$\text{var}\, T_B \text{ is much smaller than var}\, T_A \tag{8.5.5}$$

and

$$\text{the times } T_A \text{ and } T_B \text{ are independent.} \tag{8.5.6}$$

An answer to this question was searched for by referring to a sister scheme.

*

Analogously as in the problems discussed in Sections 8.2 – 8.4 a sister system supplies much better possibilities of investigations on the influence of the given factor on the compartments A and B in the cell cycle than does a system based on examination of single cells, but formulation of the problem becomes very complex.

The further remarks will be limited to a discussion of the choice of the model in a control situation (without the factor), since this is the first step of fundamental importance for studying the influence of the factor.

We shall, therefore, take into consideration a population of randomly ordered sister cells. Times T_A and T_B are denoted by symbols X_1 and X_2 for the first sister in the randomly ordered pair and symbols Y_1 and Y_2 for the second sister. Therefore, we get

$$\begin{aligned} U_1^* &= X_1+X_2, \\ U_2^* &= Y_1+Y_2, \end{aligned} \tag{8.5.7}$$

where U_1^* and U_2^* are the generation times for the first and second sister cell from the randomly ordered pair.

The model of the cell cycle for sisters is, thus, determined by specifying the family of joint distributions for unobservable times

$$X_1, \quad X_2, \quad Y_1, \quad Y_2. \tag{8.5.8}$$

We assume that both (X_1, Y_1) and (X_2, Y_2) are positively dependent and that the distributions of the particular variables X_1, X_2, Y_1, Y_2 are continuous and right-sided skew. This assumption is natural for the life times. If we assume particularly that the constant transition probability model is valid, we must assume that X_1 and Y_1 have identical exponential distributions. We then say that we are dealing with a constant transition probability model for sister cells.

Further natural constraints on the joint distribution of (X_1, X_2, Y_1, Y_2) result from the assumptions concerning the joint distribution of (U_1^*, U_2^*) the values of which are observable. It seems (cf. Sec. 8.4) that for many types of cells a regular linear dependence may be assumed between U_1^* and U_2^*:

$$(U_1^*, U_2^*) \in \text{LRLF}^*, \tag{8.5.9}$$

the correlation coefficient showing a high positive value (greater than 0.75).

An analogue to condition (8.5.5) is the assumption that

$$\operatorname{var} X_1 = \operatorname{var} Y_1 > \operatorname{var} X_2 = \operatorname{var} Y_2,$$

and an analogue to condition (8.5.6) is the assumption that the pairs (X_1, Y_1) and (X_2, Y_2) are independent. These assumptions are included into the model according to the investigator's discretion.

In agreement with assumption (8.5.5), it may be considered in the sister system that the values of X_2 and Y_2 do not differ much. In an extreme approach the assumption of constancy of T_B is generalised with the assumption that

$$X_2 = Y_2, \tag{8.5.10}$$

then

$$|U_1^* - U_2^*| = |X_1 - Y_1|. \tag{8.5.11}$$

The distribution of the observable variable $|U_1^* - U_2^*|$ depends then only on the distribution of the pair (X_1, Y_1), this allowing verification of the hypotheses concerning this distribution. Let us assume for illustration that we are dealing with a constant transition probability model and that condition (8.5.10) holds. We wish to check in this model whether the hypothesis of independence of X_1 and Y_1 is correct. The variable $|U_1^* - U_2^*|$ has, according to this hypothesis, an exponential distribution identical to that of X_1 and Y_1. However, verification of this hypothesis based on an estimator of the survival function of $|U_1^* - U_2^*|$ is difficult in practice because of the high variability of the results from one sample to another.

A better instrument for verification of this hypothesis is the correlation coefficient of (U_1^*, U_2^*) under the assumption of independence of the pairs (X_1, Y_1) and (X_2, Y_2). This coefficient is only dependent on the first and second moments of the variables X_1, X_2, Y_1, Y_2:

$$\operatorname{cor}(U_1^*, U_2^*) = \frac{w}{w+1}\operatorname{cor}(X_1, Y_1) + \frac{1}{w+1}\operatorname{cor}(X_2, Y_2),$$

where

$$w = \frac{\operatorname{var} X_1}{\operatorname{var} X_2},$$

hence

$$\operatorname{cor}(X_1, Y_1) = \frac{(w+1)\operatorname{cor}(U_1^*, U_2^*) - \operatorname{cor}(X_2, Y_2)}{w} \geqslant \frac{(w+1)\operatorname{cor}(U_1^*, U_2^*) - 1}{w}.$$

As already stated, $\operatorname{cor}(U_1^*, U_2^*)$ is close to 1, and w is larger than 1 by virtue of (8.5.5). By specifying the lower bounds for $\operatorname{cor}(U_1^*, U_2^*)$ and w we find the lower boundary for $\operatorname{cor}(X_1, Y_1)$ (e.g. for $\operatorname{cor}(U_1^*, U_2^*) = 0.8$ and $w = 2$ we have $\operatorname{cor}(X_1, Y_1) \geqslant 0.7$). We can, thus, state that X_1 and Y_1 must be strongly positively dependent.

The foregoing examples illustrate the way in which a cell cycle sister model can be built and checked. Extensive empirical and theoretical investigations will be necessary in the future, based especially on information supplied by the sample correlation coefficient, the sample monotone dependence function and the sample survival function of $|U_1^* - U_2^*|$. It is hoped these studies will allow to establish whether and under what additional assumptions, the transition probability model constant can be taken into consideration.

8.6. Final remarks

Practical application of statistical methods in biological investigations evokes criticism both of biologists and statisticians. Baserga (1978) expressed his reservations as follows: "We biologists often use mathematics and numbers like a drunken man uses the lamppost when he returns home on a Saturday night, that is, more for support than for illumination".

The reservations mainly concern the application of too primitive statistical instruments in view of the degree of complexity of the problem. Especially empirical distributions are as a rule described by means of arithmetical means and standard deviations, notwithstanding the distribution type and the measurement scale on which the variable is measured and studied. Nachtwey and Cameron (1968) call attention to these difficulties. Sometimes one begins by discarding outliers from the data (e.g. particularly long GTs), treating them as "untypical" (Sisken, 1963, p. 162). Neither is attention paid to the fact that the correlation coefficient is an appropriate measure of dependence only in the case of regular linear dependence, and that its symmetrized version should be used in the case of codependence.

The influence of a given factor is usually investigated in two-sample significance tests based on sample means. Rejection of the hypothesis of equality of distributions, for cells subjected to the influence of the factor and those free of it, is often treated uncritically as the assertion that both distributions differ significantly from the biological point of view. Numerous authors do not seem to notice the dangers connected with literal treatment of statements supplied by significance tests. Investigators are in general interested rather in the hypothesis that the distributions of the studied feature *do not differ much* from one another in both populations, than in the hypothesis that they are *equal.* When testing the latter hypothesis they risk that it will be almost certainly rejected even when the differences in distribution are very small, if only the data are sufficiently abundant. The result of testing the hypothesis of identity of distributions may, thus, be treated at most as a certain description of data. The danger connected with uncritical interpretation of "a significant result" consisting in rejecting the equality of distributions has been suggestively expressed by Klecka (1981, p. 67) who opposed statistical significance to "factual" significance.

Significance tests may, thus, be described as statistical instruments too primitive for investigation of the cell cycle.

Cell cycle sister models with compartments A and B are also open to criticism, especially the way of checking them. In the pertinent literature the term "transition probability model" is used without specifying which family of distributions of (X_1, X_2, Y_1, Y_2) is taken into account. The results of experiments are as a rule presented by means of what is known as "alpha and beta curves". The alpha curve is a plot of the percentage of cells from the sample which survived over period t; such a curve is shown in Fig. 8.5.4 as the plot of function $\hat{\alpha}(t)$ introduced on page 216. The beta curve gives the percentage of pairs of randomly ordered cells in the sample for which the absolute value of the difference $|U_1^* - U_2^*|$ is larger than t. For both curves the logarithmic scale is applied on the axis of ordinates. As stated on page 216, the alpha curve in the TP model should lie close to the dashed line as in Fig. 8.5.3. Since in the TP model with additional assumption (8.5.3) the expression $|U_1^* - U_2^*|$, equal to $|X_1 - Y_1|$, has the exponential distribution with parameter λ, the beta curve should be close to the plot of $\alpha_{0,\lambda}(t)$, i.e., it should be shifted to the left in relation to the alpha curve approximately by the value t_0 (cf. Fig. 8.5.3). The description of alpha and beta curves, verification of the TP model on this basis and evaluation of the parameters λ and t_0 occupies much space in the literature (cf. Shields, 1977, 1978). The authors of such papers are interested above all, whether one may consider that the kink in the alpha curve is distinct and whether the beta curve and the descending part of the alpha curve lie more or less on parallel lines. Such conclusions are, however, unreliable and unprecise. First of all it is not clear, how can t_0 and λ be determined from Fig. 8.5.4. In our opinion these discussions are vague, above all because departure from assumption (8.5.3) is postulated in them without giving precise assumptions concerning the model of (X_1, X_2, Y_1, Y_2). They also are unconvincing because the obtained information from data is in general limited to the alpha and beta curves only. Among others Green (1980), Castor (1980) and Koch (1980) criticized such an approach.

In our opinion there exist very interesting possibilities of studying the model of (X_1, X_2, Y_1, Y_2) if we assume that (X_1, Y_1), (X_2, Y_2), $(U_1^*, U_2^*) \in$ LRLF* and that (X_1, Y_1) and (X_2, Y_2) are independent. Then the pairs of random variables from the family LRLF* should be considered which may be presented as sums of independent random pairs from LRLF*. The question remains unsolved whether such an assumption will not rule out the constant transition probability model. Such a presentation can be given, on the other hand, for distributions of the family $\mathscr{T}_2$ defined in Section 5.3, the marginal distributions in this family being continuous and right-sidedly skew. Therefore, the fit of this model with the results of experiments is worth checking and the eventual biological interpretation of such a distribution for the pair (X_1, Y_1) should be considered. It would seem that the gamma distribution with suitable parameters may be interpreted as connected with the existence of more than one control point in the cycle.

*

The greatest difficulties in utilisation of the literature concerning sister cells is caused by the lack of specification according to which experimental variants (in the sense of diagram 8.3.1) of the investigations were performed. Some authors use the system of a "control sister" and an "experimental sister" basing solely on the similarity of the sisters accepted "at a guess" (Wiese, 1965; Frankel, 1965; Shepard, 1965; Parson and Rustad, 1968; Stone and Miller, 1965). In other papers information is lacking whether ordering of the sister cells was observable or not. It seems that in most cases ordering was not observed, nevertheless, statistics were commonly used which are not suitable for this case. It is, therefore, difficult to ascertain which of the experimental variants was chosen and whether dependence or codependence was considered (Gill and Hanson, 1968; Kaczanowska *et al.*, 1976). Only in few papers were these problems taken into account (e.g. Kimball *et al.*, 1959). Moreover, nearly in all investigations evaluation refers solely to the strength of dependence or codependence and not to its nature.

In any sister system it is also important to ascertain whether the ordering is irrelevant to the examined feature, and this is not considered in the biological literature. Attention was called to this by Ćwik *et al.* (1982).

Many authors do not seem to notice that the results of experiments conducted on whole populations, on isolated cell pairs and on isolated sisters should be treated differently. Usually the results are compared with the remark that dependence (codependence) is stronger in some cells and weaker in others or that it does not exists at all. These three "levels" are, however, not equivalent. Cells in the population undergo intercellular influences and the dependence or codependence found may be of exogenous character. It will be dependent on the composition and number of cells in the population. This is also true to some extent in a situation in which sister cells are cultured together (interactions between sister cells are known). On the other hand, when sisters are cultured separately the existence of dependence or of codependence should be interpreted as due to endogenous causes. As mentioned in Section 8.3, this distinction is important in the construction of cell cycle models, and particularly in adopting assumptions on the dependent or independent development of cells in a population.

Chapter 9

Survival analysis for censored data

Key words: *grouped data, censored data, subdistribution function, Kaplan–Meyer estimator, survival function, hazard function, influence curve, martingale, bootstrap method, random censorship model, proportional hazards model, multiple decrement model, two-sample problem, point process.*

9.1. Introduction

In this chapter we consider first of all the problem of estimating the survival function and its parameters when observability is limited. This problem often appears in medical research.

Any inference concerning the survival of patients is greatly hindered by the necessity of adapting the plan of an experiment to the specific circumstances of collecting data. Moreover, the diversity and complexity of the situations encountered in practice is enormous. Here we discuss only the simplest models of data censoring without taking into account the specific features of the individual practical problems. Stress is laid upon the formal consequences of the limitations of observability in data censoring.

Consider the following simplified model of a program of implantations of heart pacemakers. Suppose that the program starts on January 1st 1970. Patients with cardiac defects arriving at the hospital after this date are examined by a medical board which decides whether the implantation of a heart pacemaker is indicated or not. If the decision is positive the patient undergoes the operation, and afterwards is examined once a week. The parameter on which we want to base our opinion whether the use of heart pacemakers is advisable is the average postoperation survival time in a population under study.

We assume that the population of patients with implanted heart pacemakers is homogeneous with respect to the postoperation survival time in the sense that the random variable denoting this time is well defined for all patients in that population. We assume that a partial analysis of the program is carried out four years after the program started, i.e., on January 1st 1974. For a long lasting program we aim at obtaining an estimate of the parameter in question as soon as possible. For patients with implanted heart pacemakers who died before January 1st 1974 the survival time is given by the difference between the time of death and the time of the operation; for patients who are still alive on January 1st 1974 we know only that their true survival time will exceed the time which elapsed from the moment of the operation. Hence, for a certain part of the population only censored information is available; what we know is that $x > t$, where x denotes the servival time, and t is the length of the time interval between the moment of operation and January 1st 1974.

This is an example of problems with limited observability. Such problems arise in medical research whenever an analysis has to be carried out before the research program is completed. In these problems, as in problems with full observability, we are interested in estimating the survival function at any moment s. In other words, we want to estimate the probability that the survival time of an individual is longer than s.

Censoring of information may also be caused by other factors. Suppose that in a given group of patients we investigate deaths caused by a specific disease. For the patients who have died of other reasons the information we need is censored; we only know that the potential moment of death caused by the disease in question would not be prior to the real moment of death. It also happens that a patient leaves the hospital at his (her) own request or ceases to attend the periodical check-up examinations. Then, as before, we only know how long that patient was certainly alive after the operation but we do not know the real survival time.

Counting from January 1st 1970 to January 1st 1974, the history of the program can be represented by Fig. 9.1(a) or—with a certain loss of information—by Fig.

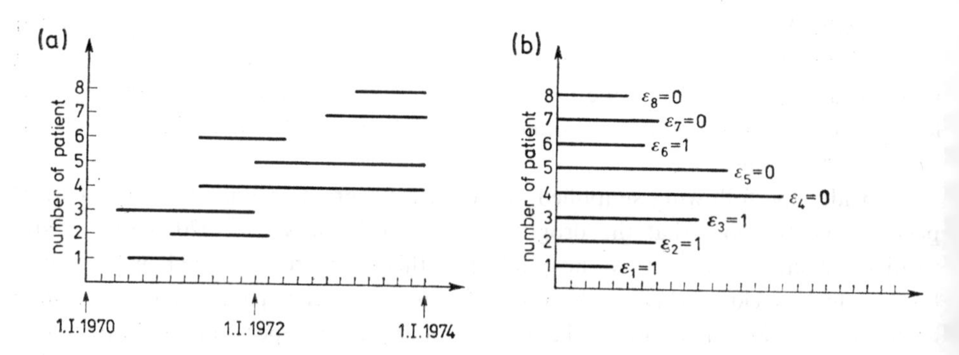

FIG. 9.1.1. Graphical representation of patient survival time (a) with respect to real time, (b) with espect to the moment of operation.

9.1(b). The difference is that in Fig. 9.1(a) the zero moment is the beginning of the investigation while in Fig. 9.1(b) the zero moment is the moment of operation (it can be seen that this corresponds to shifting the starting point of each vector representing the individual survival time to the y-axis). For example, patient N° 1 operated six months after the beginning of the investigations died 8 months after the operation. Patient N° 8, who is still alive on January 1st 1974, was operated on 10 months before that date. Variables $\varepsilon_1, \ldots, \varepsilon_8$ for patients N° 1, ..., N° 8 take on the value 0 in the case of death and the value 1 in the case of censoring.

The problem presented here has a great deal in common with problems of estimating the lifetime of items in reliability theory. The following is a typical experiment in investigating this kind of problems. All the items of a group are tested simultaneously for their reliability time; observations are continued either

(a) until all the elements have failed

or

(b) during a fixed period of time T

or

(c) until the failure of the k-th consecutive element.

In scheme (a) the lifetime of all the items is observable; in scheme (b) it may be possible to say about some of the elements only that their lifetime is longer than T; in scheme (c) full observability is available only for the k elements with the shortest failure-free time. Thus, in cases (b) and (c) we deal with censored information.

In the above example, illustrating the reliability problems, the censoring mechanism is fully known to us and is introduced intentionally to reduce the costs of the experiments. In contrast to this in implantation experiments patients arrive randomly at a priori unknown moments and it is almost impossible to gather a group of patients to be operated on the same time.

However, the two types of problems are identical from the formal point of view in spite of the differences concerning observation schemes: essentially, in both cases the problem is to estimate the survival function when, for some of the elements examined, the only information available is that their survival time is greater than a given period of time.

In Section 9.2 we give a certain convenient representation of the survival function and a model pertaining to implantation programs. In Section 9.3 we define the Kaplan–Meier (K–M) estimator which is a natural estimator of the survival function in that model. Next we show that the K–M estimator is an maximum likelihood (ML) estimator and that it is the weak limit of a sequence of survival function estimators based on grouped data. In Section 9.4 we present some asymptotic properties of the K–M estimator and in Section 9.5 we discuss the problem of comparing two therapies when observability is limited. To construct the appropriate test statistics discussed in Section 9.5 we make use of the results of the preceding sections. Some possible generalizations of the problems presented and a survey of current trends in this domain can be found in Section 9.6.

9.2. A MODEL OF RANDOM CENSORSHIP

Let X be a random variable assuming positive values and representing the length of the survival period (for instance, in the example concerning the implantation of heart pacemakers the variable X is the length of the time interval from the moment of implantation until death).

Let F be the distribution function of X. Hence, $F(0) = 0$. Assume that F (together with all the other distribution functions in this chapter) is a convex combination of an absolutely continuous distribution function and a discrete distribution function (i.e. the singular part of F is zero). In the sequel a representation of F will be given by means of a function λ defined as

$$\lambda(s) = \lim_{t \to 0^+} \frac{P(X \in [s, s+t) \mid X \geqslant s)}{t} = \left.\frac{\mathrm{d}F}{\mathrm{d}t}\right|_{t=s} \cdot (1-F(s^-))^{-1} \tag{9.2.1}$$

if F is continuous at s, and

$$\lambda(s) = \frac{F(s)-F(s^-)}{1-F(s^-)} = \frac{P(X = s)}{P(X \geqslant s)} = P(X = s \mid X \geqslant s) \tag{9.2.2}$$

if F has a jump at s.

The function λ, called the *hazard function*, is useful in estimating the conditional probabilities of death in a short time interval. Namely, if F is continuous in an interval containing t, then, by (9.2.1), the probability $P(X \in [t, t+dt] | X \geqslant t)$ can be estimated by $\lambda(t)dt$. The hazard function λ uniquely determines the distribution function F. This is based on the following identity proved by Peterson (1977):

$$1-F(s) = \exp(-\Lambda_F(s)), \tag{9.2.3}$$

where

$$\Lambda_F(s) = \mathop{\int^{\circ}}_0^s \lambda(t)\,\mathrm{d}t + \sum -\ln(1-\lambda(t)).$$

The symbol $\mathop{\int^{\circ}}_0^s$ means that the integration is carried out over open intervals in which F is continuous, and the summation is taken over all jump points of F which are greater than s.

The function $1-F$ appearing in (9.2.3) will be called the *survival function* and will be denoted by F'. Thus, for any continuous random variable X,

$$F'(s) = \exp\left(-\int_0^s \lambda(t)\,\mathrm{d}t\right), \tag{9.2.4}$$

and for any discrete random variable X with values $s_1, \ldots, s_n$,

$$F'(s) = \prod_{i:\, s_i \leqslant s} (1-\lambda_i), \tag{9.2.5}$$

where $\lambda_i = \lambda(s_i)$, $i = 1, \ldots, n$. Another representation of F', *for* any $\xi_0, \ldots, \xi_n$ such that $0 = \xi_0 < \xi_1 < \ldots < \xi_n = s$, is given by

$$F'(s) = \prod_{i=1}^{n-1} (1-q_i), \tag{9.2.6}$$

where $q_i = P(X \in (\xi_i, \xi_{i+1}] \,|X > \xi_i)$, $i = 1, \ldots, n-1$, denote the conditional probability of the death of the patient in the interval $(\xi_i, \xi_{i+1}]$ under the condition of his (her) survival to the beginning of the interval.

Consider now a situation which is of particular interest to us, namely where in addition to the mechanism determining the length of the survival period a censoring of information takes place. We make a somewhat artificial assumption that each element of the population in question is described by a pair (x, y), where x is the survival time and y is the length of the time interval from a certain zero moment (e.g. the moment of implantation) to the moment of information censoring. For instance, in the heart pacemaker implantation program, the information was censored on January 1st 1974 and y was the length of the time interval from the moment of implantation till the end of the investigation. Hence, we observe $\min(x, y)$ and, moreover, we know which of the elements x, y is greater. Let Y be a random variable with values equal to the length of time interval from zero moment to the moment of information censoring. The observable variables are random variables Z and ε,

$$Z(X, Y) = \min(X, Y),$$

$$\varepsilon(X, Y) = \begin{cases} 1 & \text{for} \quad X \leqslant Y, \\ 0 & \text{for} \quad X > Y, \end{cases}$$

which are functions of (X, Y). In order to avoid ambiguity in the definition of ε, we assume that no value is taken simultaneously by both of the two variables X, Y with positive probability. We consider a model in which the survival and the censoring mechanisms are not influenced by each other, i.e. we assume that X and Y are independent. This model will be called the *model of random censorship.* In the case of n patients we are given two sequences $X_1, \ldots, X_n$ and $Y_1, \ldots, Y_n$ of independent random variables with distributions equal to the distributions of X and Y, respectively. It is assumed that X_i and Y_j $(i, j = 1, 2, \ldots, n)$ are independent of each other. Note that in the reliability experiments presented in Section 9.1 the variable Y is equal to the constant T in case (b), and to the k-th order statistic (for $X_1, X_2, \ldots, X_n$) in case (c). Therefore, case (c) is not covered by the random censorship model.

This model does not include the case where some of the patients leave the hospital at their own request and further information concerning them is not available. In this case the assumption of the independence of the survival and the censoring mechanisms is violated.

Let F, G, H be the distribution functions of X, Y, Z, respectively. In the considered model the survival functions satisfy the condition

$$H'(s) = F'(s) \cdot G'(s).$$

In the sequel we also use *subdistribution functions* defined as

$$\tilde{F}_i(t) = P(Z \leqslant t, \varepsilon = i), \quad i = 0, 1.$$

Hence,

$$\begin{aligned} \tilde{F}_0(s) &= \int_0^s F'(t) \mathrm{d}G(t), \\ \tilde{F}_1(s) &= \int_0^s G'(t) \, \mathrm{d}F(t). \end{aligned} \tag{9.2.7}$$

From the formal point of view the roles of the survival and the censoring mechanisms are symmetric in our model. Therefore, any method suitable for the estimation of F can be applied also to G whenever G satisfies the regularity assumption imposed on F. The estimation of G is important when we are interested in examining the censoring mechanism (e.g. for better planning in future experiments).

9.3. The Kaplan–Meier estimator

Here we are concerned with the estimation of the survival function F' in the model of random censoring.

Let us assume that as the outcome of an experiment we have obtained a sequence $(z_1, \varepsilon_1), \ldots, (z_n, \varepsilon_n)$. Denoting by $z_{(i)}$ the i-th order statistic in the sequence $z_1, \ldots, z_n$ we assume that $z_{(1)} < z_{(2)} < , \ldots, < z_{(n)}$, i.e. that no ties appear in the z-sequence. When we estimate the value $F'(s)$, it is natural to assign the weight 1 to each $z_i > s$, and the weight 0 to each z_i such that $z_i \leqslant s$ and the corresponding concomitant ε_i is equal to 1. It is difficult, however, to assign a reasonable weight in the case where $z_i \leqslant s$ and $\varepsilon_i = 0$. In this situation the conditional probability of survival till the moment s is equal to $F'(s)/F'(z_i)$. If we have at our disposal any estimator of F', denote by $\hat{F}_1'$, then this conditional probability may be estimated by $\hat{F}_1'(s)/\hat{F}_1'(z_i)$. Thus it is intuitively reasonable to replace $\hat{F}_1$ by the estimator $\hat{F}_2$ defined as follows:

$$n \cdot \hat{F}_2'(s) = N(s) + \sum_{z_i \leqslant s, \varepsilon_i = 0} \frac{\hat{F}_1'(s)}{\hat{F}_1'(z_i)}, \tag{9.3.1}$$

where $N(s) = \#\{z_i > s\}$. It may be expected that $\hat{F}_2$ will be better than $\hat{F}_1$. Starting from any initial estimator, we can obtain in this way a sequence of estimators whose elements should approximate the function F with increasing accuracy. Note that if there exists a limit $\hat{F}'$ of such a sequence, then it satisfies

$$n \cdot \hat{F}'(s) = N(s) + \sum_{z_i \leqslant s, \varepsilon_i = 0} \frac{\hat{F}'(s)}{\hat{F}'(z_i)}. \tag{9.3.2}$$

In view of (9.3.2), no further improvement of $\hat{F}'$ is possible. It can be shown that any sequence of estimators generated by the iteration procedure (9.3.1) converges, for each s, to the unique solution of (9.3.2) given by

$$\hat{F}'(s) = \begin{cases} \prod_{i:\, z_{(i)} \leqslant 0} \left(1 - \dfrac{1}{n-i+1}\right)^{\varepsilon_{(i)}} & \text{for} \quad s \leqslant z_{(n)}, \\ 0 & \text{for} \quad s > z_{(n)}, \end{cases} \tag{9.3.3}$$

where $\varepsilon_{(i)}$ is the concomitant of the i-th order statistic $z_{(i)}$ in the sequence $(z_1, \varepsilon_1), \ldots, (z_n, \varepsilon_n)$. This estimator is called the *Kaplan–Meier estimator* (cf. Kaplan and Meier, 1958) abbreviated in the sequel to the *K–M estimator* and denoted by $\hat{F}'$ (or $\hat{F}_n'$ whenever it is necessary to indicate the size of the sample). Analogously, the function $1 - \hat{F}'$ will be called the *K–M estimator of the distribution function* F and

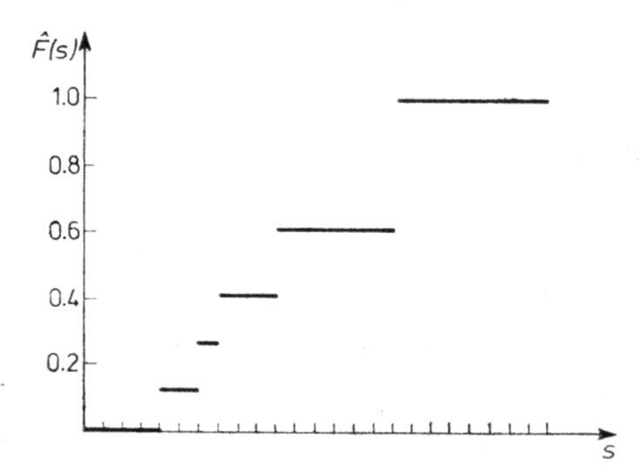

FIG. 9.3.1. Realizations of the K–M estimator of the survival mechanism distribution function corresponding to data from Fig. 9.1.1(b).

denoted by $\hat{F}$ (or by $\hat{F}_n$). Figure 9.3.1 represents the K–M estimator for data from Fig. 9.1.1(b).

It is easily observed that in the absence of censoring (i.e., when $\varepsilon_i = 1$, $i = 1, 2, \ldots, n$) the K–M estimators reduce to the empirical survival function and the empirical distribution function, respectively.

It follows from (9.2.5) and the discussion at the beginning of Section 9.2 that, for a discrete random variable X_n^* with the distribution determined by the K–M distribution function $\hat{F}_n$, the hazard function is of the form

$$\lambda_i = \lambda(z_{(i)}) = \frac{\varepsilon_{(i)}}{n-i+1},$$

where $n-i+1$ is the number of individuals who survived till the moment $z_{(i)}$. The points z_i with $\varepsilon_i = 1$ are the only points of jumps of the K–M estimator. Note that if $\varepsilon_n = 1$ (i.e. if the last observation is death) then the product $\prod_{i=0}^{n} (1-1/(n-i+1))^{\varepsilon_{(i)}}$ is equal to zero and there is no need to define $F'(s)$ additionally for $s > z_{(n)}$ as has been done in (9.3.3). With this modification the K–M estimator is left continuous (and not right continuous) at $z_{(n)}$.

The K–M estimator is often resticted to the interval $[0, z_{(n)}]$. This is motivated as follows: let $t_F = \sup\{t: F(t) < 1\}$ and let t_G be defined for G in the same way; it is intuitively clear that when $t_F > t_G$ the values of $F(s)$ cannot be reasonably estimated for $s \in (t_G, t_F)$, since all observations greater than t_G are censored with probability 1. For n tending to infinity, $z_{(n)}$ tends almost everywhere to $t^* = \min(t_F, t_G)$, and hence putting $\hat{F}'(s) = 0$ for s greater than $z_{(n)}$ may cause large estimation errors in the interval $(z_{(n)}, t_F)$. In all limit theorems presented below we make assumptions which ensure that total information censoring in this part of the support of F will not occur. In particular, we make the assumption that $t_G > t_F$.

Since the roles of F and G are formally identical in the model, we can analogously estimate the survival function of the censoring mechanism Y. The K–M estimator of G' is of the form

$$\hat{G}'(s) = \prod_{i:\, z_{(i)} \leqslant s} \left(1 - \frac{1}{n-i+1}\right)^{1-\varepsilon_{(i)}} \quad \text{for} \quad s \leqslant z_{(n)}.$$

Since $H' = F' \cdot G'$, the natural estimator of H' for $s \leqslant z_{(n)}$ is of the form

$$\hat{F}' \cdot \hat{G}'(s) = \prod_{i:\, z_{(i)} \leqslant s} \left(1 - \frac{1}{n-i+1}\right)^{\varepsilon_{(i)}} \cdot \prod_{i:\, z_{(i)} \leqslant s} \left(1 - \frac{1}{n-i+1}\right)^{1-\varepsilon_{(i)}} = \frac{N(s)}{n}.$$

We have thus obtained the empirical survival function of the random variable Z. This is not surprising in view of the observability of Z.

The values of the K–M estimator in the interval $[0, z_{(n)}]$ can be found by means of a very simple algorithm (Efron, 1967). Observe first that if $z_{(i)}$ is not censored then, according to the definition of the K–M estimator (cf. (9.3.2)), its jump at $z_{(i)}$ is equal to

$$w_i = F'(z_{(i)}^-) - F'(z_{(i)}) = \frac{1}{n-i+1} \prod_{j=1}^{i-1} \left(\frac{n-j}{n-j+1}\right)^{\varepsilon_{(j)}}. \tag{9.3.4}$$

Values of w_i are determined consecutively:

1° we put $w_i = 1/n$ for $i = 1, 2, \ldots, n$;

2° for $i = 1, \ldots, n-1$:

— if $\varepsilon_{(i)} = 0$ we add $w_i/(n-1)$ to w_j, for $j = i+1, \ldots, n$, and we put $w_i = 0$ (i.e. the mass concentrated at $z_{(i)}$ is equally distributed among $z_{(i+1)}, \ldots, z_{(n)}$);

— if $\varepsilon_{(i)} = 1$ we increase i by 1;

3° if $\varepsilon_{(n)} = 0$ we put $w_n = 0$.

The K–M estimator of the distribution function F in the interval $[0, z_{(n)}]$ is a step distribution function which is continuous on the right with jump w_i at $z_{(i)}$.

The correctness of this procedure can easily be shown. If $i_1 < i_2 < \ldots < i_l$ denote the indices of the order statistics corresponding to the successive censored observations (i.e. $\varepsilon = 0$), then, according to the algorithm, to each point $z_{(i)}$, $i \in (i_1, i_2)$, corresponds the mass $\frac{1}{n} + \frac{1}{(n-i)n}$ equal to $\frac{1}{n}\left(\frac{n-i+1}{n-1}\right)^{1-\varepsilon_{(i)}}$. Generally, for $k \leqslant n$ and $\varepsilon_{(k)} = 1$, the mass concentrated at $z_{(k)}$ is equal to product $\frac{1}{n}\prod_{i=1}^{k-1}\left(\frac{n-i+1}{n-1}\right)^{1-\varepsilon_{(i)}}$, which in view of (9.3.4) is equal to w_k.

*

Now we shall show that the K–M estimator is the maximum likelihood estimator. Let us consider the general case. Let $t_1 < t_2 < \ldots < t_k$ be the observed different moments of death. Let d_i be the number of deaths at the moment t_i, and let m_i be the number of patients for which the information censoring occurs in the time interval $[t_i, t_{i+1})$ at the moments $t_{i1}, \ldots, t_{im_i}$ $(i = 0, \ldots, k)$ where $t_0 = 0$, $t_{k+1} = +\infty$.

We have thus replaced the representation of the observable data given by $(z_1, \varepsilon_1), \ldots, (z_n, \varepsilon_n)$ by the representation $(t_1, d_1, t_{11}, \ldots, t_{1m_1}), \ldots, (t_k, d_k, t_{k1}, \ldots$ $\ldots, t_{km_k})$. The two forms are clearly equivalent but the latter is more useful in constructing the likelihood function for observable data. Since the death at the moment t_i occurs with probability $F'(t_i^-)-F'(t_i)$ and the probability of survival at the moment t_{il} is equal to $F'(t_{il})$, the likelihood function of observed data in the model of random censorship is of the form

$$L = \prod_{j=0}^{k} \Big([F'(t_j^-)-F'(t_j)]^{d_j} \cdot \prod_{i=1}^{m_j} F'(t_{jl}) \Big). \tag{9.3.5}$$

Further argumentation proceeds in two stages. First, we note that the survival function maximizing the function L must be a step function with jumps at $t_1, \ldots, t_k$. This follows from the fact that $F'(s)$ is nonincreasing, and hence the maximum of L is attained on the set of functions such that

$$F'(t_{jl}) = F'(t_j) \quad \text{for} \quad j = 1, \ldots, k,\ l = 1, \ldots, m_j,$$
$$F'(t_{0l}) = 1 \quad \text{for} \quad l = 1, \ldots, m_0.$$

Next, by representing $F'(t)$ in the form (9.2.5) and by applying standard methods of finding $\hat{\lambda}_j$ which maximize L we obtain $\hat{\lambda}_j = d_j/n_j$, where $n_j = m_j+d_j+ \ldots +$ $+m_k+d_k$. Clearly, $n_0 = n$.

If $d_i = 1$ for $i = 1, \ldots, k$, i.e. if all the recorded moments of death are different, then

$$n_i = n-i+1-\prod_{j=0}^{i-1} m_j$$

and, as can easily be seen, the ML estimator coincides with the K–M estimator.

Thus, the estimator of the survival function for the representation of data previously considered can be written in the form

$$F'(s) = \begin{cases} \prod_{i:\, z_i \leqslant s} \left(\dfrac{n-R_i}{n-R_i+1} \right)^{\varepsilon_i} & \text{for} \quad t \leqslant z_{(1)}, \\ 0 & \text{for} \quad t > z_{(n)}, \end{cases} \tag{9.3.6}$$

where R_i denotes the rank of the element $(z_i, 1-\varepsilon_i)$ in the lexicographically ordered sequence $(z_1, 1-\varepsilon_1), \ldots, (z_n, 1-\varepsilon_n)$.(1) This estimator of the survival function is called the *K–M estimator admitting ties*.

Note that we could reverse the order of presentation, first introducing the K–M estimator as an ML estimator and next showing that if ties are absent, it is the limit of the sequence of estimators such that each of them is an improved version (in the intuitive sense) of the preceding one.

---

(1) This conforms to the convention that moments of death are regarded as prior to the censored observations coinciding with them.

Let us note that there exists also another natural approach to the estimation of the survival function. Namely, for random variables X, Y with continuous distribution functions and for $s < t_H$, we have

$$\lambda_F(s) = \lim_{h\to 0^+} \frac{P(Z \in [s, s+h), \varepsilon = 1)}{h \cdot P(Z \geqslant s)} = \left.\frac{\mathrm{d}\tilde{F}_1}{\mathrm{d}t}\right|_{t=s} \cdot (H'(s))^{-1}.$$

Therefore, by (9.2.4), we obtain

$$F'(s) = \exp\Big(-\int_0^s (H'(z))^{-1}\,\mathrm{d}\tilde{F}_1(z)\Big) \tag{9.3.7}$$

for $s < t_H$ (Yang, 1977). Thus, in the absence of ties, the natural estimator of Λ_F is given by (Nelson, 1972)

$$\Lambda_n(s) = \sum_{z_{(i)}\leqslant s} \frac{\varepsilon_{(i)}}{n+1-i}. \tag{9.3.8}$$

It can be shown, however, that the resulting estimator of the survival function of the form $\exp(-\Lambda_n)$ is asymptotically equivalent to the K–M estimator. This follows from the inequality

$$0 < -\log \hat{F}_n'(t) - \Lambda_n(t) < \frac{n-N(t)}{n\cdot N(t)}.$$

Consider now the case where all data are grouped into intervals independent of the moments of death, and the available information consists of the initial number of patients and the numbers of deaths and censorings in each interval. We show that also in this case the K–M estimator emerges naturally.

Let $0 = \xi_0 < \xi_1 < \ldots < \xi_k = T$ be a partition of the interval $[0, T]$. Denote by I_k the interval $(\xi_{k-1}, \xi_k]$ and let q_k be, as previously, the conditional probability of death in the interval I_k under the condition that the patient was alive at the beginning of I_k:

$$q_k = \frac{F(\xi_k) - F(\xi_{k-1})}{F'(\xi_{k-1})}. \tag{9.3.9}$$

Put

$$\begin{aligned}
d_{1k} &= \#\{i\colon X_i \in I_k,\ Y_i > \xi_k,\ \varepsilon_i = 1\},\\
n_{1k} &= \#\{i\colon X_i > \xi_{k-1},\ Y_i > \xi_k\},\\
d_{2k} &= \#\{i\colon X_i \in I_k,\ Y_i \in I_k,\ \varepsilon_i = 1\},\\
n_{2k} &= \#\{i\colon X_i > \xi_{k-1},\ Y_i \in I_k\},\\
w_k &= \#\{i\colon Y_i \in I_k,\ \varepsilon_i = 0\},\\
d_k &= d_{1k} + d_{2k}, \qquad n_k = n_{1k} + n_{2k}.
\end{aligned}$$

In contrast with the random variables d_k, n_k, w_k, the random variables d_{ik}, n_{ik} $(i = 1, 2)$ are, in general, not observable.

In medical research two estimators of q_s are usually considered:

1° Berkson and Gage (1950) use the estimator

$$\hat{q}_k = \begin{cases} \dfrac{d_k}{n_k - \frac{1}{2} w_k} & \text{for} \quad n_k > 0, \\ 1 & \text{for} \quad n_k = 0. \end{cases} \tag{9.3.10}$$

The estimator $\hat{q}_k$ refers to recognizing the number $n_k - w_k/2$ as the number of patients exposed to the risk of death in the interval I_k. Every individual for whom the observation is censored is, by convention, treated as surviving, within the interval I_k, exactly half of it and thus subjected to the risk of death only during that period.

2° The estimator based on reduced sample defined by

$$\tilde{q}_k = \frac{d_{1k}}{n_{1k}} \tag{9.3.11}$$

is also considered. This estimator is observable only when the censoring mechanism is known. Since this condition is not always satisfied, the estimator $\hat{q}_k$ is used more frequently.

Let us now investigate the properties of the estimators $\hat{q}_1$ and $\tilde{q}_1$ of the probability of death in the first interval I_1.

The variables introduced above have a binomial distribution, where

$$\begin{aligned} E(d_{11}) &= nF(\xi_1)G'(\xi_1), \\ E(n_{11}) &= nG'(\xi_1), \\ E(d_{21}) &= n\int_0^{\xi_1} G' \mathrm{d}F, \\ E(n_{21}) &= nG(\xi_1), \\ E(w_1) &= n\int_0^{\xi_1} F' \mathrm{d}G. \end{aligned} \tag{9.3.12}$$

If $n \to \infty$, the quantities n_{11}/n and d_{11}/n tend with probability 1 to $F(\xi_1)G'(\xi_1)$ and $G'(\xi_1)$, respectively. Hence the estimator $\tilde{q}_1$ is consistent in the strong sense. By using the equality

$$\hat{q}_1 = \frac{n_{11}}{n - \frac{1}{2} w_1} \frac{d_{11}}{n_{11}} + \frac{n_{21} - \frac{1}{2} w_1}{n - \frac{1}{2} w_1} \frac{d_{21}}{n_{21} - \frac{1}{2} w_1}$$

and (9.3.12), it can be shown (Breslow and Crowley, 1974) that the condition

$$F(\xi_1) = \frac{\int_0^{\xi_1} F \,\mathrm{d}G}{G(\xi_1) - \int_0^{\xi_1} \frac{1}{2} F' \,\mathrm{d}G} \tag{9.3.13}$$

is necessary and sufficient for the consistency of the estimator $\hat{q}_1$. If the distribution function G is absolutely continuous, condition (9.3.13) is equivalent to the condition:

$$F(x) = 1 - \left(\frac{1}{1+cG(x)}\right)^{1/2}$$

for a certain $c > 0$. For example, $\hat{q}_k$ is consistent if the censoring mechanism is uniform in the interval $[0, T]$ and F is given by

$$F(x) = 1 - \left(\frac{1}{1+cx}\right)^{1/2}.$$

The estimators $\tilde{q}_k$ give rise to a natural estimator of the survival function, defined as

$$\hat{F}'(s) = \prod_{k:\, \xi_k \leqslant s} (1-\tilde{q}_k)$$

(we assume that elements censored at ξ_k are not taken into account in calculating w_k). This estimator depends on the adopted partition. If $\mathscr{K}_n$ is a normal sequence of partitions such that

$$\sup_k |\xi_{k,\mathscr{K}_n} - \xi_{k-1,\mathscr{K}_n}| \to 0,$$

then the sequence of estimators corresponding to these intervals is convergent to the K–M estimator in the weak topology of the space $D[0, T]$ (by $D[0, T]$ we denote the space of functions which are continuous on the right and have left-hand side limits on the interval $[0, T]$ with the Skorokhod metric). A similar result can be obtained if $\hat{q}_k$ is used instead of $\tilde{q}_k$ (cf. Breslow and Crowley, 1974).

9.4. Main asymptotic properties of the K–M estimator

The K–M estimator is used to estimate the survival function and its parameters in the case of censored data and is a counterpart of the empirical survival function in the case of full observability. In standard statistical analysis the following properties of the K–M estimator are particularly useful:

Property 1. *The K–M estimator is consistent in the strong sense for* $t < t_H = \sup\{t\colon H(t) < 1\}$, *i.e. for each* $t < t_H$ *we have* $\hat{F}_n(t) \to F(t)$ *with probability* 1.

This property follows from the following representation of F by means of the subdistribution functions $\tilde{F}_0$ and $\tilde{F}_1$ defined in Section 9.2 (cf. Peterson, 1977): *for any* $t < t_H$ *we have*

$$F(s) = \Phi(\tilde{F}_0, \tilde{F}_1)(s) \quad for \quad s \leqslant t, \tag{9.4.1}$$

where Φ *is a function which maps* $B[0, t] \times B[0, t]$ *into* $B[0, t]$, *and* $B[0, t]$ *is the space of all bounded functions on* $[0, t]$ *with the* sup *metric.*

Observe that the estimators

$$\tilde{F}_{1n}(t) = \frac{\#\{i\colon z_i \leqslant t,\, \varepsilon_i = 1\}}{n},$$

$$\tilde{F}_{0n}(t) = \frac{\#\{i\colon z_i \leqslant t,\, \varepsilon_i = 0\}}{n}$$

are maximum likelihood estimators of $\tilde{F}_1$ and $\tilde{F}_0$, respectively. Consequently, $\Phi(F_{0n}, F_{1n})$ is an ML estimator of F. Since in Section 9.3 we have shown that the K–M estimator is an ML estimator, we obtain

$$\hat{F}_n(t) = \tilde{\Phi}(\tilde{F}_{0n}, \tilde{F}_{1n})(t). \tag{9.4.2}$$

In this representation the K–M estimator is a function of natural estimators of subdistribution functions $\tilde{F}_0, \tilde{F}_1$. By the Gliwienko–Cantelli theorem and by the continuity of Φ (Gill, 1981), we obtain Property 1.

For F and G such that $t_F < t_G \leqslant +\infty$ we have a stronger property, namely

PROPERTY 2 (Főldes and Rejtő, 1981).

$$P\left(\sup_{u\in R} |\hat{F}_n(u) - F(u)| = O\left(\frac{\log\log n}{n}\right)^{1/2}\right) = 1. \tag{9.4.3}$$

The analogous property for the empirical distribution function is as follows:

$$P\left(\overline{\lim_{n\to\infty}} \left(\frac{n}{2\log\log n}\right)^{1/2} \sup_{u\in R} |\hat{F}_n(u) - F(u)| = c(F)\right) = 1,$$

where

$$c(F) = \sup_{u\in R} \{F(u)(1-F(u))\}^{1/2}.$$

PROPERTY 3. *For $t < t_H$, the K–M estimator of the survival function at point t is an asymptotically unbiased estimator of $F'(t)$.*

This follows from the inequality

$$1 \geqslant \frac{E\hat{F}(s)}{F'(s)} \geqslant 1 - e^{-nH'(s)}. \tag{9.4.4}$$

Hence, for $t < t_H$, the bias of the estimator decreases exponentially with the increasing sample size.

PROPERTY 4 (Breslow and Crowley, 1974). *For continuous distribution functions F, G and for $T < t_H$, $n^{1/2}(\hat{F}_n - F)$ is weakly convergent in $D[0, T]$ to a Gaussian process with the mean 0 and the covariance function of the form*

$$\Gamma(s, t) = F'(s)F'(t)\int_0^s (H')^{-2} \mathrm{d}F_1 \quad \textit{for} \quad s \leqslant t. \tag{9.4.5}$$

This result can be rewritten in the following form: *for $s \in [0, t_H)$ the ratio*

$$\frac{n^{1/2}(\hat{F}_n(a^{-1}(s)) - F(a^{-1}(s)))}{F(a^{-1}(s))}, \tag{9.4.6}$$

where $a(s) = \int_0^s (H')^{-2}\, \mathrm{d}\tilde{F}_1$ is weakly convergent to the Wiener process $W(s)$ (Efron, 1967). As an immediate corollary we obtain

PROPERTY 5 (Gillespie and Fisher, 1979). *For each* $d_1, d_2 \in \boldsymbol{R}$, *and any* $c_1 < 0$, $c_2 > 0$,

$$P\left(\frac{\hat{F}_n(t)}{1+(c_2+d_2 a_n(t))n^{-1/2}} \leqslant F(t) \leqslant \frac{\hat{F}_n(t)}{1+(c_1+d_1 a_n(t))n^{-1/2}}; 0 \leqslant t \leqslant T\right) \to$$
$$\to P(c_1+d_1 t \leqslant W(t) \leqslant c_2+d_2 t; \quad 0 \leqslant t \leqslant a(T)), \tag{9.4.7}$$

where $a_n(t)$ *is an estimator of* $a(t)$ *obtained by substituting the empirical distribution functions in the formula defining* $a(t)$.

Another theorem concerning symmetric asymptotic confidence bands for F has been proved by Hall and Wellner (1980).

PROPERTY 6. *Under the assumptions of Property* 4 *and for* $\eta > 0$, *we have*

$$P(\hat{F}_n(t)-\eta D_n(t) \leqslant F(t) \leqslant \hat{F}_n(t)+\eta D_n(t); \; 0 \leqslant t \leqslant T) \to G_l(\eta), \tag{9.4.8}$$

where

$$D_n(t) = n^{-1/2}\hat{F}_n(t)(K_n(t))^{-1},$$

$$K_n(t) = \left\{1+n \sum_{i:\, z_i<t} (n-i)^{-1}(n-i+1)^{-1}\varepsilon_i\right\}^{-1},$$

$$l = a(T)\{1+a(T)\}^{-1}.$$

The function $G_l(\eta)$ is tabulated in Hall and Wellner (1980).

Using Property 4 we can derive the asymptotic variance (as var) for asymptotically normal estimators of distribution parameters pertaining to the K–M estimator:

$$M = \int_0^\infty x \, d\hat{F}_n(x),$$

$$M_b = \int_0^b x \, d\hat{F}_n(x).$$

Namely, we have

$$\text{as var}\,\hat{M} = \frac{1}{n}\int_0^\infty (H'(s))^{-2}\left[\int_s^\infty F'(t)\,dt\right]^2 d\tilde{F}_1(s),$$
$$\text{as var}\,\hat{M}_b = \frac{1}{n}\int_0^b (H'(s))^{-2}\left[\int_s^b F'(t)\,dt\right]^2 d\tilde{F}_1(s). \tag{9.4.9}$$

Finally, for any distribution function F, the Kaplan–Meier estimator of the median m of F is of the form:

$$\hat{m} = \inf\{m: \; \hat{F}(m) \geqslant \tfrac{1}{2}\}.$$

The asymptotic variance of $\hat{m}$ for absolutely continuous F with density f is given by (Reid, 1981a)

$$\text{as var}\,\hat{m} = \frac{1}{n}\,[f(m)]^{-2}\int_0^m [H'(s)]^2\,\mathrm{d}\tilde{F}_1(s).$$

The properties listed above permit estimating parameters of F and constructing confidence intervals. It should be mentioned, however, that for samples of small size the properties of this method are not known sufficiently well.

*

Finally we shall discuss the estimation of distribution parameters for censored data by means of sample randomization. First, let us recall the method which in the case of full observability is called the *bootstrap method.* This method seems to meet the following postulate formulated by Kendall (1971):

> *Evidently therefore a method to be really satisfactory must allow for the possibility of more or less independent repeat analyses, so that we can see whether a surprising feature of one "solution" is in fact common to all "solutions", or at any rate, to most of them.*

For an arbitrary distribution function F, we want to estimate a parameter $\theta(F)$ by using a statistic $\theta(F_{\mathrm{emp}})$ which is a function of the empirical distribution function F_{emp}. Let $(x_1, x_2, \ldots, x_n)$ be a sample of size n from F. By $(x_1^*, x_2^*, \ldots, x_n^*)$ we denote a sample of size n from F_{emp} (i.e. we perform n drawings with replacements from the set $\{x_1, \ldots, x_n\}$). Let F_{emp}^* be the empirical distribution function of this sample and let $\hat{\theta}^* = \theta(F_{\mathrm{emp}}^*)$. By repeating this operation N times we obtain the sequence $\theta^{*1}, \ldots, \theta^{*N}$. As the estimator of $\theta(F)$ and the measure of its precision we take, respectively, the mean and the variance from this sample. Modifications of this procedure consist in generating the sample $(x_1^*, \ldots, x_n^*)$ not from F_{emp} directly, but from its smoothed counterpart, e.g. obtained by the kernel method.

Applying the above method to censored data consists in regarding the sequence of points $(z_1, \varepsilon_1), \ldots, (z_n, \varepsilon_n)$ as a subset of a plane and constructing on its basis a bivariate empirical distribution function. The value of the estimator $\theta(F_{\mathrm{emp}}^*)$ is replaced by $\theta(\hat{F}_{\mathrm{emp}})$, where $\hat{F}_{\mathrm{emp}}$ is the K–M estimator pertaining to the sample drawn from F_{emp}. The only difference with respect to the classical method lies in basing the computation on the K–M estimator and not on the empirical distribution function. By the Peterson theorem (cf. (9.4.2)) the properties of the bootstrap method thus modified can be derived from the respective properties of the original bootstrap method for uncensored data.

A survey of similar techniques applicable to censored data can be found in Efron (1981a, 1981b).

9.5. A TWO-SAMPLE PROBLEM FOR CENSORED DATA

In medicine the problem of comparison of two survival functions appears quite naturally, for example in comparing traditional and new treatments. This problem, known in a more general formulation as a *two-sample problem*, is widely discussed in the literature. It becomes more complicated if for such a comparison we have to use data part of which are censored. The tests applied in this case are usually modifications of tests used in the corresponding classical two-sample problems for uncensored data. The K–M estimator is then a very useful tool.

The formulation of the two-sample problem for censored data is similar to its formulation in the case of full observability. We assume that two independent samples, called *experimental* and *control data*, are available. The experimental sample corresponds to the experimental medical treatment and the control sample to the classical treatment. In both samples part of the data are censored. This means that the control sample is of the form $\big((z_{1i}, \varepsilon_{1i}), i = 1, \ldots, n\big)$ and the experimental sample is of the form $\big((z_{2j}, \varepsilon_{2j}), j = 1, \ldots, m\big)$, where $z_{1i} = \min(x_{1i}, y_{1i})$, $z_{2j} = \min(x_{2j}, y_{2j})$, and y_{1i}, y_{2j} are the times of censoring for the i-th and the j-th patient in the first and the second sample, respectively. The distribution functions of the survival and the censoring mechanisms are F_1, G_1 in the first case, and F_2, G_2 in the second case. We want to test the hypothesis H_0: $F_1 = F_2$ versus the alternative hypothesis H_a: $F_1 < F_2$. In other words, we want to find out whether the experimental treatment is better than the classical treatment, if data censoring occurs in both samples. If we know that the censoring mechanism is the same for both treatments, then the null hypothesis implies the equality $F_1'G_1' = F_2'G_2'$, which can be tested by the Wilcoxon test. However, we do not then use all the available information; moreover, the assumption that the two censoring mechanisms coincide seriously restricts the area of possible applications of this model. Therefore, in the general case the censoring mechanisms are assumed to be different. Posing the problem in this way is certainly indicated if the experiments involving classical therapy and experimental therapy are distant in time.

We shall now discuss two examples of testing H_0 versus H_a. Both tests are modifications of the Wilcoxon test. If the quantities x_{1i}, x_{2i} were observable, then the Wilcoxon test function applied to these data would be the normalized sum of expressions equal to 1 if $x_{1i} \geqslant x_{2j}$, and equal to 0 if $x_{1i} < x_{2j}$, for $i = 1, 2, \ldots, n$, $j = 1, \ldots, m$. Note that

$$\begin{aligned} z_{1i} > z_{2j} \ \wedge \ \varepsilon_{2i} = 1 \quad &\Rightarrow \quad x_{1i} > x_{2j}, \\ z_{1i} < z_{2j} \ \wedge \ \varepsilon_{1i} = 1 \quad &\Rightarrow \quad x_{1i} < x_{2j}. \end{aligned} \tag{9.5.1}$$

The first modification of the Wilcoxon test for censored data, proposed by Gehan (1965), is based on the statistic W_G which is the sum of the scores u_{ij}, equal to 1 or 0 if the respective premises in implications (9.5.1) hold, and equal to $\frac{1}{2}$ otherwise. This refers to the fact that whenever the expression u_{ij} in the original Wilcoxon test is unobservable, it could be replaced by the probability that the i-th element of the control sample will survive longer than the j-th element of the experimental sample; this probability is approximated by $\frac{1}{2}$. For any survival functions F', G', denote by $P(F' \geqslant G')$ the integral $-\int_0^\infty F'(t)\,dG'(t)$. Under relatively weak assumptions we can obtain the asymptotic distribution of modified Wilcoxon statistic W_G. Namely, if the distribution functions F_i, G_i $(i = 1, 2)$, are continuous and $m/(m+n)$ tends to a positive constant η, then, under hypothesis H_0, $(m+n)^{1/2}(W_G - \frac{1}{2})$ is weakly convergent to $N\left(0, \frac{1}{12}\left(\frac{1}{\eta}\sigma_1^2 + \frac{1}{1-\eta}\sigma_2^2\right)\right)$, where $\sigma_1^2 = P\big(G_1'(G_2')^2 > (F')^3\big)$ and $\sigma_2^2 = P\big((G_1')^2 G_2' > (F')^3\big)$. The quantities σ_1^2, σ_2^2 can be estimated by their sample counterparts.

The test based on statistic W_G is consistent.

Setting $u_{ij} = \frac{1}{2}$ for dubious cases is sometimes highly inadequate, e.g. if the two censoring mechanisms differ from each other. It may result in the power of the test being small, especially for samples of small size. This intuition has been confirmed in a large number of case studies (cf. Prentice and Marek, 1979).

The second modification of the Wilcoxon test differs from the first in the definition of the scores u_{ij} in dubious cases. Namely, in such cases, the scores u_{ij} is defined as the value of a certain estimator of the probability that the i-th element from the first sample lives longer than the j-th element from the second sample. For example, if $z_{1i} < z_{2j}$ and $(\varepsilon_{1i}, \varepsilon_{2j}) = (0, 1)$, this probability can be estimated by means of $F_1'(z_{2j})/F_1'(z_{1j})$.

TABLE 9.5.1.
Scores u_{ij} in the Wilcoxon test for censored data.

$(\varepsilon_{1i}, \varepsilon_{2j})$	$z_{1i} \geqslant z_{2j}$	$z_{1i} < z_{2j}$
(1, 1)	1	0
(0, 1)	1	$\dfrac{\hat{F}_1'(z_{2j})}{\hat{F}_1'(z_{1i})}$
(1, 0)	$1 - \dfrac{\hat{F}_2'(z_{1i})}{\hat{F}_2'(z_{2j})}$	0
(0, 0)	$-\int_{z_{1i}}^{\infty} \dfrac{F_1'(s)\,dF_2'(s)}{F_1'(z_{1i})F_2'(z_{2j})}$	$1 - \dfrac{F_2'(z_{1i})}{F_1'(z_{2j})} - \int_{z_{1i}}^{\infty} \dfrac{F_1'(S)\,dF_2'(s)}{F_1'(z_{1i})F_2'(z_{2j})}$

Table 9.5.1 shows the values of u_{ij} in all the cases. In Efron (1967) it is shown that the expression $\sum u_{ij}/(m \cdot n)$ can be transformed into the form $P(\hat{F}_1' \geqslant \hat{F}_2')$ and so can be regarded as an adaptation of the Wilcoxon test. However, computational difficulties make the use of this test cumbersome, especially if the number of censored observations is large in both samples. The properties of the above tests in some families of distributions of survival functions are given in Efron (1967); see also Gill (1980).

In practice, the construction of a control sample is sometimes difficult and nontypical. For example, consider an experiment similar to that described in Section 9.1. The experiment concerns heart transplantations; each patient classified to undergo the operation waits for a suitable donor. In effect, at the time of the analysis (January 1st, 1974) both the patients who have not been operated and those who have are considered. In both groups there are patients who are no longer alive at the moment of the analysis. The patients from the first group are characterized by the time interval from the moment of being qualified for the treatment to the moment of their death or to the date of the analysis. The patients from the second group are characterized by the waiting time before the operation and the survival time after the operation (till the moment of death or the moment of the analysis). In certain ill-planned investigations it happened that only those patients from the first group who died were included in the control group. This of course lead to a rather tendentious presentation of the experimental therapy in a good light.

A nonstandard method proposed by Turnball *et al.* (1974) estimates the survival functions of the patients who did not undergo the operation on the basis of both groups. For the patients of the second group the waiting time for transplantation is identified with the censored survival time, and the moment of transplantation with the moment of censoring. These data are then processed for testing the null hypothesis by means of the K–M estimator.

9.6. Final remarks

Let us now discuss some generalizations and modifications of the problem of estimation of the survival function in the case of censored data. We shall also make some remarks on the corresponding two-sample problem.

The model of random censorship can be generalized in the following way. Let (X_i), $i = 1, \ldots, n$, be a sequence of independent random variables with distribution functions (F_i), $i = 1, \ldots, n$. Assume that the random variable $Z = \min(X_1, \ldots, X_n)$ having the distribution function H and the variables $\varepsilon_i = \chi_{\{X_i = Z\}}$ are observable. This model corresponds to situations where it is reasonable to assume that death may be caused by n independently acting risks. The random variable X_i is then interpreted as the potential survival time in a hypothetical situation in which the i-th risk is the only operating risk. The function $W_i'(t) = P(Z > t, Z = X_i)$ is interpreted as the probability that the survival time caused by the i-th risk in the presence of all other risks is longer than t. According to this interpretation, F_i' is called the *i-th net survival function*, and W_i' is called the *i-th crude survival function.* Observe that the moment of censoring can be interpreted formally as the moment of death caused by the risk which is independent of the operating risks. Hence, the model of random censorship given in Section 9.2 is a particular case of the model presented here, which is called a *model of independent competing risks.*

The most questionable assumption in this model is the assumption of independence of risks and therefore models allowing interdependence of risks are considered and independence of risks is regarded as a hypothesis to be tested. However, it is shown by Tsiatis (1975) that this hypothesis is not observably reducible (in the sense of the definition in Section 2.2) since, for given crude survival functions $W_1', \ldots, W_n'$, we can always define independent random variables $Y_1, \ldots, Y_n$ such that the probability $P(Z > t, Z = Y_i)$ is equal to $W_i'(t)$, for $i = 1, \ldots, n$.

In recent years intensive studies have been devoted to successive generalizations of this model, which do not assume independence of risks. The following model, called the *multiple decrement model*, has been introduced by Aalen (1976). Let M be a Markov chain with one distinguished state 0 and n absorbing states $1, 2, \ldots, n$. Denote by $P_i(t)$ the probability that at the moment t the chain is in the state i and put $P_0(0) = 1$. Then the functions λ_i,

$$\lambda_i(s) = \left.\frac{\mathrm{d}P_i}{\mathrm{d}t}\right|_{t=s} \cdot (P_0(s))^{-1}, \tag{9.6.1}$$

define the infinitesimal conditional probability of transition from state 0 to state i at a given moment s. Assume that we observe realizations of the phenomenon in k such models. On the basis of this observation we want to estimate the net transition probability

$$p_i(t) = 1-\exp\Big(-\int_0^t \lambda_i(s)\,\mathrm{d}s\Big), \quad i = 1, 2, \ldots, n, \tag{9.6.2}$$

which is interpreted as the probability of transition from 0 to i in the time interval $[0, t]$ provided $\lambda_j = 0$ for $j \neq i$. Observe that the model of n independent competing risks is a particular case of the multiple decrement model. The deaths caused by the i-th risk are then interpreted as absorbing states, and it is assumed that $P_i(t) = W_i(t)$. As can easily be seen, the parameter $\lambda_i(t)$ is then equal to the hazard function of X_i and the net transition probability $p_i(t)$ is equal to the distribution function $F_i(t)$. The K–M estimator used in the multiple decrement model for estimating $p_i(t)$ has similar asymptotic properties as in the random censorship model (cf. Aalen, 1978a).

A much wider generalization than the multiple decrement model is the approach based on the theory of point processes. Let $N(t) = (N_i(t), i = 1, 2, \ldots, r)$ be an r-dimensional process with nondecreasing trajectories $N_1(\cdot), \ldots, N_r(\cdot)$ which are continuous on the right and assume only natural values. However, the jumps of trajectories are assumed to be equal to 1 and there are no common points of jumps. Let $\mathscr{F}_t$ be a family of σ-fields related to this process and let $\Lambda(t)$ denote a stochastic process defined by the formula

$$\Lambda(t) = \lim_{h\to 0^+} \frac{1}{h} E\big(N(t+h)-N(t) \mid \mathscr{F}_t\big). \tag{9.6.3}$$

The proposed approach (Aalen, 1978b) consists in investigating the properties of the process N by using the properties of Λ. The main fact exploited in this investigations is that the process $M(t)$,

$$M(t) = N(t) - \int_0^t \Lambda(s)\,\mathrm{d}s, \tag{9.6.4}$$

is a square integrable martingale (this follows from the Doob decomposition of a submartingale). A specific form of this model is obtained by putting

$$\Lambda(t) = \alpha(t) \cdot Y(t), \tag{9.6.5}$$

where $\alpha(t)$ is a positive (and deterministic) function and $Y(t)$ is an r-dimensional process observed together with the process $N(t)$.

In particular, in the random censorship model we take as $Y(t)$ the number of elements observed up to the moment t. By using this approach it can be shown that in an interval $[0, T]$ such that $F(T) < 1$ the process

$$W(t) = \frac{\hat{F}'(\min(t, \tau))}{F'(\min(t, \tau))}, \tag{9.6.6}$$

where $\tau = \inf\{t\colon Y(t) = 0\}$ is a martingale. A detailed survey of these results and their applications to the two-sample problem can be found in Gill (1980).

Another trend in modifications and generalizations is to consider censoring on the left and analyse situations where both right and left censoring occurs. A model of this type was used for estimating the mean time at which a certain kind of monkeys come down from the trees in the morning (Wagner and Altman 1973). Left-hand data censoring resulted from the researchers arriving at the observation point too late, namely after the monkeys has already come down from the trees.

A model in which the moments of right-hand censoring were fixed was first considered by Meier (1975). Later this situation was described in terms of the theory of point processes (Gill, 1980). It is an important problem in these particular cases when we cannot assume that the moments of censoring are generated by a random mechanism.

The assumptions are often modified in practice by taking instead of the hazard function $\lambda(t)$ a function $f(x, \lambda(t))$, where x is an observable vector of parameters such as the age of the patient, his blood pressure, or the sugar level in his blood. This is a more realistic approach than that presented in the preceding sections but it requires specifying the form of dependence of the hazard function on the parameter vector. The most common model of this type is the model of proportional hazards,

$$f(x, \lambda(t)) = \lambda(t) \cdot e^{x \cdot \beta}. \tag{9.6.7}$$

In this model the value of the vector β is first estimated by using adapted ML methods and treating $\lambda(t)$ as a nuisance parameter. Next, the resulting value $\hat{\beta}$ is substituted in the estimator of $\lambda(t)$ obtained for a fixed β. These problems are considered in Kalbfleisch and Prentice (1980).

There also exist models in which a certain special form of the hazard functions λ_F, λ_G is assumed. In particular, these functions are assumed to be step functions or to have a certain parametric form. A survey of such methods is given in the book of Kalbfleisch and Prentice (1980).

In the bayesian approach to the random censorship model it is assumed that the survival function F' is generated by the Dirichlet process with a parameter α, where α is an arbitrary positive measure on $[0, +\infty)$. Under this assumption, in the case of a quadratic loss function, the bayesian estimator $F'_{n,\alpha}(u)$ is of the form:

$$F'_{n,\alpha}(u) = \begin{cases} \dfrac{N(u)+\alpha(u)}{n+\alpha(0)} & \text{for} \quad 0 < u \leqslant z_{(1)}, \\ \dfrac{N(u)+\alpha(u)}{n+\alpha(u)} \cdot \prod_{j=1}^{i} \left(\dfrac{N(z_{(j)})+\alpha[z_{(j)})+1}{N(z_{(j)})+\alpha[z_{(j)})} \right)^{1-\varepsilon_{(j)}} & \\ & \text{for } z_{(j)} \leqslant u < z_{(i+1)}, \\ 0 & \text{for } u > z_{(n)}, \end{cases} \tag{9.6.8}$$

where $\alpha(t) = \alpha((t, +\infty))$ and $\alpha[t) = \alpha([t, +\infty))$ (cf. Susarla and Van Ryzin, 1978). Földes and Rejtö (1981) proved that if the distribution functions F and G are continuous, $t_F < t_G < +\infty$ and $\alpha[T_F) > 0$, then

$$P\left(\sup_{u \in R} |F'_{n,\alpha}(u) - F'(u)|\right) = O\left(\left(\frac{\log \log n}{n}\right)^{1/2}\right) = 1. \tag{9.6.9}$$

If it is not the distribution function F but the parameter $\theta(F)$ which is of interest then the statistic $\theta(\hat{F})$ is most frequently used. We do this in point and interval estimation by making use of the asymptotic normality of the statistic $\theta(\hat{F})$ and of the fact that asymptotic variance of $\theta(\hat{F})$ may be calculated with the aid of the so-called delta method (cf. Rao, 1965). Moreover, if the influence function IC(θ, F, t) is known, then we can directly apply the equality proved by Reid (1981b),

$$\text{as var}\, \theta(\hat{F}) = \frac{1}{n} \int_0^\infty \frac{1}{[H'(s)]^2} \left(\int_s^\infty F'(t) g(t)\, dt \right)^2 d\tilde{F}_1(s), \tag{9.6.10}$$

where

$$g(t_0) = \frac{d}{dt} \text{IC}(\theta, F, t)_{|t=t_0}.$$

Using this equality we obtain the estimator by replacing the unknown quantities with their estimates. In particular, as the estimate of the variance of the K–M estimator at point t we consider (for the case without ties) the following expression:

$$[F'(t)]^2 \sum_{z_i < t} \frac{\varepsilon_i}{N(z_i)\big(N(z_i)+1\big)} \tag{9.6.11}$$

known in the analysis of lifetables as the *Greenwood formula* (cf. Elandt-Johnson and Johnson, 1980). This approach can be used, for example, in the case of the mean value and the quantiles. Let us consider the problem of the estimation of the density function f of a random variable X, availing ourselves of the classical methods applied in the case of complete observability.

In the classical case the kernel estimators f_n are often used

$$f_n(x) = \frac{1}{h(n)} \int_R K\left(\frac{x-y}{h(n)}\right) dF_n(y), \tag{9.6.12}$$

where $h(n)$ is a sequence of numbers such that $h(n) \to 0$, and $nh(n) \to \infty$, K is a fixed density function, and F_n is an empirical distribution function.

Instead of deterministic bandwidth $h(n)$, also a random bandwidth $R(n, x)$ is considered defined as the distance from x to the $k(n)$-th nearest neighbour, where $k(n)$ is a fixed sequence of natural numbers tending to infinity.

If the data are censored analogous estimators are defined as

$$
\begin{aligned}
f_n(x) &= \frac{1}{h(n)} \int_R K\left(\frac{x-y}{h(n)}\right) d\hat{F}_n(y), \\
\hat{f}_n(x) &= \frac{1}{R_1(n)} \int_R K\left(\frac{x-y}{R_1(n)}\right) d\hat{F}_n(y),
\end{aligned}
\tag{9.6.13}
$$

where $R_1(n)$ denotes the distance to the $k(n)$-th uncensored neighbour. It has been shown (Mielniczuk, 1986) that, under some conditions on F, G, $k(n)$, and $h(n)$

$$
\begin{aligned}
&[nh(n)]^{1/2}(f_n(x)-f(x)) \xrightarrow{\mathscr{L}} N\left(0, \frac{f(x)}{G'(x)} \int_R K^2(y)\, dy\right), \\
&[k(n)]^{1/2}(\hat{f}_n(x)-f(x)) \xrightarrow{\mathscr{L}} N\left(0, 2f^2(x) \int_R K^2(y)\, dy\right).
\end{aligned}
\tag{9.6.14}
$$

In the literature many particular forms of the two-sample problem discussed in Section 9.4 are considered. The test proposed by Mantel (1966) is the most powerful test if the censoring mechanisms in both samples are the same and the alternative hypothesis states that the ratio of the hazard functions is constant and different from 1. Fleming *et al.* (1980) consider a test which is a modification of the Kolmogorov–Smirnov test in the case where the alternative hypothesis states that the plots of the hazard functions intersect.

The two-sample problem becomes greatly simplified if the survival hazard functions is assumed to be known in the control sample and the censoring hazard function is assumed to differ from it only by a constant. For this situation a test has been constructed on the basis of a modification of the von Mises test statistic (cf. Koziol and Green, 1976). In the literature, also the situation where the population consists of several subpopulations is dealt with. For data censoring, a survey of test statistics for two-sample problems in the case of such populations was given by Lininger *et al.* (1979). They evaluate the power of most commonly used tests on the basis of simulation experiments.

Testing the independence of the random variables X_1 and X_2 if the value of at least one of them is censored has been investigated by Brown *et al.* (1974), Oakes (1982), Cuzick (1982), and Dąbrowska (1984). Just as for two-sample problems, suggested test statistics are adaptations of the corresponding statistics used in the classical case. The last of the works cited above gives the form of the rank tests which are locally most powerful in a certain family of alternatives.

Chapter 10

Latent variables in experimental psychology

Key words: *latent variable, subject ability, item difficulty, item discriminating power, response strength, item threshold level.*

10.1. Introduction

In social sciences, primarily in psychology and sociology, researchers are interested in measurement of such quantities as abilities, predispositions, attitudes, preferences and artistic likings. These variables serve to describe human groups for solving various problems involving groups or particular individuals. They are commonly called *latent variables* (as distinct from *manifest variables*, such as stature, body weight or eye pigment).

It is assumed that latent variables influence types of human behaviour which lend themselves to observation. Consequently, information on latent variables is derived from research findings concerning human behaviour in suitably performed experiments.

The theoretical foundations for this field of study should be furnished by the measurement theory together with statistics. However, the current relationships of the measurement theory with statistics are loose rather and for this reason, latent variables are studied in psychometry either in reference to measurement theory or to statistics.

The study of latent variables by statistical methods not involving the measurement theory has a long tradition in social sciences. In such investigations it is necessary to start out from the assumption that there exist certain unobservable latent variables having real or vector values. The next step is to choose an experimental scheme which specifies the technique of observation of the behaviour of a human group selected out of a population. Further, it is assumed that the observable results of behaviour have a certain distribution dependent on the values of latent variables and possibly also on parameters of the experimental scheme, unknown to the investigator. This constitutes a formal description of the phenomenon under consideration. The purpose of investigations may be to get information on the values of unobservable latent variables, on the distribution of these variables in a population or on the relationships of latent variables with the results of experiments.

Most research is devoted to experimental schemes in which each individual in the selected group is subjected to stimuli which cause a response observable in two states. Most frequently, a subject is given test items such that each of them may only be answered correctly or incorrectly. We assume that each subject is represented by a value η of some latent trait called his *ability*. Also, each item is represented by a value α of some latent trait called item *difficulty*. It should be emphasized that "ability" and "difficulty" are just conventional terms, corresponding to various latent variables appearing in concrete experiments. Likewise, we should understand conventionally the answering of test items.

In this chapter, we present statistical problems concerning investigations of the subject ability and item difficulty most frequently found in literature. In dealing with these problems, it is usually assumed that abilities and difficulties are real-valued. A situation is also admitted, in which two or more components describing various types of ability are distinguished, and then ability $\boldsymbol{\eta}$ is a vector with m components (for a fixed m). Similarly, item difficulty $\boldsymbol{\alpha}$ can be a vector. Traditionally, the number of components of vectors $\boldsymbol{\eta}$ and $\boldsymbol{\alpha}$ is taken to be identical.

The descriptions of experimental schemes in particular problems differ in that sometimes the subject set is a simple random sample from a population (the ability distribution in this population being usually treated as known) and sometimes we are concerned with a concrete subject set without referring it to any particular population. We distinguish therefore models with a random subject group and with a fixed subject group. Similarly we could speak of a random or fixed item set, but usually the item set is treated as fixed.

The descriptions of experiments differ also in the assumptions concerning the distribution of subject responses. Note that in experiments of this type, there is no suitable heuristics to determine the links between the observable results of experiments and the latent trait values. Therefore parametric models are rather seldom appropriate here.

Given in Section 10.2 is a systematic survey of models describing the experiments. Also discussed are the goals of investigations, which may be specified in various less or more detailed ways. Sometimes we want to know the abilities of subjects and the difficulties of items, and sometimes we are interested only in ordering the items according to increasing difficulty. We may also be interested in other item parameters bearing on the responses. We take into account the fact that the less information we have on the phenomenon, in other words, the more general is its description, the less particular should be the information sought by us, since otherwise we can not expect reliable statements. Models may differ from one another, yet it may happen that phenomenon descriptions corresponding to different explanations of the subject's behaviour are formally equivalent.

Models in which the concept of response strength of the subject is introduced are described at the end of Section 10.2. We assume here that an item will be answered correctly when and only when the response strength exceeds the threshold level specific to this item. The goal of investigations in threshold models are threshold levels and parameters of the joint distribution of ability and response strength.

It appears that it is possible to distinguish a class of parametric threshold problems and a class of parametric ability-difficulty problems, such that the two classes are equivalent. This is demonstrated in Section 10.2. The equivalence of the two classes is used to solve parametric ability-difficulty problems (Sec. 10.3). Estimation of item ordering in some non-parametric models is described in Section 10.4. Selected literature items on latent trait analysis and some problems relating to measurement scales are discussed in Section 10.5.

10.2. Typical experimental schemes

Taking part in an experiment are n subjects and each of them answers all items from a certain k-element item set. Let us denote the subject group by $\boldsymbol{O}$ and the item group by $\boldsymbol{Z}$.

We assume that each subject $o \in \boldsymbol{O}$ can be described by an m-dimensional ability vector $\boldsymbol{\eta} \in \boldsymbol{R}^m$ and each item $z \in \boldsymbol{Z}$ by an m-dimensional difficulty vector $\boldsymbol{\alpha} \in \boldsymbol{R}^m$. Therefore, the set $\boldsymbol{O} \cup \boldsymbol{Z}$ is represented by the sequence $(\boldsymbol{\eta}_1, \ldots, \boldsymbol{\eta}_n, \boldsymbol{\alpha}_1, \ldots, \boldsymbol{\alpha}_k)$ and the realization of the whole subject-item experiment is

$$\omega = (\boldsymbol{\eta}_1, \ldots, \boldsymbol{\eta}_n, \boldsymbol{\alpha}_1, \ldots, \boldsymbol{\alpha}_k, \boldsymbol{x}_1, \ldots, \boldsymbol{x}_n), \tag{10.2.1}$$

where $\boldsymbol{\eta}_i = (\eta_{i1}, \ldots, \eta_{im})^T$, $i = 1, \ldots, n$, is the ability vector, $\boldsymbol{\alpha}_j = (\alpha_{j1}, \ldots, \alpha_{jm})^T$, $j = 1, \ldots, k$, is the difficulty and $\boldsymbol{x}_i = (x_{i1}, \ldots, x_{ik})^T$, $i = 1, \ldots, n$, is a zero–one vector, such that for each $j = 1, \ldots, k$, $x_{ij} = 1$ when i-th subject answers correctly the j-th item, and $x_{ij} = 0$ in the opposite case. Observable experimental data are $\boldsymbol{x}_1, \ldots, \boldsymbol{x}_n$.

Let $\tilde{\boldsymbol{\eta}}_i = (\tilde{\eta}_{i1}, \ldots, \eta_{im})^T$, $\tilde{\boldsymbol{\alpha}}_j = (\tilde{\alpha}_{j1}, \ldots, \tilde{\alpha}_{jm})^T$, $\boldsymbol{X}_i = (X_{i1}, \ldots, X_{ik})^T$ denote random vectors with values $\boldsymbol{\eta}_i, \boldsymbol{\alpha}_j, \boldsymbol{x}_i$. Obviously, $\tilde{\boldsymbol{\eta}}_i$ are degenerated (they assume a constant value with probability 1) in models with a fixed subject group, and $\tilde{\boldsymbol{\alpha}}_j$ are degenerated in models with a fixed item set.

The manner in which subjects answer test items is formalized by a conditional distribution of vectors $\boldsymbol{X}_1, \ldots, \boldsymbol{X}_n$ with given abilities and difficulties. Let us denote

$$p_{ij} = P(X_{ij} = 1 | \tilde{\boldsymbol{\eta}}_i = \boldsymbol{\eta}_i, \tilde{\boldsymbol{\alpha}}_j = \boldsymbol{\alpha}_j), \quad i = 1, \ldots, n, \quad j = 1, \ldots, k.$$

Probability p_{ij} depends on the properties of both the item and the subject. In the simplest case, the probability is taken to depend only on subject ability and item difficulty according to a certain function p: $\boldsymbol{R}^{2m} \to [0, 1]$. Then for each pair (i, j)

$$p_{ij} = p(\boldsymbol{\eta}_i, \boldsymbol{\alpha}_j). \tag{10.2.3}$$

This assumption can be weakened, if we admit that answering of test items depends not only on the subject's ability and item difficulty, but also on other properties of the item. In place of a single function p we have then k functions p_j: $\boldsymbol{R}^m \to [0, 1]$, $j = 1, \ldots, k$, and we replace (10.2.2) by

$$p_{ij} = p_j(\boldsymbol{\eta}_i), \quad i = 1, \ldots, n, \quad j = 1, \ldots, k. \tag{10.2.4}$$

In the sequel we confine ourselves to considering models, in which assumptions (10.2.3) or (10.2.4) hold. We shall also assume that the responses of subject o to item z and of subject o' to item z' are independent if $(o, z) \neq (o', z')$, i.e.,

$$P(X_1 = x_1, \ldots, X_n = x_n \mid \tilde{\eta}_1 = \eta_1, \ldots, \tilde{\eta}_n = \eta_n, \tilde{\alpha}_1 = \alpha_1, \ldots, \tilde{\alpha}_k = \alpha_k)$$

$$= \prod_{i=1}^{n} \prod_{j=1}^{k} P(X_{ij} = x_{ij} \mid \tilde{\eta}_i = \eta_i, \tilde{\alpha}_j = \alpha_j)$$

$$= \prod_{i=1}^{n} \prod_{j=1}^{k} [p_j(\eta_i)]^{x_{ij}} (1 - p_j(\eta_i))^{1-x_{ij}}. \tag{10.2.5}$$

Furthermore, we confine ourselves to a situation, in which the item set is treated as fixed. As for subject ability, we shall admit both a situation with a fixed subject set $\boldsymbol{O}$ and a situation with a random subject group. In the latter case, the random variables $\tilde{\eta}_1, \ldots, \tilde{\eta}_n$ are independent and identically distributed. This distribution is taken to be continuous with a density denoted by h. The joint marginal distribution of vectors $X_1, \ldots, X_n$ has then the form

$$P(X_1 = x_1, \ldots, X_n = x_n) = \prod_{i=1}^{n} \int_{R^m} \prod_{j=1}^{k} [p_j(t)]^{x_{ij}} [1 - p_j(t)]^{1-x_{ij}} h(t) \, dt. \tag{10.2.6}$$

In models with a fixed subject group, the joint marginal distribution of random vectors $X_1, \ldots, X_n$ is exactly the same as the conditional distribution (10.2.5).

To sum up, the distribution on the set of phenomenon realizations of the form (10.2.1) is specified by functions $p_1, \ldots, p_k$, item difficulties $\alpha_1, \ldots, \alpha_k$ and moreover by density h in the case of a random subject group or by abilities $\eta_1, \ldots, \eta_n$ in the case of a fixed subject group. We present below different variants of assumptions concerning functions $p_1, \ldots, p_k$ and number m.

In the first of the considered variants we assume that $m = 1$ and that

$$\eta_1 < \eta_2 \quad \Rightarrow \quad p_j(\eta_1) \leqslant p_j(\eta_2) \qquad \forall \eta_1, \eta_2 \in \boldsymbol{R};\ j = 1, \ldots, k, \tag{10.2.7}$$

$$\alpha_s < \alpha_t \quad \Leftrightarrow \quad p_s \succcurlyeq p_t, \qquad s, t = 1, \ldots, k, \tag{10.2.8}$$

where $p_s \succcurlyeq p_t$ means that for each η, $p_s(\eta) \geqslant p_t(\eta)$ and that the inequality is strict on a set of positive Lebesgue measure. Hence, the set of functions $p_1, \ldots, p_k$ is ordered in the sense that the graphs of functions $p_1, \ldots, p_k$ do not intersect. We additionally assume that

$$p_j(\alpha_j) = \tfrac{1}{2}, \quad j = 1, \ldots, k. \tag{10.2.9}$$

Thus, item difficulties are standardized in a way with respect to subject abilities.

Presented in Fig. 10.2.1 are the graphs of several functions satisfying (10.2.7), (10.2.8) and (10.2.9).

In the second variant of assumptions for $p_1, \ldots, p_k$ we take for $j = 1, \ldots, k$

$$p_j(\eta) = F(\beta_j^T(\eta - \alpha_j)), \tag{10.2.10}$$

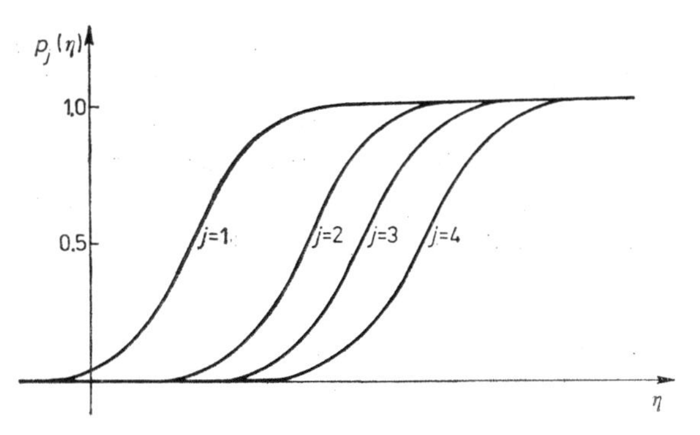

FIG. 10.2.1. Functions p_j ($j = 1, 2, 3, 4$) in the nonparametric model which satisfy (10.2.7), (10.2.8), (10.2.9).

where F is the distribution function of a certain symmetric and continuous standardized univariate distribution, and $\boldsymbol{\beta}_j = (\beta_{j1}, \ldots, \beta_{jm})^T$ is an element of a certain subset $\boldsymbol{B}$ of $\boldsymbol{R}^m$. We therefore admit different cases according to the choice of m, F and $\boldsymbol{B}$. Specifically, we shall consider the following possibilities:

$$m \geqslant 1, \quad F = F_{N(0,1)}, \quad \boldsymbol{B} = \boldsymbol{R}, \tag{10.2.11}$$

$$m = 1, \quad F = F_{N(0,1)}, \quad \boldsymbol{B} = \boldsymbol{R}^+, \tag{10.2.12}$$

$$m = 1, \quad F = F_{L(0,1)}, \quad \boldsymbol{B} = \boldsymbol{R}^+, \tag{10.2.13}$$

$$m = 1, \quad F = F_{L(0,1)}, \quad \boldsymbol{B} = \left\{\frac{\sqrt{3}}{\pi}\right\}. \tag{10.2.14}$$

We call a model *normal*, if the condition (10.2.11) or (10.2.12) is satisfied, *logistic*, if (10.2.13) is satisfied, and *Rash*, if (10.2.14) is satisfied. In the Rash model

$$p_j(\eta_i) = p(\eta_i, \alpha_j) = \frac{\exp(\eta_i - \alpha_j)}{1 + \exp(\eta_i - \alpha_j)}.$$

Conditions (10.2.7) and (10.2.9) are satisfied in the normal model specified by (10.2.12) and in the logistic model. Condition (10.2.8) is not always fulfilled (it is fulfilled if $\boldsymbol{\beta}_1 = \ldots = \boldsymbol{\beta}_k$).

With $m = 1$ and $\beta_j \in \boldsymbol{R}^+$, the expression $F(\beta_j(\eta_i - \alpha_j))$ can be interpreted as follows: the item is correctly answered only when ability η_i is greater than the sum of α_j and the random error with mean zero and standard deviation $1/\beta_j$. The standardized error has the distribution function F.

If $\beta_j \to 0$, then for any number η $p_j(\eta) \to \frac{1}{2}$, irrespective of the value α_j. A model in which

$$p_j(\eta) = \tfrac{1}{2}, \quad j = 1, \ldots, k, \quad \eta \in \boldsymbol{R}, \tag{10.2.15}$$

will be called here a *random response model*. On the other hand, if $\beta_j \to \infty$, then $p_j(\eta) \to 1$ for $\eta > \alpha_j$, and $p_j(\eta) \to 0$ for $\eta < \alpha_j$. A model, in which

$$p_j(\eta) = \begin{cases} 0 & \text{for} \quad \eta < \alpha_j, \\ \frac{1}{2} & \text{for} \quad \eta = \alpha_j, \\ 1 & \text{for} \quad \eta > \alpha_j, \end{cases} \quad j = 1, \ldots, k, \tag{10.2.16}$$

is the opposite of the random response model, since for $\eta_i \neq \alpha_j$ the response is already determined by the sign of $\eta_i - \alpha_j$. The greater is β_j the more accurately can we forecast the response of a subject with ability η_i to the j-th item, basing on the sign of $\eta_i - \alpha_j$. Parameter β_j is called the *item discriminating power*. Shown in Fig. 10.2.2 are the graphs of functions $p_1, \ldots, p_4$ in the normal model specified by (10.2.12) with $\beta_1 = 0$, $\beta_2 = 0.01$, $\beta_3 = 2$, $\beta_4 = 100$.

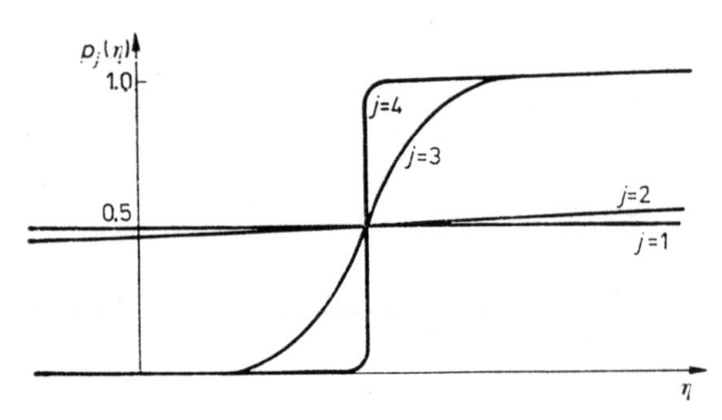

FIG. 10.2.2. Functions p_j ($j = 1, 2, 3, 4$) in the normal model for $m = 1$, for discriminating powers β_j equal to 0, 0.01, 2, 100, respectively, and for identical item difficulties α_j.

Models (10.2.15) and (10.2.16) are extremal for models (10.2.12), (10.2.13) and (10.2.14) what means that affinity between X_{ij} and $(\eta_i - \alpha_j)$ is in these models weakest or strongest, respectively. Model (10.2.16) satisfies assumptions (10.2.7), (10.2.8) and (10.2.9); model (10.2.15) satisfies (10.2.7) and (10.2.9) but does not satisfy (10.2.8).

Sometimes the model (10.2.16) is being modified by taking $p_j(\alpha_j) = 0$, $j = 1, \ldots, k$. Such a model is called the *Guttman model*.

In the sequel we consider certain special parametric and nonparametric models for random or fixed subject group. In parametric models of both types we admit models, in which condition (10.2.10) is satisfied for given m, F, $\boldsymbol{B}$ (including normal, logistic and Rash models). In the case of parametric models with a random subject group, we assume that density h is known completely or up to a finite number of real-valued parameters. In nonparametric models of both types, we take $m = 1$ and assume that $p_1, \ldots, p_k$ satisfy (10.2.7), (10.2.8) and (10.2.9). Regarding nonparametric models with a random subject group we assume h to be the density of an arbitrary continuous real-valued random variable.

In some models, indexing of a family of distributions is not an injection, i.e. the same distribution can correspond to different indices. Therefore, whenever the value of a function of the index is the goal of investigations, it may happen that this goal is unidentifiable. In such a case, it becomes necessary to modify the goal of investigations or to narrow the model by imposing further constraints on the index set.

We now present another way of modelling the subject's behaviour. We introduce what is called *response strength of the i-th subject*, denoted by

$$\boldsymbol{Y}_i = (Y_{i1}, \ldots, Y_{ik})^T,$$

which depends on his ability vector $\tilde{\boldsymbol{\eta}}_i$. We take the response strengths Y_{ij} to be standardized random variables. We assign to item z_j ($j = 1, \ldots, k$) number γ_j, called the *item threshold level*. We assume that the i-th subject correctly answers j-th item if and only if $Y_{ij} \geqslant \gamma_j$. We shall consider only certain special threshold models for a random subject group, in which the density of vector $\tilde{\boldsymbol{\eta}}_i$ is known up to the covariance matrix $\boldsymbol{\Phi}$ (i.e. a family (h_{Φ}) of density functions indexed by an arbitrary covariance matrix $\boldsymbol{\Phi}$ is admitted for consideration). We shall assume additionally that

(i) $m < k$;

(ii) the response strength of the i-th subject depends linearly on the ability up to the random error $\tilde{\boldsymbol{e}}_i$:

$$Y_i = \boldsymbol{\Lambda}\tilde{\boldsymbol{\eta}}_i + \tilde{\boldsymbol{e}}_i, \quad i = 1, \ldots, n, \tag{10.2.17}$$

where $\tilde{\boldsymbol{\eta}}_i = (\tilde{\eta}_{i1}, \ldots, \tilde{\eta}_{im})^T$, $\tilde{e}_i = (\tilde{e}_{i1}, \ldots, \tilde{e}_{ik})^T$, $\Lambda = [\lambda_{jt}]$ is a $k \times m$ matrix of rank m, and $\lambda_{jt} \in \boldsymbol{R}$;

(iii) random errors are mutually independent and do not depend on ability;

(iv) there exist positive numbers $\psi_1, \ldots, \psi_k$ such that

$$\frac{\tilde{e}_{ij}}{\sqrt{\psi_j}} \sim F, \quad i = 1, \ldots, n, \quad j = 1, \ldots, k,$$

where F is the distribution function of a symmetric and continuous standardized univariate distribution.

Let $\boldsymbol{\Psi}$ denote the covariance matrix of $\tilde{\boldsymbol{e}}_i$ (i.e., a diagonal matrix with elements of the diagonal equal to $\psi_1, \ldots, \psi_k$) and let $\Sigma = [\sigma_{st}]$ denote the covariance matrix of $\boldsymbol{Y}_i$. Hence

$$\Sigma = \boldsymbol{\Lambda}\boldsymbol{\Phi}\boldsymbol{\Lambda}^T + \boldsymbol{\Psi}, \tag{10.2.18}$$

and since by assumption $\sigma_{ss} = 1$, $s = 1, \ldots, m$, therefore

$$\psi_j = 1 - \boldsymbol{\lambda}_j \boldsymbol{\Phi} \boldsymbol{\lambda}_j^T, \quad j = 1, \ldots, k, \tag{10.2.19}$$

where $\boldsymbol{\lambda}_j = (\lambda_{j1}, \ldots, \lambda_{jm})$.

Models in which Y_i satisfying (10.2.17) are considered, are called *factor analysis models*. Vector $\tilde{\boldsymbol{\eta}}_i$ in these models is called the *vector of principal factors*, vector $\tilde{\boldsymbol{e}}_i$, the *vector of specific factors* and the matrix Λ, *factor loading matrix*. The described threshold models, in which the response strengths satisfy (10.2.17) will henceforth be called *factor threshold models*.

In such models the phenomenon realization is a sequence of values of random vectors

$$(\tilde{\boldsymbol{\eta}}_1, \ldots, \tilde{\boldsymbol{\eta}}_n, \tilde{\boldsymbol{e}}_1, \ldots, \tilde{\boldsymbol{e}}_n, \boldsymbol{Y}_1, \ldots, \boldsymbol{Y}_n, \boldsymbol{X}_1, \ldots, \boldsymbol{X}_n), \tag{10.2.20}$$

where as previously $\boldsymbol{X}_i = (X_{i1}, \ldots, X_{ik})$ are the responses of i-th individual to k items.

With fixed n, k, m, F and with a fixed family of densities (h_{Φ}), the family of distributions in a factor threshold model is of the form $(P_\theta, \theta \in \boldsymbol{\Theta})$, where $\boldsymbol{\Theta}$ is the set of all triplets

$$\theta = (\gamma, \Lambda, \Phi) \tag{10.2.21}$$

for any $\gamma = (\gamma_1, \ldots, \gamma_k)^T$.

From (10.2.17), the distribution of observable data for $\theta \in \Theta$ is given by the formula

$$P_\theta(X_1 = x_1, \ldots, X_n = x_n) =$$

$$= \prod_{i=1}^{n} \int_{R^m} \prod_{j=1}^{k} \left[F\left(\frac{\lambda_j t - \gamma_j}{\sqrt{\psi_j}}\right)\right]^{x_{ij}} \left[1 - F\left(\frac{\lambda_j t - \gamma_j}{\sqrt{\psi_j}}\right)\right]^{1-x_{ij}} h_\Phi(t)\,\mathrm{d}t. \tag{10.2.22}$$

Let us compare now the family of distributions (10.2.22) with the family of distributions of observable data in the previously considered parametric ability-difficulty model with a random subject group. We shall consider therefore ability-difficulty models, in which for fixed n, k, m, F (10.2.10) is valid for any α_j and β_j belonging to R^m, and vector $\tilde{\eta}$ has a density h_Φ known up to covariance matrix Φ. Let

$$\theta' = (\alpha_1, \ldots, \alpha_k, \beta_1, \ldots, \beta_k, \Phi), \tag{10.2.23}$$

and let Θ' be the set of all such sequences for any $\alpha_1, \ldots, \alpha_k$, $\beta_1, \ldots, \beta_k, \Phi$. The distribution of observable data for $\theta' \in \Theta'$ is given by the formula

$$P'_{\theta'}(X_1 = x_1, \ldots, X_n = x_n) =$$

$$= \prod_{i=1}^{n} \int_{R^m} \prod_{j=1}^{k} [F(\beta_j^T(t-\alpha_j))]^{x_{ij}} [1 - F(\beta_j^T(t-\alpha_j))]^{1-x_{ij}} h_\Phi(t)\,\mathrm{d}t. \tag{10.2.24}$$

We now seek a transformation $\tau\colon\ \Theta \to \Theta'$ such that for any n, k, m, h, F and any vectors $x_1, \ldots, x_n$ for $m < k$

$$P'_{\tau(\theta)}(X_1 = x_1, \ldots, X_n = x_n) = P_\theta(X_1 = x_1, \ldots, X_n = x_n). \tag{10.2.24}$$

It follows from the comparison of (10.2.24) and (10.2.22) that corresponding to any θ is θ' given by the equalities

$$\alpha_j = \frac{\lambda_j^T \gamma_j}{\lambda_j \lambda_j^T}, \quad \beta_j = \frac{\lambda_j^T}{\sqrt{1 - \lambda_j \Phi \lambda_j^T}}. \tag{10.2.26}$$

The transformation τ is an injection, and

$$\Theta'' = \tau(\Theta) = \{(\alpha_1, \ldots, \alpha_k, \beta_1, \ldots, \beta_k, \Phi) \in \Theta'\colon\ \alpha_j = d_j \beta_j \text{ for some number } d_j, j = 1, \ldots, k\}. \tag{10.2.27}$$

For $m = 1$, $\Theta'' = \Theta'$, and for $m > 1$, Θ'' is the proper subset of Θ'.

If we restrict the set of indices to Θ'', then the corresponding family of distributions in the ability-difficulty model is identical with the family of distributions in the factor threshold model.

Going further, if δ is an estimator of θ in the factor threshold model, that is if some $\theta \in \Theta$ is assigned by δ to each sequence $(x_1, \ldots, x_n)$ then there exists an estimator δ' of θ' in the ability-difficulty model (10.2.10), namely

$$\delta'([x_{ij}]) = \tau(\delta([x_{ij}])).$$

The same applies of course to any function of θ. Hence, every estimation problem in the factor threshold model can be equivalently presented as some estimation problem in a suitably restricted ability-difficulty model, and vice versa.

10.3. Inference in parametric models

We shall discuss successively parameter estimation in parametric models with a random subject group and with a fixed subject group.

In the case of a random subject group, we usually consider a factor threshold models and estimate $\theta = (\Lambda, \gamma, \Phi)$. Then, we can use (10.2.26) to estimate θ' in the ability-difficulty model (10.2.10), if we assume that $\theta' \in \Theta''$, where Θ'' is defined by (10.2.27).

We shall confine ourselves to factor threshold models, in which $F = F_{N(0,1)}$ and $\tilde{\eta}_i \sim N_m(0, \Phi)$, $i = 1, \ldots, n$; $m \geqslant 1$.

*

Most commonly in use are estimators by the maximum likelihood method and estimators by the method of least squares. The use of the maximum likelihood method is restricted in practice by numerical problems.

Less complicated numerically is estimation by the method of least squares. We denote

$$\pi_j = P(X_{ij} = 1), \quad i = 1, \ldots, n; \quad j = 1, \ldots, k, \tag{10.3.1}$$

$$\pi_{st} = P(X_{is} = 1, X_{it} = 1),$$
$$i = 1, \ldots, n; \quad s, t = 1, \ldots, k, \quad s \neq t. \tag{10.3.2}$$

In the considered factor threshold model

$$\pi_j = \frac{1}{\sqrt{2\pi}} \int_{\gamma_j}^{\infty} \exp(-\tfrac{1}{2} z^2)\, dz, \tag{10.3.3}$$

$$\pi_{st} = \frac{1}{2\pi |\Sigma_{st}|^{1/2}} \int_{\gamma_s}^{\infty} \int_{\gamma_t}^{\infty} \exp[-\tfrac{1}{2}(z_1, z_2) \Sigma_{st}^{-1} (z_1, z_2)^T]\, dz_1\, dz_2, \tag{10.3.4}$$

where

$$\Sigma_{st} = \begin{bmatrix} 1 & \sigma_{st} \\ \sigma_{st} & 1 \end{bmatrix}, \tag{10.3.5}$$

$$\sigma_{st} = \sum_{u=1}^{m} \sum_{v=1}^{m} \lambda_{su} \lambda_{tv} \varphi_{uv}, \tag{10.3.6}$$

where $\lambda_{su}, \lambda_{tv}, \varphi_{uv}$ are the elements of matrices Λ and Φ respectively. As estimators of probabilities π_j, π_{st}, $j = 1, \dots, k$, $s, t = 1, \dots k$, $s \neq t$, we take

$$\hat{\pi}_j = \frac{1}{n}\sum_{i=1}^{n} X_{ij}, \tag{10.3.7}$$

$$\hat{\pi}_{st} = \frac{1}{n}\sum_{i=1}^{n} X_{is}X_{it}. \tag{10.3.8}$$

Let

$$\xi = (\pi_1, \dots, \pi_k, \pi_{12}, \dots, \pi_{1k}, \pi_{23}, \dots, \pi_{2k}, \dots \pi_{k-1\,k})^T. \tag{10.3.9}$$

$$\hat{\xi} = (\hat{\pi}_1, \dots, \hat{\pi}_k, \hat{\pi}_{12}, \dots, \hat{\pi}_{1k}, \hat{\pi}_{23}, \dots, \hat{\pi}_{2k}, \dots, \hat{\pi}_{k-1\,k})^T, \tag{10.3.10}$$

The difference $\hat{\xi}-\xi$ is a random vector with mean vector $\mathbf{0}$ and a covariance matrix $\operatorname{cov}\hat{\xi}$. Vector ξ and matrix $\operatorname{cov}\hat{\xi}$ are functions of the estimated parameter $\theta = (\Lambda, \gamma, \Phi)$. For example, for $k = 2$

$$\operatorname{cov}\hat{\xi} = \begin{bmatrix} \frac{1}{n}\pi_1(1-\pi_1) & \frac{1}{n}(\pi_{12}-\pi_1\pi_2) & \frac{1}{n}(1-\pi_1)\pi_{12} \\ \frac{1}{n}(\pi_{12}-\pi_1\pi_2) & \frac{1}{n}\pi_2(1-\pi_2) & \frac{1}{n}(1-\pi_2)\pi_{12} \\ \frac{1}{n}(1-\pi_1)\pi_{12} & \frac{1}{n}(1-\pi_2)\pi_{12} & \frac{1}{n}\pi_{12}(1-\pi_{12}) \end{bmatrix}.$$

For fixed θ the expression

$$G(\theta, \hat{\xi}) = (\hat{\xi}-\xi)^T[\operatorname{cov}\hat{\xi}]^{-1}(\hat{\xi}-\xi)$$

is a certain measure of departure of the observed values $\hat{\xi}$ from $\xi(\theta)$. But this is so complicated that it is difficult to find the value $\theta \in \Theta$ for which $G(\theta, \hat{\xi})$ attains the minimal value. For this reason, instead of the function $G(\theta, \hat{\xi})$ we consider the function $\hat{G}(\theta, \hat{\xi})$ replacing π_j, π_{st}, $j = 1, \dots, k$, $s = 1, \dots, k-1$, $t = s+1, \dots, k$ in matrix $\operatorname{cov}\hat{\xi}$ by their estimators $\hat{\pi}_j$ and $\hat{\pi}_{st}$. We estimate parameter θ determining for the observed sequence $x_1, \dots, x_n$ an element of the Θ for which the expression $\hat{G}(\theta, \hat{\xi})$ reaches a minimum.

Such an estimator of θ is consistent and asymptotically efficient in the class of estimators which are functions of vector $\hat{\xi}$.

The computational difficulties in the least squares method can be reduced by taking advantage of the fact that vector ξ given by (10.3.6) depends on $\theta = (\Lambda, \gamma, \Phi)$ only through γ and σ_{st} (defined by (10.3.6)). Let $\varepsilon = (\gamma, (\sigma_{st});\ s = 1, \dots, k-1,\ t = s+1, \dots, k)$ and let $\hat{\varepsilon}$ be the estimator of ε obtained from equations (10.3.3) and (10.3.4) modified by replacing π_j in (10.3.3) and π_{st} in (10.3.4) by their estimators $\hat{\pi}_j$ and $\hat{\pi}_{st}$, defined above. We seek then the estimators of $\theta = (\Lambda, \gamma, \Phi)$ for which $(\hat{\varepsilon}-\varepsilon)^T W^{-1}(\hat{\varepsilon}-\varepsilon)$ reaches a minimum, where W is a suitably chosen matrix dependent on $\hat{\xi}$.

The least squares methods just described can be generalized by adding to ξ the probabilities

$$\pi_{s,t,u} = P(X_{is} = X_{it} = X_{iu} = 1), \quad i = 1, \ldots, n, \ s, t, u = 1, \ldots k, \quad s \neq t \neq u.$$

This would cause, however, considerable numerical complications.

*

In parametric ability-difficulty models with a fixed subject group, we shall confine ourselves to the case when $m = 1$. The unknown parameter is

$$(\eta_1, \ldots, \eta_n, \alpha_1, \ldots, \alpha_k, \beta_1, \ldots, \beta_k). \tag{10.3.11}$$

The number of components of this parameter increases with the number of subjects. We are therefore confronted with a more difficult situation than in the case of simple random samples, where with increasing size of sample, a better estimation of parameters becomes possible, e.g., in estimation by the maximum likelihood method. The asymptotic properties of maximum likelihood estimators of parameter (10.3.11) with $n \to \infty$ are unsatisfactory (e.g., in the Rash model with $k = 1$, such estimators of item difficulty are inconsistent). On the other hand, numerical problems grow with n.

Consequently other estimating methods than the maximum likelihood method have to be sought.

Suppose first that we want to estimate only $\alpha_1, \ldots, \alpha_k, \beta_1, \ldots, \beta_k$ with $\eta_1, \ldots, \eta_n$ treated as nuisance parameters. In models, in which there exists a sufficient statistic for nuisance parameters, (α_j) and (β_j) can be estimated by the method of conditional maximum likelihood given in Section 2.3. But a sufficient statistic for $\eta_1, \ldots, \eta_n$ exists only in the Rash model (defined by (10.2.14)). The likelihood function in this model can be written in the form

$$L(\alpha_1, \ldots, \alpha_k, \eta_1, \ldots, \eta_n, \boldsymbol{x}_1, \ldots, \boldsymbol{x}_n) = \frac{\exp\left(\sum_{i=1}^{n} \eta_i x_{i\cdot}\right)}{\prod_{i=1}^{n} \prod_{j=1}^{k} (1+\exp(\eta_i - \alpha_j))} \exp\left(-\sum_{j=1}^{n} \alpha_j x_{\cdot j}\right), \tag{10.3.12}$$

where $x_{i\cdot} = \sum_{j-1}^{k} x_{ij}$, $x_{\cdot j} = \sum_{i=1}^{n} x_{ij}$. The likelihood function will not change, when we add an arbitrary constant to each η_i and each α_j, $i = 1, \ldots, n$, $j = 1, \ldots, k$. Parameters $\alpha_1, \ldots, \alpha_k$ and $\eta_1, \ldots, \eta_n$ become identifiable when a linear constraint is imposed on them. We usually take

$$\sum_{j=1}^{k} \alpha_j = 0. \tag{10.3.13}$$

It is seen from (10.3.12) (on the basis of the factorization theorem) that a vector $T = (X_{1\cdot}, \ldots, X_{n\cdot})$ is sufficient for the vector of nuisance parameters $(\eta_1, \ldots, \eta_n)$. The conditional likelihood function has the form

$$\begin{aligned} & L_W(\alpha_1, \ldots, \alpha_k, t_1, \ldots, t_n, \boldsymbol{x}_1, \ldots, \boldsymbol{x}_n) \\ & = P(\boldsymbol{X}_1 = \boldsymbol{x}_1, \ldots, \boldsymbol{X}_n = \boldsymbol{x}_n \mid T_1 = t_1, \ldots, T_n = t_n) \\ & = \frac{\prod_{i=1}^{n} \exp\left(-\sum_{j=1}^{k} \alpha_j x_{ij}\right)}{\sum_{\{x_{ij}:\, x_{i\cdot}=t_i\}} \left(-\sum_{j=1}^{k} \alpha_j x_{ij}\right)} \end{aligned} \qquad (10.3.14)$$

We obtain the conditional maximum likelihood estimators by maximizing (10.3.14) with respect to $\alpha_1, \ldots, \alpha_k$, taking into account (10.3.13). The likelihood equations are relatively simply solved by approximate methods. The estimators of parameters $\alpha_1, \ldots, \alpha_k$ thus obtained are consistent and asymptotically normal. Next, with known $\alpha_1, \ldots, \alpha_k$, we can estimate (η_i) by the maximum likelihood method seeking the maximum value of function (10.3.12), i.e., an equivalently maximum value of the function

$$P(T_1 = t_1, \ldots, T_n = t_n) = \frac{\prod_{i=1}^{n} \exp(\eta_i t_i) \sum_{\{x_{ij}:\, x_{i\cdot}=t_i\}} \left(-\sum_{j=1}^{k} \alpha_j x_{ij}\right)}{\prod_{j=1}^{k} \left(1+\exp(\eta_i - \alpha_j)\right)} \qquad (10.3.15)$$

for $(\eta_1, \ldots, \eta_n) \in \boldsymbol{R}^n$. With a fixed number of subjects and with the number of items tending to infinity, such estimators are consistent and asymptotically normal. The properties of thus adapted estimators of parameters (η_i) are unknown.

10.4. Inference in nonparametric models

In ability and difficulty testing the choice of the parametric model is usually based on tradition or is related to the actual possibilities of carrying out the calculations. More suitable in practice appears to be a nonparametric description of the phenomenon. In keeping with Section 10.2, we shall consider models, in which $m = 1$ and the functions $p_1, \ldots, p_k$ satisfy (10.2.7), (10.2.8) and (10.2.9), with $\alpha_s \neq \alpha_t$ for $s \neq t$, and the subject group is a simple random sample, drawn from a population according to density h. We seek information about item difficulties $\alpha_1, \ldots, \alpha_k$. However, the vector $(\alpha_1, \ldots, \alpha_k)$ cannot be taken for the goal of investigations, since under so general assumptions concerning $p_1, \ldots, p_k$ it is unidentifiable. Consequently, the goal has to be suitably modified.

We shall consider problems, in which the goal of investigations is the ordering of difficulties according to increasing values. In accordance with the definition of identifiability given by (1.2.1) the goal thus stated is identifiable if for any different $(\alpha_1, \ldots, \alpha_k, p_1, \ldots, p_k, h)$ and $(\alpha_1', \ldots, \alpha_k', p_1', \ldots, p_k', h')$ the equality of distributions of $X_1, \ldots, X_n$ implies that item difficulties are identically ordered in both cases.

But obviously, following from the equality of two distributions of observable data is the equality of the corresponding vectors $\boldsymbol{\pi} = (\pi_1, \ldots, \pi_k)$ and $\boldsymbol{\pi}' = (\pi_1', \ldots, \pi_k')$ where

$$\pi_j = \int_{-\infty}^{+\infty} p_j(z)h(z)\,dz,$$
$$\pi_j' = \int_{-\infty}^{+\infty} p_j'(z)h'(z)\,dz \tag{10.4.1}$$

are the probabilities of a correct answer to j-th item, for the first and the second distribution respectively. The ordering of the coordinates of vectors $\boldsymbol{\pi}$ and $\boldsymbol{\pi}'$ is therefore identical, hence, by (10.4.1) and (10.2.8), also identical is the ordering of functions $(p_1, \ldots, p_k)$ and $(p_1', \ldots, p_k')$. Therefore, the ordering of difficulties $(\alpha_1, \ldots, \alpha_k)$ and $(\alpha_1', \ldots, \alpha_k')$ is identical, too.

The ordering of items according to increasing difficulties is equivalent to the ordering of probabilities $\pi_1, \ldots, \pi_k$ according to decreasing values. Hence, a natural estimation of the first named ordering is based on $\hat{\pi}_1, \ldots, \hat{\pi}_k$ (cf. (10.3.7)). If for some permutation $j_1, \ldots, j_k$ of $1, \ldots, k$

$$\hat{\pi}_{j_1} > \ldots > \hat{\pi}_{j_k},$$

then we state that $\alpha_{j_1} < \ldots < \alpha_{j_k}$, otherwise our decision is suspended. We now present the properties of this estimator for $k = 2$. Let

$$Z^{(n)} = \hat{\pi}_1 - \hat{\pi}_2. \tag{10.4.2}$$

The estimator of the ordering assigns the decision "$\alpha_1 < \alpha_2$" when $Z^{(n)} > 0$, decision "$\alpha_1 > \alpha_2$", when $Z^{(n)} < 0$ and suspends the decision, when $Z^{(n)} = 0$. We evaluate the quality of the estimator by the probability of a wrong decision, equal to $P(Z^{(n)} < 0)$ when $\alpha_1 < \alpha_2$ or to $P(Z^{(n)} > 0)$ when $\alpha_1 > \alpha_2$ and by the probability of a suspended decision $P(Z^{(n)} = 0)$.

Let us assume that $\alpha_1 < \alpha_2$ and denote by A, B, C the probabilities of wrong, correct and suspended decision, respectively, in the case $n = 1$:

$$A = P(Z^{(1)} < 0) = \pi_2 - \pi_{12}, \tag{10.4.3}$$
$$B = P(Z^{(1)} > 0) = \pi_1 - \pi_{12}, \tag{10.4.4}$$
$$C = P(Z^{(1)} = 0) = 1 - \pi_1 - \pi_2 + 2\pi_{12}, \tag{10.4.5}$$

where

$$\pi_{12} = \int_{-\infty}^{+\infty} p_1(z)p_2(z)h(z)\,dz. \tag{10.4.6}$$

Then

$$P(Z^{(n)} < 0) = \sum_{s=1}^{n} \binom{n}{s} C^{n-s} \sum_{t=0}^{[(s-1)/2]} \binom{s}{t} A^{s-t}B^t, \tag{10.4.7}$$

$$P(Z^{(n)}) > 0) = \sum_{s=1}^{n} \binom{n}{s} C^{n-s} \sum_{t=[s/2]+1}^{s} \binom{s}{t} A^{s-t} B^t, \tag{10.4.8}$$

$$P(Z^{(n)} = 0) = \sum_{s=0}^{n} \binom{n}{s} C^{n-s} \cdot f(s, A, B), \tag{10.4.9}$$

where

$$f(s, A, B) = \begin{cases} \binom{s}{s/2} A^{s/2} B^{s/2} & \text{for even } s, \\ 0 & \text{for odd } s. \end{cases}$$

Hence, for any n, the probabilities of a wrong, correct and suspended decision depend only on probabilities A, B and C.

With $n = 1, 2, 3$, we can demonstrate that for arbitrary fixed π_1 and π_2, the probability of a wrong decision (cf. (10.4.7)) is a decreasing function of π_{12}. For $n > 3$, there is no proof as yet of the probability of a wrong decision being monotonic but this has been verified numerically on many examples.

Note that if (10.2.7) and (10.2.8) occur and α_1 and α_2, then

$$\pi_1 \pi_2 < \pi_{12} \leqslant \pi_2. \tag{10.4.10}$$

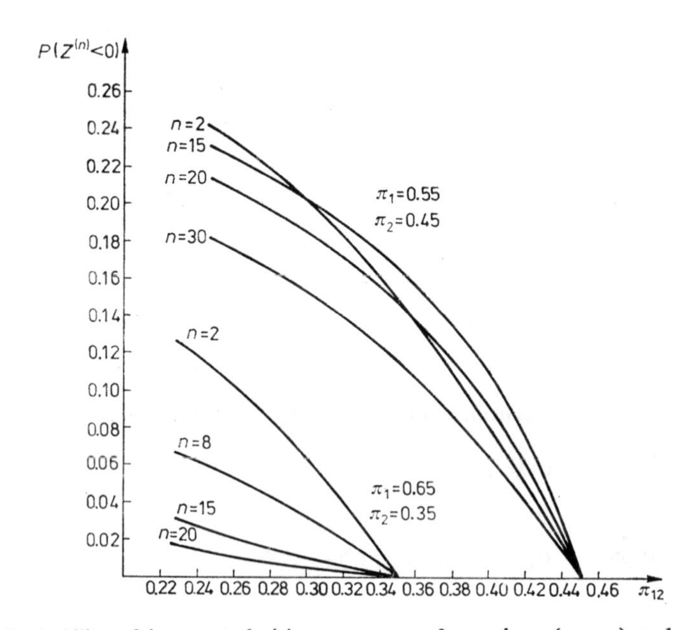

FIG. 10.4.1. Probability of incorrect decision versus π_{12} for various (π_1, π_2) and various n.

The graphs of the probability of wrong decision as a function of π_{21} are presented in Fig. 10.4.1 for some pairs (π_1, π_2) and some n.

If we retain assumptions (10.2.7) and (10.2.9) but weaken assumption (10.2.8) in the manner given below

$$\alpha_s < \alpha_t \iff p_s(\eta) \geqslant p_t(\eta) \text{ for each } \eta \in \boldsymbol{R}, \tag{10.4.11}$$

then

$$\pi_1 \pi_2 \leqslant \pi_{12} \leqslant \pi_2 \tag{10.4.12}$$

and the equality $\pi_1 \pi_2 = \pi_{12}$ occurs if and only if we deal with a random response model, i.e.,

$$p_1(\eta) = p_2(\eta) \equiv \tfrac{1}{2}.$$

Let us consider the other extreme situation in the inequalities (10.4.10) and (10.4.12):

$$\pi_{12} = \pi_2 .$$

Under (10.2.7), (10.2.8) and (10.2.9), this equality occurs if and only if the sets $\{\eta\colon 0 < p_1(\eta) < 1\}$ and $\{\eta\colon 0 < p_2(\eta) < 1\}$ are disjoint (cf. Fig. 10.4.1). In these models, which we shall call here *modified Guttman models*, it never happens that the more difficult item is answered correctly while the less difficult one is answered incorrectly. Note that in the Guttman model defined in Section 10.2, the equality $\pi_{12} = \pi_2$ holds but the model is left out of the considerations in this section because the assumption (10.2.9) is not satisfied.

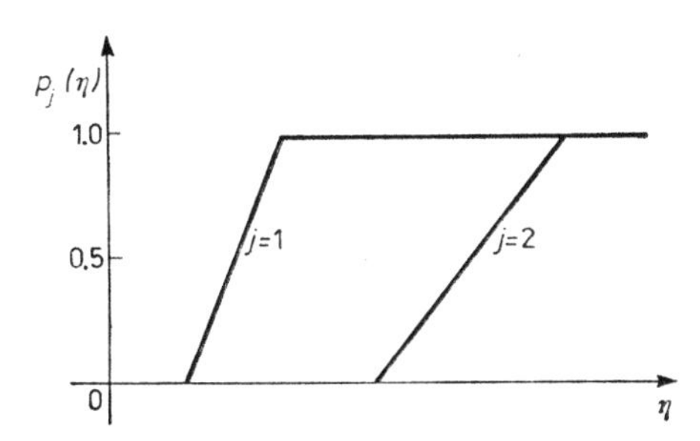

FIG. 10.4.2. Functions p_1, p_2 in Guttman modified model.

In the random response model, the probability of a wrong decision and the probability of a correct decision tend with $n \to \infty$ to $\frac{1}{2}$ (and the probability of a suspended decision tends to zero). In the modified Guttman model, the probability of a wrong decision is equal to 0 and the probability of a correct decision tends to 1 with $n \to \infty$. In the remaining cases ($\pi_1 \pi_2 < \pi_{12} < \pi_2$), the probability of a wrong decision tends to 0 and the probability of a correct decision tends to 1.

For $k > 2$, the situation is more complicated. We shall consider only the case when $\alpha_1 < \ldots < \alpha_k$, since for other permutations $j_1, \ldots, j_k$ of indices $1, \ldots, k$ the considerations are analogous.

The probability of a correct decision equals

$$P(\hat{\pi}_1 > \ldots > \hat{\pi}_k),$$

the probability of a suspended decision equals

$$P\big(\hat{\pi}_1 \geqslant \ldots \geqslant \hat{\pi}_k \wedge \sim(\hat{\pi}_1 > \ldots > \hat{\pi}_k)\big),$$

and the probability of a wrong decision equals

$$P\big((\hat{\pi}_1 < \hat{\pi}_2) \vee (\hat{\pi}_2 < \hat{\pi}_3) \vee \ldots \vee (\hat{\pi}_{k-1} < \hat{\pi}_k)\big).$$

Note that to order k items, at least $k-1$ subjects are needed, since otherwise the probability of a correct decision equals zero.

For any $k \geqslant 2$ and $n \geqslant k-1$ the probabilities of a wrong, correct and suspended decision depend on $\pi_{j_1, \dots, j_r}$, $1 \leqslant r \leqslant k$, $j_1 < \dots < j_r$, where

$$\pi_{j_1, \dots, j_r} = \int_{-\infty}^{+\infty} p_{j_1}(z) p_{j_2}(z) \dots p_{j_r}(z) h(z) \, dz .$$

Yet, for any n and k, the formulas expressing these probabilities are complicated owing to the fact that:

(i) n subjects can answer k items in 2^{nk} ways,

(ii) $\hat{\pi}_1, \dots, \hat{\pi}_k$ are not independent.

In the random response model, the probabilities of a correct and a wrong decision tend, with $n \to \infty$, to $1/k!$ and to $1-1/k!$ respectively, and the probability of a suspended decision tends to 0. For an arbitrary model satisfying (10.2.7), (10.2.8) and (10.2.9), the asymptotic properties of the considered probabilities are exactly the same as for $k = 2$.

10.5. Final remarks

We have confined ourselves in this chapter to the study of latent variables using items with two possible answers: correct or wrong. Models, in which the correctness of an answer is evaluated at more than two stages, are discussed, for example, by Andersen (1972), (1973a), (1973b), (1973c), (1980), Bock (1972) and Rash (1961).

In Section 10.2, we consider ability-difficulty models, in which p_j is of the form (10.2.10). The normal model was introduced by Lawley (1943) and furthered by Lord (1952), (1953). The logistic model (10.2.13) was proposed by Birnbaum (1968) and the model (10.2.14) by Rash (1961). The interpretation of parameters in these models seems to be natural and in harmony with intuition. In the literature, we meet models, in which an item is described by several parameters lacking an interpretation of this type. For example, Lazarsfeld (1950, p. 371) considered models with functions $p_1, \dots, p_k$ in the form of polynomials of a known degree for these η, for which the polynomial value belongs to interval (0, 1). The polynomial coefficients are the unknown parameters describing the item. Lazarsfeld (1950, p. 410) and Torgerson (1958, p. 374) considered in place of the polynomial the function $g_j(\eta) = \alpha_{0j} - \alpha_{1j}(1 - 2\chi_{\{\eta - \alpha_{2j} \geqslant 0\}})$ and Lazarsfeld (1959) considered the function $g_j(\eta) = \alpha_{0j} - \alpha_{1j} \exp(\alpha_{2j} \eta)$, where $\alpha_{0j}, \alpha_{1j}, \alpha_{2j}$ are the unknown parameters of the item.

If any item is answered so that the subject selects one of several admissible answers, amongst which exactly one is correct, then the expression (10.2.10) is usually modified as follows

$$p_j(\eta) = c_j + (1 - c_j) F(\beta_j^T(\eta - \alpha_j)),$$

where c_j is the reciprocal of the number of possible answers to j-th item (Birnbaum, 1968; Kolakowski and Bock, 1970; Samejima 1973).

Samejima (1974) discusses the relationships between ability-difficulty models and factor threshold models. But it does not follow clearly from this study, which problems can be treated as equivalent for the two models.

The estimation of parameters in parametric models was illustrated in Section 10.3 with the use of several commonly employed methods for a random or fixed subject group. In the earliest contributions concerned with random subject groups (Lord, 1952; Indow *et al.*, 1962), factor threshold models with $m = 1$ were considered. Attempts were made to use factor analysis methods, with the covariance matrix estimated by means of the sample tetrachoric correlation matrix (Sec. 5.6). However, such an estimator of the covariance matrix is unsatisfactory for several reasons (Gourlay, 1951). Maximum likelihood estimation for $m = 1$ (and the related numerical difficulties) were discussed by Bock and Lieberman (1970). A method of least squares for $m = 1$ was given by Christoffersson (1975); a modification of it was proposed by Muthén (1978). The maximum likelihood method in models with a fixed subject group for (10.2.12) is discussed by Tucker (1951) and Lord (1953), and also by Birnbaum (1968) for (10.2.13). Such estimators may have unsatisfactory asymptotic properties, as confirmed by the example given by Andersen (1973a), in which it is demonstrated that in item difficulty estimation in the Rash model, maximum likelihood estimators with $k = 2$ are inconsistent. On the other hand, Haberman (1977) proved that if certain assumptions are satisfied and the number of items and the number of subjects grow to infinity, then the maximum likelihood estimators of item difficulty are consistent. The conditional maximum likelihood method is described in the studies of Andersen (1970), (1972), (1973a).

Meriting attention are the estimation methods with deferred decision considered in some contributions (Bock and Lieberman, 1970; Christoffersson, 1975; Andersen, 1973b). Generally speaking, they are applicable to a situation, in which the phenomenon cannot be described by the considered parametric model, but departures from this model are not very large. Statistic is being sought to measure the departure from the parametric model and a certain threshold level (constant or random) is fixed. If this threshold is exceeded, no estimation is carried out, otherwise some estimator recognized as suitable for the considered parametric model is used.

The nonparametric models presented in Section 10.4 are discussed by Mokken (1971), Stokman and van Schuur (1980) and Gáfriková (1987). Specifically, the parameter

$$H_{1,2}(\alpha_1, \alpha_2, p_1, p_2, h) = \begin{cases} \dfrac{\pi_{12}-\pi_1\pi_2}{(1-\pi_1)\pi_2} & \text{if } \alpha_1 < \alpha_2, \\ \dfrac{\pi_{12}-\pi_1\pi_2}{(1-\pi_2)\pi_1} & \text{if } \alpha_1 > \alpha_2, \end{cases} \tag{10.5.1}$$

is considered in Mokken's study. Under the assumptions (10.2.7), (10.2.8) and (10.4.11), it assumes values in the interval [0, 1] and it equals 0 if and only if $\pi_{12} = \pi_1\pi_2$ (i.e., in a random response model) and equals 1 if and only if $\pi_{12} = \pi_2$ (i.e., in modified Guttman models). For any fixed π_1, π_2, the ordering of models according to the value of $H_{1,2}$ is equivalent to the ordering of these models according to the probability of a wrong decision. Parameter $H_{1,2}$ can therefore be treated then as a measure of departure from the random response model.

Parameter $H_{1,2}$ cannot be estimated directly from a sample by means of $\hat{\pi}_1$, $\hat{\pi}_2$, $\hat{\pi}_{12}$ because the ordering of difficulties α_1 and α_2 is unknown. It is, therefore, necessary first to order the difficulties on the basis of $Z^{(n)}$ and only then define

$$\hat{H}_{1,2} = \begin{cases} \dfrac{\hat{\pi}_{12}-\hat{\pi}_1\hat{\pi}_2}{(1-\hat{\pi}_1)\hat{\pi}_2}, & \text{if } \hat{\pi}_1 > \hat{\pi}_2, \\ 0, & \text{if } \hat{\pi}_1 = \hat{\pi}_2, \\ \dfrac{\hat{\pi}_{12}-\hat{\pi}_1\hat{\pi}_2}{(1-\hat{\pi}_2)\hat{\pi}_1}, & \text{if } \hat{\pi}_1 < \hat{\pi}_2. \end{cases}$$

Mokken finds an asymptotic distribution of the expression $\hat{\pi}_{12}-\hat{\pi}_1\hat{\pi}_2$ and proposes to suspend the decision of the ordering of items, when there are no grounds to reject the hypothesis that $\pi_{12} = \pi_1\pi_2$.

For k items with $k > 2$, we consider parameter $H_{1,\dots,k}$ which for $\alpha_1 < \dots < \alpha_k$ is of the form

$$H_{1,\dots,k}(\alpha_1, \dots, \alpha_k, p_1, \dots, p_k, h) = \frac{\sum_{s=1}^{k-1}\sum_{t=s+1}^{k}[\pi_{st}-\pi_s\pi_t]}{\sum_{s=1}^{k-1}\sum_{t=s+1}^{k}\pi_s(1-\pi_t)} \tag{10.5.2}$$

and for any other ordering of item difficulties, it is defined by an obvious modification of (10.5.2). Under the assumptions (10.2.7), (10.2.8) and (10.4.11), parameter $H_{1,\dots,k}$ assumes value in the interval $[0, 1]$, with $H_{1,\dots,k} = 0$ if and only if we deal with a random response model, and $H_{1,\dots,k} = 1$ if and only if we deal with a modified Guttman model.

For $k > 2$ and for fixed $\pi_1, \dots, \pi_k$ the ordering of models according to the increasing values of $H_{1,\dots,k}$ is not equivalent to the ordering of models according to the increasing probabilities of a wrong decision. Parameter $H_{1,\dots,k}$ for fixed $\pi_1, \dots, \pi_k$ has the same value for different sets $\{\pi_{st}: s < t,\ s, t = 1, \dots, k\}$ and moreover the probability of a wrong decision for $n \geqslant k-1$ depends on $\pi_{j_1,\dots,j_r}$, $2 \leqslant r \leqslant k$, $j_1 < \dots < j_r$. It follows that $H_{1,\dots,k}$ does not measure the departure of the model from the random response model even for fixed $\pi_1, \dots, \pi_k$.

*

The present chapter though it covers only part of the investigations concerned with latent variables in experimental psychology, illustrates the procedures used in such investigations. The descriptions of many psychological investigations would be, it seems, more lucid, if we consistently separated out within them a description of the experiment itself, a description of the goal of investigations and a description of the inferring method used. For it should be said that the links of psychological investigations with statistics are full of reticences and frequently psychologists tend to make statistics a fetish, which hinders proper planning of the experiment and proper evaluation of the results. Present-day textbooks of methodology in psychological research are proof of the great importance attached to statistics in this field (e.g. Brzeziński, 1984; Nowakowska, 1975).

Design of research instruments holds a special place in psychological research methodology, especially in psychological tests (intelligence, special abilities, personality, knowledge). This theme has altogether been left out in this chapter. No reference was made to the design of the test items to be answered, the merit criteria of tests (accuracy, reliability, etc.) or lastly the item answering process itself. Besides, only binary tests were considered.

Psychological research methodology is linked with the measurement theory, as emphasized in contemporary monographs of this theory (e.g. Pfanzagl, 1968; Roberts, 1978; Krantz *et al.*, 1971). Commonly used in textbooks of experimental psychology are measurement scales, which were introduced by Stevens (1951), namely nominal, ordinal, interval and ratio scales. However, the precision with which the measurement scale and other concepts of theory of measurement are defined in psychological literature leaves much to be desired. The psychologist must usually fall back on intuitive definitions.

Measurement scales are of paramount importance in defining latent variables. Unfortunately, only some very special problems are considered in the literature on latent trait analysis. As regards subject ability and item difficulty testing (for items which can be answered either correctly or incorrectly), the Guttman scale is proposed (Guttman, 1944; Stouffer *et al.*, 1950; Ducamp and Falmagne, 1969). Let us recall the definition of the Guttman scale, considering a somewhat more general situation.

Let us start from introducing an empirical relational structure for subject item groups jointly, since what we aim at is to suitably interrelate abilities and difficulties. Therefore, the set of interest to us is $\boldsymbol{O} \cup \boldsymbol{Z}$, where $\boldsymbol{O}$ is the subject group and $\boldsymbol{Z}$ the item group. Defined on the product $\boldsymbol{O} \times \boldsymbol{Z}$ is a function q with values in the interval [0, 1] whose value $q(o, z)$ we interpret as the probability of item z being correctly answered by subject o.

In the deterministic case, q assumes only one of two values, 0 or 1.

Let $R_{\succ}$ be a subset of the set $\boldsymbol{O} \times \boldsymbol{Z}$, such that

$$R_{\succ} = \{(o, z) \in \boldsymbol{O} \times \boldsymbol{Z} \colon q(o, z) > \tfrac{1}{2}\}.$$

By inequality $q(o, z) > \frac{1}{2}$ we understand that subject o has predispositions to correct approach to z. We consider the relational structure $(\boldsymbol{R}^2, R_{>})$, where

$$R_{\succ} = \{(v_1, v_2) \in \boldsymbol{R}^2 \colon v_1 > v_2\}$$

and seek a homomorphism $(f_1, f_2)\colon (\boldsymbol{O} \times \boldsymbol{Z}, R_{\succ}) \to (\boldsymbol{R}^2, R_{>})$. Function f_1 assigns to subject o the number $\eta = f_1(o)$ interpreted as the ability of subject o; function f_2 assigns to item z the number $\alpha = f_2(z)$ interpreted as the difficulty of this item. In the deterministic case, the homomorphism (f_1, f_2) is called the *Guttman scale.*

Moreover, let us assume that items from set $\boldsymbol{Z}$ are compatibly answered by subject from set $\boldsymbol{O}$ in the following sense:

$$\begin{aligned} \exists z \in \boldsymbol{Z} \ q(o,z) > q(o',z) &\Rightarrow \forall z \in \boldsymbol{Z} \ q(o,z) \geqslant q(o',z), \\ \exists o \in \boldsymbol{O} \ q(o,z) < q(o,z') &\Rightarrow \forall o \in \boldsymbol{O} \ q(o,z) \leqslant q(o,z'). \end{aligned} \tag{10.5.3}$$

By means of q we determine the following binary relation $\boldsymbol{R}_0$ in $\boldsymbol{O} \cup \boldsymbol{Z}$:

$$\begin{aligned} oR_0 z &\Leftrightarrow q(o,z) > \tfrac{1}{2}, \\ zR_0 o &\Leftrightarrow q(o,z) < \tfrac{1}{2}, \\ oR_0 o' &\Leftrightarrow \exists z \in \boldsymbol{Z} \ q(o,z) > q(o',z), \\ zR_0 z' &\Leftrightarrow \exists o \in \boldsymbol{O} \ q(o,z) < q(o,z'). \end{aligned} \tag{10.5.4}$$

For relational structures $\mathscr{R}_0 = (\boldsymbol{O} \cup \boldsymbol{Z}, \boldsymbol{R}_0)$, $\mathscr{R} = (\boldsymbol{R}, >)$, we seek the homomorphism $f: \mathscr{R}_0 \to \mathscr{R}$. This means that we want suitably to interrelate the measurement of ability and difficulty. Let f_1 and f_2 be the restrictions of f in respect to $\boldsymbol{O}$ and $\boldsymbol{Z}$, and $\boldsymbol{R}_{01}$ and $\boldsymbol{R}_{02}$ be the restrictions of $\boldsymbol{R}_0$ to these sets. It follows from the definition of $\boldsymbol{R}_0$ (cf. (10.5.4)) that if f is a homomorphism from $\mathscr{R}_0$ to $\mathscr{R}$ then f_1 is a homomorphism from $(\boldsymbol{O}, \boldsymbol{R}_{01})$ to $(\boldsymbol{R}, >)$ and f_2 is a homomorphism from $(\boldsymbol{Z}, \boldsymbol{R}_{02})$ to $(\boldsymbol{R}, >)$ and the pair (f_1, f_2) is a homomorphism from $(\boldsymbol{O} \times \boldsymbol{Z}, \boldsymbol{R}_{\succ})$ to $(\boldsymbol{R}^2, \boldsymbol{R}_{>})$.

In view of (10.5.3), $\boldsymbol{R}_0$ is asymmetric and negatively transitive, i.e., for any $a, b \in \boldsymbol{O} \cup \boldsymbol{Z}$ $aR_0 b \Rightarrow \sim bR_0 a$ and for any $a, b, c \in \boldsymbol{O} \cup \boldsymbol{Z}$ $\sim aR_0 b \wedge \sim bR_0 c \Rightarrow \sim aR_0 c$. Hence (Roberts, 1979, Theorem 3.1), there exists the homomorphism $f: \mathscr{R}_0 \to \mathscr{R}$ and it is of the form $f(a) = \#\{b \in \boldsymbol{O} \cup \boldsymbol{Z}: aR_0 b\}$ and f_1, f_2 are ordinal scales.

If the answer is deterministically related to any pair (o, z) then f_1 assigns to each subject $o \in \boldsymbol{O}$ the sum of the number of items answered correctly by subject o and of the number of subjects who answered correctly less items than did subject o, and function f_2 assigns to each item $z \in \boldsymbol{Z}$ the sum of the number of subjects who answered it correctly and of the number of items which were answered correctly by a lesser number of subjects than was item z. Such definitions of functions f_1 and f_2 seem to be somewhat artificial in practice; however, they make it possible to link the measurement of subject ability and item difficulty. When we consider separately the set $\boldsymbol{O}$, we may define the function f_1' assigning to each subject $o \in \boldsymbol{O}$ the number of items correctly answered by him. Analogously for the set $\boldsymbol{Z}$ we may define function f_2' assigning to each item $z \in \boldsymbol{Z}$ the number of subjects who answered it correctly. Function f_1' is a homomorphism from $(\boldsymbol{O}, \boldsymbol{R}_{01})$ to $(\boldsymbol{R}, >)$ and function f_2' is a homomorphism from $(\boldsymbol{Z}, \boldsymbol{R}_{02})$ onto $(\boldsymbol{R}, >)$, but the pair (f_1', f_2') is not a homomorphism from $(\boldsymbol{O} \times \boldsymbol{Z}, \boldsymbol{R}_{\succ})$ to $(\boldsymbol{R}^2, \boldsymbol{R}_{>})$.

It follows from the foregoing considerations that ability and difficulty measurement on ordinal scale can be introduced in a natural way. However, the parametric models discussed in Section 10.3 cannot be considered with the ordinal scales for abilities and difficulties. Admissible on the other hand, are the nonparametric models discussed in Section 10.4. Intuitively, it is quite clear that measurement scales should play a fundamental role in formulating statistical problems involving ability and difficulty. But this aspect is but mentioned in the literature.

Finally, let us mention screening and classification problems which concern a latent trait measured on at least ordinal scale and exploit parallel opinions by different experts. More precisely, we aim at classifying an object to two classes (in one of them the values of the latent trait are smaller than a chosen quantile of the latent trait distribution, in the other one they are greater). Learning sample consists of the sequence of opinions about the value of the latent trait, given by two or more expert. It is usually assumed that the experts deliver their opinions independently of each other according to the same distribution. Thus, the opinions are conditionally independent given the value of the latent trait. Under these assumptions the type of dependence between opinion of one of the experts and the latent trait in question implies the type of dependence between parallel opinions of both experts. Holland and Rosenbaum (1986) gave an overview of this kind of dependence problems.

Chapter 11

Queueing models of computer systems

Key words: *computer system performance evaluation, queueing models, queueing theory, multi-access computer systems, think time, service time, response time, stretch factor.*

11.1. INTRODUCTION

Suppose that a new computer centre is being planned with the aim of providing service in the field of data processing for a certain group of customers. The core of such a centre is a computer device, i.e. a set of hardware and software resources. The job of the designer is to develope a computer system which would satisfy requirements meeting the tasks facing the centre. There are various possible performance evaluation criteria, e.g.:

(i) user convenience,
(ii) reliability,
(iii) responsiveness,
(iv) efficiency.

We are interested here in operational criteria, i.e. criteria related to specified operating conditions and reflecting the massive aspect of processing, in contrast to technical criteria, which are essentially independent of the system load created by the customers (with the exception of the most general qualitative specification of the load). Among the above mentioned criteria, (i) and (ii) represent technical criteria, and (iii) and (iv) represent operational ones.

Evaluation criteria may take the form of constraints (feasibility conditions) or objective functions (representing the measure of efficiency of the system). A typical problem for the analyst usually lies somewhere between verifying whether the planned or already existing system satisfies the feasibility conditions and finding a solution which would optimize the efficiency of the system. In particular, it may turn out that the imposed constraints are inconsistent and there exists no feasible solution. Thus the problem of computer system analysis consists in evaluating several variants and may necessitate complex investigations revealing the relations between various parameters of the system and the population of its users. Since the behaviour of the users is seldom known with sufficient accuracy, in particular before a given system is put into operation, and moreover since it may vary considerably in time, the evaluation is often performed for the already existing system with the aim of modifying (tuning) it.

The basic methods of evaluating computer systems are based on observation and modelling. By a model we understand here a precise description of the relationship between the quantities sought for and the set of parameters representing the properties of the system and its users which is essential for the investigations in question. Every model is a certain idealization; its construction is inherently associated with the following dilemma: simplicity versus accuracy. A model must be sufficiently simple to enable one to determine the values of the unknowns (an important factor here is the amount of computation required to obtain the solution) but also it must be sufficiently accurate to reflect the relevant features of the system. Solving the above dilemma is always a matter of a tradeoff and requires a thorough knowledge of the system in question as well as the methods to be applied.

Among the different types of models used to describe computer systems stochastic models play the major role. In a stochastic model the behaviour of the users and processing data are regarded as certain random processes and the evaluation criteria are formulated in terms of certain characteristics of those processes. In some cases the quantities sought for can be derived analytically (possibly with some help of numerical techniques). If analytical methods fail, then the simulation approach must be used. The essence of this approach is the construction of an algorithm (in the form of a computer program) able to generate pseudorealizations of the random process considered. The data generated by such an algorithm are then subjected to an appropriate statistical treatment in order to estimate the required characteristics.

According to the type of the method applied, stochastic models are classified as analytical or simulation models. Most analytical models are based on the queueing theory which has proved particularly useful for describing the processes taking place in computer systems.

In addition to modelling, we should also stress the role of the observations of real systems, which is indispensable for assessing the input parameters of a model by means of the relevant statistical procedures.

This chapter is addressed to the reader interested in applications of probability tools and statistical methods to computer system modelling, and its aim is to present some basic concepts and methods in this field. Because of the limited space, the author has confined himself to the presentation of a simple classical model of a multi-access computer system.

In the next three sections of this chapter the following problems are discussed: the construction of a queueing model, the determination of an analytical solution, and finally the estimation of the input parameters and an assessment of sensitivity of the solution to estimation errors. The last section is devoted to the problem of adequacy of such models and to a brief overview of recent results and new directions of research in the field of queueing models.

11.2. A QUEUEING MODEL

Multi-access computer systems are designed for use by many (several to several hundred) customers simultaneously in interactive (conversational) mode at remote access points called *terminals*. This kind of cooperation with a computer is mainly used in centralized data base systems (such as airline reservation systems, scientific information systems, banking systems, etc.) and in some computer centres (research institutes, universities).

A rough description of the operation of a multi-access system is as follows (cf. Fig. 11.2.1). Assume a certain number of independent customers, each of them served

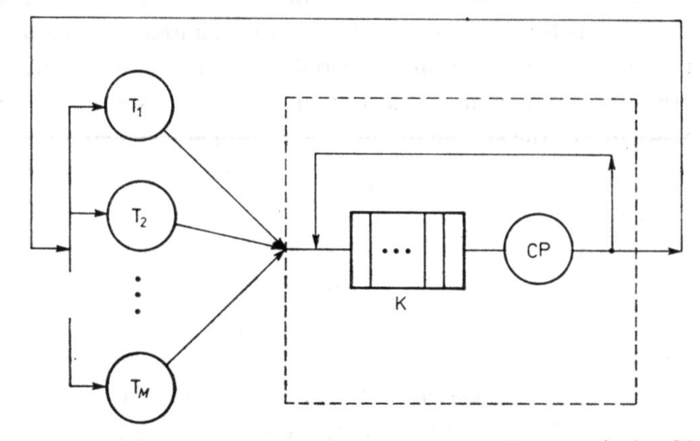

FIG. 11.2.1. Simplified scheme of multi-access system: $T_1, \ldots, T_M$—terminals, CP—central unit, K—queue of requests. The central part of the system is circumscribed by a dotted line. Arrows show the directions of the flow of customer requests and system responses.

by the system via a separate terminal. At any given time a customer may be in one of two possible states: thinking or waiting. In the *thinking state* the customer can make a request and send it by remote control to the system (e.g. by typing with the help of the keyboard). After sending in the request the customer enters the *waiting state* and remains in it until he receives an answer from the system (e.g. on the screen). Then he reenters the thinking state and the cycle is repeated. The time spent by a customer in the thinking state during one cycle will be called the *think time*.

The requests generated at the terminals are sent via the transmission lines to the central part of the system, where they are registered and put into a queue to wait for the execution by the central processor. We assume the following queueing discipline: (i) the request arriving at the system is put at the end of the queue, (ii) the request at the beginning of the queue is continuously processed up to a given time quantum limit. If the execution of the request is completed before the current quantum elapses, the request leaves the queue; otherwise it is put at the end of the queue. In both cases the next request assumes the first position in the queue and the next time quantum starts immediately(1).

(1) In general scheduling algorithm is one of the parameters of the computer system. The method described above is called *round-robin scheduling*.

After executing the request the system sends the customer the answer (whose content depends on the request itself and on the result of its execution). The time spent by the customer in the waiting state during one cycle will be called the *response time* of the system. The response time thus includes two components: the time required to process the request (depending only on the request and the speed of the processor), which will be called *service time*, and the time spent on waiting for the allocation of the processor (transmission time in both directions can be neglected). During the waiting state the customer is barred from making any other requests.

It should be stressed that the above description is simplified and neglects several important points relating to storage management and other functions of the system. We tacitly assume that they are inessential for the functioning of the system.

It is reasonable to consider think time and service time as random variables and ascribe to them certain probability distributions. In the sequel, we assume that these distributions are exponential (these assumptions will be discussed in the final section of this chapter). Hence, the above model of a computer system (together with its customers) is characterized by three input parameters: mean think time ($\bar{\sigma}$), mean service time ($\bar{\tau}$) and the number of active terminals (M).

One of the most important operational characteristic of a multi-access system is the stretch factor (r). It will be defined precisely in the next section; intuitively we define it as the ratio of the mean response time ($\bar{R}$) to the mean service time ($\bar{\tau}$):

$$r = \frac{\bar{R}}{\bar{\tau}}. \tag{11.2.1}$$

We shall now consider the functional relation between the stretch factor and the input parameters of the model. Accordingly, we shall complete the construction of our model in order to be able to construct the probability space in which the necessary random processes may be defined.

Let us assume that the system operates in stable conditions for an infinite time starting at the moment $t = 0$. Hence, we assume that each of the M customers generates an infinite stream of requests. Each request is represented in the model by a two dimensional random variable of the form

$$X_k^{(m)} = (\sigma_k^{(m)}, \tau_k^{(m)}), \quad m = 1, 2, \ldots, M, \quad k = 1, 2, \ldots,$$

representing the think time needed to generate the request and its service time. We make some additional assumptions, namely

(i) for each request the think time and the service time are independent random variables,

(ii) the random variables $X_k^{(m)}$ for $k = 1, 2, \ldots$ and $m = 1, \ldots, M$ are mutually independent,

(iii) all think times have the same exponential distribution with the parameter $\lambda = 1/\bar{\sigma}$,

(iv) all service times have the same exponential distribution with the parameter $\mu = 1/\bar{\tau}$.

Under the above assumptions each of the M streams $X^{(m)} = (X_k^{(m)}, k = 1, 2, \ldots)$ is described by the same probability space. A joint stream of requests is represented by an M-element family of sequences $(X^{(m)}, m = 1, 2, \ldots, M)$. The probability space $(\boldsymbol{\Omega}, \mathscr{A}, P)$ which represents the joint stream is the product of M spaces for separate streams.

In the space $(\boldsymbol{\Omega}, \mathscr{A}, P)$ we now define a random sequence $(t_n, n = 1, 2, \ldots)$ describing the arrival moments of customer generated requests (the requests are numbered in the order of their arrival). This sequence is determined uniquely (up to a subset of realizations of measure zero) by the joint stream of requests. Next, we consider the processes $(T_n, n = 0, 1, \ldots)$, $(R_n, n = 1, 2, \ldots)$, $(L_t, t \geqslant 0)$, where

$$T_n = \begin{cases} t_1 & \text{for} \quad n = 0, \\ t_{n+1} - t_n & \text{for} \quad n > 0 \end{cases}$$

is the interarrival time of two consecutive requests (t_1 is equal to the arrival time of the first request), R_n is the response of the system to the n-th request and L_t is the number of requests in the queue at the moment t.

It can be shown that all three processes $(T_n, n = 0, 1, \ldots)$, $(R_n, n = 1, 2, \ldots)$, $(L_t, t \geqslant 0)$ are renewal processes and for all of them there exist limiting distributions. Therefore we may use the notions of the limiting interarrival times (T_∞), the limiting response time (R_∞) and the limiting length of the queue (L_∞), in the sense that

$$T_n \xrightarrow[n \to \infty]{\mathscr{L}} T_\infty, \qquad R_n \xrightarrow[n \to \infty]{\mathscr{L}} R_\infty, \qquad L_t \xrightarrow[t \to \infty]{\mathscr{L}} L_\infty .$$

11.3. Analysis of the model

In the previous section we have defined the stretch factor of a system as the ratio of the mean response time to the mean service time without, however, giving a precise definition of the first of these two quantities. Now, since we are interested in long term prediction of system behaviour, we can now define the mean response time as the mean value of the limiting response time:

$$\bar{R} = E(R_\infty). \tag{11.3.1}$$

Let

$$\bar{T} = E(T_\infty),$$
$$\bar{L} = E(L_\infty).$$

It is known from queueing theory (cf. Stidham, 1972) that the following relation (called *Little's formula*) holds:

$$\bar{R} = \bar{L}\bar{T}. \tag{11.3.2}$$

We shall now determine $\bar{L}$ and $\bar{T}$. Note that $(L_t, t \geqslant 0)$ is equivalent to the birth-and-death process with the parameters

$$\lambda_k = \begin{cases} \lambda(M-k) & \text{for} \quad 0 \leqslant k \leqslant M-1, \\ 0 & \text{for} \quad k \geqslant M, \end{cases}$$

$$\mu_k = \begin{cases} \mu & \text{for} \quad 1 \leqslant k \leqslant M, \\ 0 & \text{for} \quad k \geqslant M+1. \end{cases}$$

Indeed, the properties of the exponential distribution and the mutual independence of requests imply that if at the moment t the queue contains k requests (i.e. $M-k$ customers are in the waiting state), then the probability of the arrival of a new request in the time interval $(t, t+h)$ $(h > 0)$ for small h is equal to $\lambda(M-k)h+$ $+o(h)$ $(0 \leqslant k \leqslant M-1)$ and the probability that the servicing of any of the requests is completed in the same interval is equal to $\mu h+o(h)$ $(1 \leqslant k \leqslant M)$.

The limiting distribution $\pi_k = \lim_{t\to\infty} P(L_t = k)$ can easily be determined from the general solution for birth-and-death processes, namely (cf. e.g. Kleinrock, 1975, p. 106)

$$\pi_k = \begin{cases} \pi_0 \left(\dfrac{\lambda}{\mu}\right)^k \dfrac{M!}{(M-k)!} & \text{for} \quad 0 \leqslant k \leqslant M, \\ 0 & \text{for} \quad k > M, \end{cases} \tag{11.3.3}$$

where

$$\pi_0 = \left(\sum_{k=0}^{M} \left(\frac{\lambda}{\mu}\right)^k \frac{M!}{(M-k)!}\right)^{-1}. \tag{11.3.4}$$

In this way we obtain

$$\bar{L} = \sum_{k=0}^{M} k\pi_k. \tag{11.3.5}$$

In the limit as $n \to \infty$ the expected number of requests arriving at the queue in a time unit must be equal to the expected number of requests leaving the queue in the same unit of time; the latter is equal to $(1-\pi_0)\mu$. Hence, the limiting value of the interarrival time is equal to

$$\bar{T} = \frac{\bar{\tau}}{1-\pi_0}. \tag{11.3.6}$$

From equations (11.3.2), (11.3.5), and (11.3.6) we obtain

$$\bar{R} = \frac{\bar{\tau}}{1-\pi_0} \sum_{k=0}^{M} k\pi_k,$$

where π_k $(k = 0, \ldots, M)$ are given by the formulae (11.3.3)–(11.3.4). Using a bit of algebra we finally obtain the relation

$$\bar{R} = \frac{M\bar{\tau}}{1-\pi_0} - \bar{\sigma} \tag{11.3.7}$$

which implies that the stretch factor is

$$r = \frac{M}{1-\pi_0} - \frac{\bar{\sigma}}{\bar{\tau}}. \tag{11.3.8}$$

The above relation may be used to determine the maximum number of active terminals if the upper bound for the stretch factor of the system is given.

It will be interesting to investigate the behaviour of the stretch factor as a function of the variables M and $\varrho = \bar{\tau}/\bar{\sigma}$ (cf. Fig. 11.3.1, and 11.3.2). For small ϱ we may use the approximation

$$r(M, \varrho) = 1+(M-1)\varrho+(M-1)(M-3)\varrho^2+o(\varrho^2). \tag{11.3.9}$$

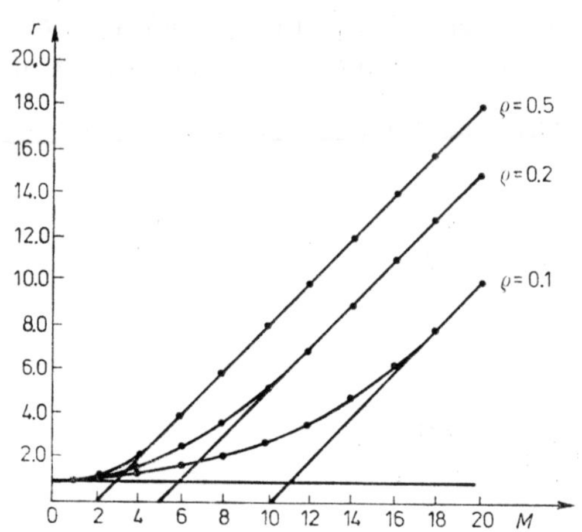

FIG. 11.3.1. The stretch factor r as a function of the number of active terminals M, for $\varrho = 0.1$, 0.2, 0.3. For the sake of readability the points of the graphs are connected by continuous lines.

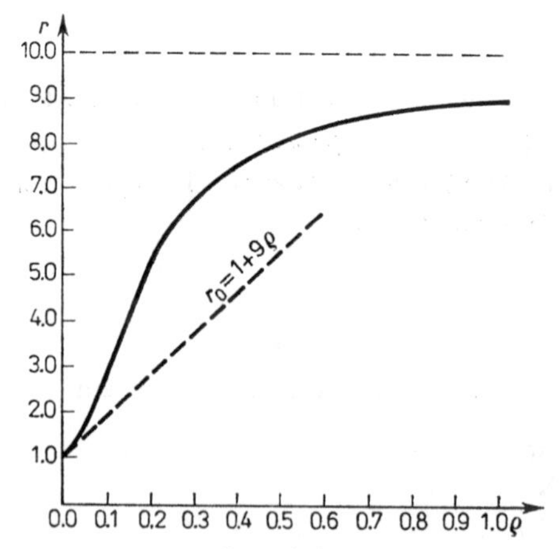

FIG. 11.3.2. The stretch factor r as a function of the parameter ϱ, for $M = 10$. The tangent line at $\varrho = 0$ is marked by a dotted line.

Moreover,

$$\lim_{\varrho\to+\infty} r(M, \varrho) = M, \quad r(1, \varrho) = 1.$$

and, as $M \to \infty$, $r(M, \varrho)$ tends to the asymptote

$$r_{\mathrm{as}}(M, \varrho) = M - \frac{1}{\varrho}.$$

Hence, for large values of ϱ the stretch factor of the system is equal to the number of active terminals, and for large values of M each new terminal increases the stretch factor by 1.

An efficient algorithm for computing the stretch factor may be found in the book by Hellermann and Conroy (1975, p. 140).

11.4. System parameter estimation

In practice to use formula (11.3.8) for assessing the stretch factor of a real system we must estimate the input parameters $\bar{\sigma}$ and $\bar{\tau}$. These quantities can be estimated from measurements performed on the stream of requests in the system with a fixed number of terminals (e.g. during experimental tests or at the initial stage of the installation of the system when only a part of the planned configuration is active). We can also use data obtained from similar existing systems. The following scheme can be proposed.

Let us assume that we observe a system with a fixed number of active terminals, say M_0. Let $N_m(t)$ be the number of requests which have been generated by the m-th customer and served by the system in the interval $[0, t]$ ($m = 1, 2, \ldots, M_0$). For a given number N we determine an observation time $t^{(N)}$ in such a way that the following inequalities hold:

$$N_m(t^{(N)}) \geqslant N, \quad m = 1, \ldots, M_0.$$

Consider now the observable initial part of a realization of the joint stream of requests, composed of M sequences

$$(\sigma_1^{(m)}, \tau_1^{(m)}),\ (\sigma_2^{(m)}, \tau_2^{(m)}), \ldots, (\sigma_N^{(m)}, \tau_N^{(m)}), \quad m = 1, 2, \ldots, M_0.$$

We introduce the statistics

$$\sigma_N = \frac{1}{NM_0} \sum_{m=1}^{M_0} \sum_{n=1}^{N} \sigma_n^{(m)},$$

$$\hat{\tau}_N = \frac{1}{NM_0} \sum_{m=1}^{M_0} \sum_{n=1}^{N} \tau_n^{(m)}.$$

It follows from the assumptions of the model that, for each N, $\bar{\sigma}_N$ and $\bar{\tau}_N$ are unbiased estimators of $\bar{\sigma}$ and $\bar{\tau}$ respectively and that

$$\sqrt{NM_0}(\hat{\sigma}_N - \bar{\sigma}) \xrightarrow{\mathscr{L}} N(0, \bar{\sigma}^2),$$

$$\sqrt{NM_0}(\hat{\tau}_N - \bar{\tau}) \xrightarrow{\mathscr{L}} N(0, \bar{\tau}^2).$$

To estimate the stretch factor we introduce

$$\hat{r}_N = r^*(\hat{\sigma}_N, \hat{\tau}_N).$$

where r^* is defined as

$$r^*(x_1, x_2) = r\left(M_0, \frac{x_2}{x_1}\right).$$

As a consequence we have

$$\sqrt{NM_0}\,(r^*(\hat{\sigma}_N, \hat{\tau}_N) - r^*(\bar{\sigma}, \bar{\tau})) \xrightarrow{\mathscr{L}} N(0, v^2),$$

where

$$v^2 = \bar{\sigma}^2 \left(\frac{\partial r^*}{\partial x_1}\right)^2 \Bigg|_{(x_1, x_2) = (\bar{\sigma}, \bar{\tau})} + \bar{\tau}^2 \left(\frac{\partial r^*}{\partial x_2}\right)^2 \Bigg|_{(x_1, x_2) = (\bar{\sigma}, \bar{\tau})}.$$

If the value of $\bar{\tau}/\bar{\sigma}$ is sufficiently small, then in view of (11.3.9) we can use the approximation

$$\tilde{r}^*(x_1, x_2) = 1 + (M_0 - 1)\frac{x_2}{x_1}$$

and then

$$\tilde{v}^2 = 2\frac{\bar{\tau}^2}{\bar{\sigma}^2}(M_0 - 1)^2$$

is an approximate value of v^2.

These results can be used to determine a confidence interval for the stretch factor for any proposed number M of terminals.

11.5. Final remarks

Models always give rise to deviations from reality whose impact on the solution of the problem in hand can hardly be evaluated in advance. This concerns postulates of stochastic nature (e.g. independence of random variables, specific forms of probability distributions) as well as structural simplifications caused by omission of some internal factors which may influence the functioning of the system. For example, in our model we have neglected the problems of storage management, data transmission between the internal store and the peripheral units, system management, etc. Therefore, some verification of the model is indispensable. If a model is constructed for an existing system accessible to experiments, then a kind of direct verification is possible since the solution calculated may be compared with the estimate obtained from empirical measurements. Such measurements usually require special equipment and rather complex procedures which often interfere with the normal functioning of the system; this topic is treated in Ferrari *et al.* (1983). If a model of the system in question does not exist or is not available for tests, verification can be performed indirectly by replacing the real system by a possibly accurate simulator. Statistical methods suitable for the latter approach are described, e.g. in Lavenberg (1983), Chapter 6.

A slightly different method of verification of the model consists in investigating its robustness. To this aim, experiments are performed on an appropriate simulation model for different probability distributions and different levels of accuracy of the description of the system. The analysis of the results of such experiments permits us to determine approximately the range of applicability of the model. It is worth mentioning that many queueing models demonstrate surprising robustness with respect to the deviations from the stochastic assumptions adopted in a given model.

Investigations of various multi-access systems show that empirical distributions of the think time can be fairly well approximated by exponential distributions whereas service time distributions are rather hyperexponential. Some recent analytical studies show that (for the queueing discipline described above) the stretch factor is fairly insensitive to the type of service time distributions. This partly explains the fact that for many real systems the model considered, despite its simplicity, gave a fairly accurate estimates of the stretch factor. This model was used for the first time by Scherr to evaluate the Compatible Time Sharing System at the M.I.T. (cf. Scherr, 1967). The problems of evaluation and the results of the investigation of multi-access systems are presented in some detail in the already mentioned book by Hellermann and Conroy, Chapter 9.

Recently, analytical methods for modelling computer systems have been significantly developed. Queueing network models have now become an important tool for system evaluation. They are significant extensions of simple queueing models, thus allowing a more accurate description of the internal structure of computer systems. A network model corresponds to a system composed of many parallel service stations with a separate queue at each station. The request which is being served moves according to fixed rules (generally probabilistic) from one station to another until servicing is completed.

Various performance measures for computer systems may be formally defined in terms of the network model, as has been done for single queueing models. For a certain interesting class of network models analytical solutions have been found. However, in most cases we must resort to approximate methods or simulation. It should be added that specialized packages of programs have been developed for computer system evaluation, making use of analytical as well as simulation methods.

The basic analytical results for queueing models, methods for an approximate analysis of network models, and simulation methodology are collected in Lavenberg (1983), where an exhaustive bibliography of the subject is included.

Closing remarks

The raison d'être of statistics is its applicability to solving practical problems. On the basis of the problems discussed in the present book some remarks can be made on the practical use of statistics and—more generally—on the relationships between theory and practice in statistics.

It must have struck the reader that the process of constructing a formal statistical problem—as presented in this book—is somewhat arbitrary and artificial. One of the reasons—as emphasized in Chapter 1—is that the generation of the realizations of a phenomenon is simulated by sampling performed according to some arbitrarily chosen probability distribution. In Part Three it was shown that the rational foundations of such an approach to modelling are in practice supplemented by assumptions which are motivated by tradition as well as by the subjective views of the researchers, and the proportions of these elements vary according to the judgement of the researcher. For example, genetic models adhere quite closely to the knowledge based on experiments, while in the cell cycle or survival models rational assumptions are often combined with other ones, made only for the sake of simplicity. Sometimes even, as in the case of psychometry, models of the mechanisms of human behaviour are based to large extent on the subjective opinions of the researchers.

Many users of statistics believe that possible negative consequences of freedom in choosing a model can be effectively neutralized by its experimental verification. This opinion, however, often leads to misunderstandings. Verification of a statistical model is tantamount to testing the hypothesis that the probability distribution which generates the realizations of a phenomenon belongs to the family of distributions specified by the model. As a rule the alternative hypothesis is only roughly defined which makes it impossible to construct a sufficiently reliable test to verify the model. We often use tests which only allow us either to reject a model or to state that there is not enough information to reject it. Unfortunately, it happens in practice that the latter conclusion is interpreted as a positive verification of the model. This interpretation which is clearly false is particularly risky when the data are scarce since those data usually lead to the conclusion that there is not enough information to reject the model. In this way it would be possible to verify positively any model, even the most absurd one.

Arbitrariness in choosing a model describing a given phenomenon goes together with arbitrariness in choosing the goal of research. The construction of a parameter adequately reflecting the intuitive meaning attached to a research goal and having a natural interpretation is sometimes difficult. A formalization of intuitions is not easy either; this can be exemplified by the formalization of the strength of the stochastic dependence of random variables presented in Chapter 5. In Chapter 6 we considered the simplest static genetic models whose parameters had an obvious interpretation. In more complex dynamic models, however, the interpretation of parameters is open to discussion.

Arbitrariness in choosing a formal description of the research goal can be found also in other chapters of Part Three. For instance, in Section 8.2 we discussed the problem how to choose the parameters evaluating the examined influence of a given factor in various schemes of experiments. In Chapter 10 the formalization of personal abilities and of test difficulties depends heavily on the chosen model. Similarly, in Chapter 11 arbitrariness in evaluating the efficiency of a given computer system can be noted.

A frequent deformation consists in narrowing down the research goals without sufficient motivation. Usually we are not interested in checking whether a certain idealized goal is accomplished but rather in checking whether the deviations from it are sufficiently small. The possible consequences of an improper formalization of the hypothesis under investigation have been pointed out several times throughout the book (e.g. in Section 5.1, when testing of the independence hypothesis was discussed).

Artificiality and arbitrariness in modelling appear also in formalizations of the requirements imposed on inference procedures. Artificiality results above all from the typical statistical approach to accessing verisimilitudity of conclusions. Namely, statistical methods guarantee, in general, that if the investigations were repeated for a sufficiently large number of independently generated sets of data, the probabilities of the occurrence of certain types of errors would be sufficiently small. The magnitudes of these probabilities are controlled by the user through confidence levels, significance levels, etc. In the case of repetitive investigations, e.g. in the quality control of products or in public opinions polls, the above approach can be accepted. However, in nonrepetitive investigations the credibility of conclusions cannot be validated by such a method. No wonder that the choice of confidence and significance levels arose so many controverses.

Another pitfall is too much belief in optimal solutions. First of all, a solution which is optimal with respect to one criterion can be inadmissible with respect to another criterion. However, our critical remark applies even if no objection is raised against the chosen criterion of optimality. For instance, minimal (and hence optimal) value of the risk may exceed a certain acceptable level and then the solution obtained is simply a wrong solution. This can be illustrated by Tables

3.2.1 and 3.2.2 from which it is possible to read how large can be the probability of misclassification under the rule which minimizes it.

Too strong requirements imposed on inference rules, on the other hand, may result in nonexistence of a solution. That is why inference rules in which, in dubious cases, the decision is deferred are of particular importance (cf. e.g. Chapter 3).

A pragmatic approach—the choice of requirements which simply allow us to solve the problem—often results in investigating only the asymptotic properties of the decision rules. In practice, however, sample sizes are usually too small to justify the conclusion that the properties of the rules based on the sample in question are sufficiently close to asymptotic properties. Thus, in modern statistics, theoretical investigations of asymptotic properties are usually suplemented by simulation experiments. This allows us to evaluate the validity of asymptotic results, as has been done in the investigation of the proof-reading process (cf. Introduction). What is more, by simulation methods one can investigate such properties of decision rules which we are not able to investigate analytically. This was done in Section 4.4, where the size and the composition of sublots obtained by screening were evaluated simultaneously.

Simulation experiments are usually expensive and time consuming, and, moreover, they can be performed only after the distribution and decision rules involved have been specified. The planning of simulation experiments is often based on asymptotic results, which correspond to large families of distributions and rules. Thus, an informal inference based on simulation experiments is interlaced with a formal analytic investigation of asymptotic properties.

In fact, the interlacing of formal and informal considerations is typical for statistical practice. Adaptation of analytical solutions may serve as an example. Adaptation of solutions was discussed in Chapters 3 and 4 in the context of discriminant and screening problems and also in Chapter 11 on the efficiency of computer systems. The properties of adapted rules are usually investigated analytically (asymptotic properties) and by simulation methods.

In most practical problems the design of the experiments is particularly important. But the troubles encountered in the formalization of a statistical problem are growing in importance if we have to choose at the same time the plan of research and the inference rule which solves the problem under this plan. Thus formal considerations are usually combined with informal ones. This is illustrated particularly in Chapter 8, dealing with cell cycle identification, in which the search of a suitable experimental scheme is a question of primary order.

Feature selection is a particular case of planning experiments. Theoretical and practical approaches to this problem are entirely different. A theoretical approach usually refers to the reduction of that part of the information which is not essential for a given problem (cf. Chapter 2). However, such a reduction by means of sufficient or prediction sufficient statistics is seldom possible in practice. Instead, various adaptations of such statistics are used. In Chapter 3 and 4 we discuss various adaptations of discriminant and screening functions and Section 3.4 deals with the choice of features with respect to a given class separability measure. Chapter 7 on father recognition discusses the construction of features with a sufficiently large discrimination power of discerning the father among the men sued. It is an interesting example of constructing real-valued statistics which reduce the rough information from the point of view of the problem to be solved. For their evaluation one may apply multistage investigation performed on subsidiary experimental data.

In theoretical investigations the problem of measurement scales is tackled only occasionally. In some domains of applications, especially in sociometry and psychometry, we make use, in rather imprecise and inconsistent way, of concepts developed in the measurement theory. In Chapter 10 we mention how to take into consideration the measurement scales in formalization of some psychometric problems.

The divergencies existing between theory and practice in statistics are thus large and still growing. Theory is developed fast without sufficient consideration for what is really needed in practice while the methodology of practical research is often stagnant. Moreover, the integration of theory and practice is often hampered by different traditions existing in the various domains of statistical applications. This lack of integration finds illustration for example in the theory of indices mentioned in Section 3.6, which is developing parallel to discriminant analysis. Also the relationships between discriminant analysis and screening are not sufficiently recognized yet. The troubles caused by the lack of unified terminology were discussed in the concluding remarks to Chapters 1 and 2.

An effective use of statistical methods is possible only when based on a conscious and sensible cooperation of statisticians and specialists in a given domain of application. In Part Three we pointed to the difficulties which may hinder this cooperation. We shall now recall some of them in reference to the particular chapters of Part Three.

In Chapter 6 we underline the difficulties arising in modelling genetic processes of population growth and in choosing stable parameters of these processes. We also mention the difficulties involved in choosing an appropriate scheme of experiments which would permit to collect data representative for the process and to find a sensible estimation of the chosen parameters. The second part of the chapter is devoted to inference in the case of a simple random sample drawn from an idealized population in a certain state of equilibrium. Therefore, we do not consider drawing from a dynamically growing population which might in practice render quite useless the results obtained by methods derived for simple random samples drawn from static populations. Applying these methods, we should have to admit various schemes of data collecting, compare results, and check the robustness of the solutions with regard to deviations from idealized assumptions. In fixing a plan of the experiment, the collaborating geneticists and statisticians should interpret together all the simplifications made in the formal description of the process of population growth on the one hand, and the formalized statistical decisions on the other hand.

The problem of establishing paternity (cf. note p. 160) presented in Chapter 7 illustrates the type of divergence existing between theoretical discriminant analysis and its practical applications. In theoretical discriminant analysis stress is laid on choosing a suitable discriminant rule according to requirements imposed upon its performance. In proving fatherhood we find that the properties of the accepted rule are less important than the choice and synthetic description of the information on the child, the mother, and the putative father. An important role is played here by a synthesis of the information on the child, the mother, and the putative father supplied by an anthropology expert's subjective evaluation. Such a synthetic evaluation, though excellent in some cases, involves the complex problems of subjective classification and, in particular, of the consistency of evaluations made by different experts. The problem of proving fatherhood differs from the typical problems of discriminant analysis also in the circumstance that learning samples are formed in a specific way. There are three kinds of learning samples: sets of families, sets of triples child–mother–randomly chosen man, sets of summons formally examined by the court. Each of these samples allows us to estimate a different parameter of the model and has its own specific observability. This causes difficulties in the organization of the research, in the unification of sets of features considered by various researchers, and in the analysis of results. Other difficulties arise in combining the judicature with statistical methods and in interpreting the results obtained from that in individual cases.

The whole of Chapter 8 is devoted to the initial phase of the cooperation of biologists and statisticians consisting in formalizing the cell cycle problems and finding appropriate plan for the experiments. It is shown how the results of biological experiments help in formulating new hypotheses and in experiment planning.

Starting a cooperation with biologists, the statistician should get acquainted with the statistical reports on former experiments and the models adopted in them. In particular, it is important to know which of the assumptions made reflect the specific biological aspects of the problem in question, and which are due to simplifications, necessity, convenience or tradition. A typical example of the influence of tradition on the choice of model is the considerable number of papers on the estimation and verification of various versions of the constant transition probability model.

A completely different type of cooperation between statisticians and specialists in empirical research is realized in problems of data censoring which are discussed in Chapter 9. The first stage of this cooperation consisted in perceiving by statisticians an interesting simple experimental model (namely, the model of random censoring of survival times) of considerable importance from the practical point of view. In the next stage the properties of various estimators of the survival function were investigated with no reference to practical applications. This field of study (in particular, the asymptotic properties of the survival function estimators) has proved to be very interesting from the mathematical point of view. However, simulation studies show that some asymptotic properties of particular interest to statisticians are not valid for samples of small or even medium sizes encountered in practice.

In investigating the survival function estimators under data censoring a considerable use is made of modern techniques of mathematical statistics, owing to, among others, the fact that this topic has been intensively studied only since the middle seventies. Chapter 10, on the other hand, illustrates how strongly the study of psychological and sociological problems is affected by tradition, which determines the choice of the model and the method. In the literature on psychometrics and sociometrics there is no distinct dividing line between the models and no clear presentation of their relationship. Investigation of nonparametric models is at the initial stage and the most commonly used method is the maximum likelihood method. No use is made of measurement scales necessary for measuring the personal ability of the persons tested and the difficulty of the tests imposed on them.

In Chapter 11 we present difficulties when one tries to apply the statistical queueing theory in practical investigation of computer systems.

In Part Three we show the borderline between the reasonable and unreasonable use of statistical methods, and the natural limits of their applicability. The existence of these limits is a consequence of the fact that though information can be processed by statistical methods, it obviously cannot be created by them. Hence, it is naive to hope that a decisive solution of the problem can be obtained from scarce data. However, the main idea of Part Three is to convince the reader that it is indeed necessary to treat each practical problem individually and to maintain a constant cooperation between statisticians and specialists in the field in question.

This opinion is in contrast with the recent tendency in statistics to replace the statistician by a package of statistical programmes and to limit his job mainly to the preparation of such packages. Thus, we are against the scheme like this:

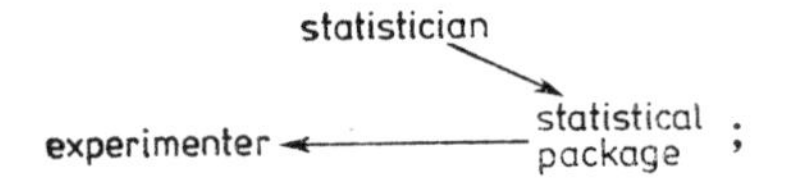

instead, we support the scheme

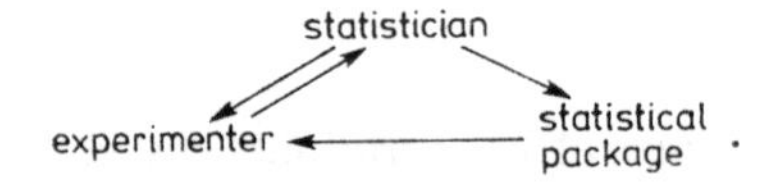

The model of a statistical problem presented in Part One should prove useful in the cooperation of statisticians with other specialists, and oblige the two parties to adjust suitably all the details of formalization of the problem dealt with. It should be stressed that the great variety of practical problems usually makes impossible to apply ready-made theoretical solutions since the specific character of any given practical problem is of paramount importance. Hence, further development of statistical theory cannot take place in separation from practical problems.

We hope that our book will fill, at least partially, the gap existing between theory and practice in statistics. We believe that statistics is a dangerous tool in the hands of incompetent or ill-willed persons. But when applied in an appropriate way, it becomes an indispensable instrument of analysing, processing and interpreting information about empirical phenomena.

Appendix

Algorithms for evaluating monotone dependence function and screening threshold

A.1. Introduction

This appendix contains fully tested algorithms for evaluating monotone dependence functions for a discrete bivariate distribution and for an empirical distribution based on a raw sample and also an algorithm for finding the screening threshold. The procedures which realize these algorithms are implemented in ISO-Fortran. In order to apply these procedures the user has to write a calling segment containing the data input, the procedure calls and the output.

The algorithms are presented according to the following scheme:

— general description;
— data format;
— numerical method;
— program structure;
— comments.

This scheme resembles that used in *Applied Statistics*, *JRSS Ser. C*, devoted, among other things, to statistical algorithms. Here, the description of application of proposed algorithms is omitted since it is described in detail in Chapters 4 and 5 of this book.

A.2. Algorithm for evaluating the monotone dependence function for a bivariate discrete distribution

— *description*

For a given discrete distribution of a pair (X, Y) of random variables, we compute the sequence of values of the monotone dependence function $\mu_{X,Y}(p_1), \ldots, \mu_{X,Y}(p_k)$, where $p_1, \ldots, p_k \in (0, 1)$.

— *data format*

A discrete distribution P of the pair (X, Y) is given by the matrix $A = [a_{ij}]$, where

$$a_{ij} = P(X = x_i, \; Y = y_j), \quad i = 1, \ldots, n, \; j = 1, \ldots, m,$$

and

$$x_1 < x_2 < \dots < x_n, \quad y_1 < y_2 < \dots < y_m.$$

A k-dimensional vector PP is a vector whose components are points $p_1, \dots, p_k$ from the interval $(0, 1)$.

— *numerical method*

The value $\mu_{X,Y}(p)$, $p \in (0, 1)$, of the monotone dependence function is computed according to the formula (cf. formulae (5.4.3), (5.4.4), (5.4.5)):

$$\mu_{X,Y}(p) = \begin{cases} \dfrac{L}{pEX - E(X; X < x_p) - x_p\big(p - P(X < x_p)\big)} & \text{for } L \geqslant 0, \\ \dfrac{L}{E(X; X > x_{1-p}) + x_{1-p}\big(p - P(X > x_{1-p})\big) - pEX} & \text{for } L < 0, \end{cases}$$

where

$$L = pEX - E(X; Y < y_p) - E(X \mid Y = y_p)\big(p - P(Y < y_p)\big).$$

For the distribution P the consecutive terms of this formula are determined in the following way:

$$EX = \sum_{i=1}^{n} x_i \mathrm{PX}(i),$$

$$E(X; Y < y_p) = \sum_{i=1}^{n} x_i \Big(\sum_{j=1}^{w(p)} a_{ij} - a_{iw(p)}\Big),$$

$$E(X \mid Y = y_p) = \sum_{i=1}^{n} \frac{x_i a_{iw(p)}}{\mathrm{PY}\big(w(p)\big)},$$

$$P(Y < y_p) = \sum_{j=1}^{w(p)} \mathrm{PY}(j) - \mathrm{PY}\big(w(p)\big),$$

$$E(X; X < x_p) = \sum_{i=1}^{v(p)} x_i \mathrm{PX}(i) - x_{v(p)} \mathrm{PX}\big(v(p)\big),$$

$$P(X < x_p) = \sum_{i=1}^{v(p)} \mathrm{PX}(i) - \mathrm{PX}\big(v(p)\big),$$

$$E(X; X > x_{1-p}) = \sum_{i=v(1-p)}^{n} x_i \mathrm{PX}(i) - x_{v(1-p)} \mathrm{PX}\big(v(1-p)\big),$$

$$P(X > x_{1-p}) = \sum_{i=v(1-p)}^{n} \mathrm{PX}(i) - PX\big(v(1-p)\big),$$

where

$$\mathrm{PX}(i) = \sum_{j=1}^{m} a_{ij},$$

$$\mathrm{PY}(j) = \sum_{i=1}^{n} a_{ij},$$

$$v(p) = \begin{cases} 1 & \text{if} \quad \mathrm{PX}(1) \geqslant p, \\ s & \text{if} \quad \sum_{i=1}^{s-1} \mathrm{PX}(i) < p \text{ and } \sum_{i=1}^{s} \mathrm{PX}(i) \geqslant p, \end{cases}$$

$$w(p) = \begin{cases} 1 & \text{if} \quad \mathrm{PY}(1) \geqslant p, \\ s & \text{if} \quad \sum_{j=1}^{s-1} \mathrm{PY}(j) < p \text{ and } \sum_{j=1}^{s} \mathrm{PY}(j) \geqslant p \end{cases}$$

(clearly, $x_{v(p)}$ and $y_{w(p)}$ are p-th quantiles x_p and y_p of the random variables X and Y, respectively).

— *program structure*

SUBROUTINE MDF (PP, FMI, K, A, N, M, NX, NY, X, PX, PY, IFAULT)

Input parameters:

PP	real array(K)	—vector containing values $p_1, \ldots, p_k$;
K	integer	—dimension of vectors PP and FMI;
A	real array(NX, NY)	—discrete distribution matrix $\mathrm{A}(i, j) = a_{ij}$;
N	integer	—number of rows of the matrix A;
M	integer	—number of columns of the matrix A;
NX	integer	—number of rows of A (as defined in the DIMENSION statement in the calling segment);
NY	integer	—number of columns of A (as declared in the DIMENSION statement in the calling segment);
X	real array(N)	—vector containing values $x_1, \ldots, x_n$.

Output parameters:

FMI	real array(K)	—vector containing function values $\mu_{X,Y}(p_1), \ldots, \mu_{X,Y}(p_k)$;
IFAULT	integer	—error indicator: = 0 no error found; = 1 at least one component of the vector PP is not from the interval (0, 1); = 2 N > NX or M > NY; = 3 at least one element of the matrix A is negative; = 4 the matrix A is degenerate, i.e., only one column (row) contains nonzero elements;

= 5 the matrix A cannot be regarded as a probability matrix, i.e.,

$$\left|\sum_{i=1}^{n}\sum_{j=1}^{m} A(i,j)-1\right| \geqslant \text{EPS};$$

= 6 there exists an l, such that

$$\text{PP}(l) \geqslant \sum_{i=1}^{n}\sum_{j=1}^{m} A(i,j);$$

= 7 the sequence $x_1, \ldots, x_n$ does not appear in increasing order.

Work-space arrays:

PX	real array(N)
PY	real array(M)

The constant EPS is the pre-set tolerance used in checking that A is a probability matrix.

— *comments*

The text of the function QUANT which computes the values v and w is affixed to the source program of MDF.

```
      SUBROUTINE MDF(PP,FMI,K,A,N,M,NX,NY,X,PX,PY,IFAULT)
      DIMENSION PP(K),FMI(K),A(NX,NY),X(N),PX(N),PY(M)
      REAL C,S,E,EX,PRB,L,P,EPS,SUM,Q
      INTEGER V,W,QUANT
      DATA EPS/0.0000001/
      IF(N.LE.0.OR.N.GT.NX.OR.M.LE.0,OR.M.GT.NY) GOTO 502
      DO 1 I=1,K
      P=PP(K)
      IF(P.LE.0.0.OR.P.GE.1.0) GOTO 501
    1 CONTINUE
      NRC=0
      SUM=0.0
      DO 3 I=1,N
      S=0.0
      DO 2 J=1,M
      C=A(I,J)
      IF(C.LT.0.0) GOTO 503
    2 S=S+C
      PX(I)=S
      IF(S.GT.0.0) NRC=NRC+1
    3 SUM=SUM+S
      IF(NRC.LE.1) GOTO 504
      IF(SUM.LT.1.0-EPS.OR.SUM.GT.1.0+EPS) GOTO 505
      NRC=0
      DO 5 J=1,M
      S=0.0
      DO 4 I=1,N
    4 S=S+A(I,J)
      PY(J)=S
      IF(S.GT.0.0) NRC=NRC+1
```

```
  5 CONTINUE
    IF(NRC.LE.1) GOTO 504
    DO 6 I=1,N-1
    IF(X(I+1).LE.X(I)) GOTO 507
  6 CONTINUE
    EX=0.0
    DO 7 I=1,N
  7 EX=EX+X(I)*PX(I)
    DO 400 NUM=1,K
    P=PP(NUM)
    IF(P.GE.SUM) GOTO 506
    E=0.0
    PRB=0.0
    S=0.0
    W=QUANT(PY,M,P)
    IF(W.EQ.1) GOTO 14
    JJ=W-1
    DO 11 I=1,N
    C=0.0
    DO 10 J=1,JJ
 10 C=C+A(I,J)
 11 E=E+C*X(I)
    DO 12 J=1,JJ
 12 PRB=PRB+PY(J)
 14 DO 15 I=1,N
 15 S=S+X(I)*A(I,W)
    L=P*EX-E-S/PY(W)*(P-PRB)
    IF(L) 23,21,16
 16 V=QUANT(PX,N,P)
    E=0.0
    PRB=0.0
    IF(V.EQ.1) GOTO 19
    JJ=V-1
    DO 18 I=1,JJ
    E=E+X(I)*PX(I)
 18 PRB=PRB+PX(I)
 19 FMI(NUM)=L/(P*EX-E-X(V)*(P-PRB))
    GOTO 400
 21 FMI(NUM)=0.0
    GOTO 400
 23 Q=1.0-P
    V=QUANT(PX,N,Q)
    E=0.0
    PRB=0.0
    IF(V.EQ.N) GOTO 25
    JJ=V+1
    DO 24 I=JJ,N
    E=E+X(I)*PX(I)
```

```
 24 PRB=PRB+PX(I)
 25 FMI(NUM)=L/(E+X(V)*(P-PRB)-P*EX)
400 CONTINUE
    IFAULT=0
    RETURN
501 IFAULT=1
    RETURN
502 IFAULT=2
    RETURN
503 IFAULT=3
    RETURN
504 IFAULT=4
    RETURN
505 IFAULT=5
    RETURN
506 IFAULT=6
    RETURN
507 IFAULT=7
    RETURN
    END


    INTEGER FUNCTION QUANT(PX,N,P)
    DIMENSION PX(N)
    REAL P,S
    K=0
    S=0.0
    DO 1 I=1,N
    K=K+1
    S=S+PX(I)
    IF(S.GE.P) GOTO 2
  1 CONTINUE
  2 QUANT=K
    RETURN
    END
```

A.3. Algorithm for evaluating the monotone dependence function for an empirical distribution based on a raw sample

— *description*

For a given sequence of points $p_1, \ldots, p_k \in (0, 1)$, we compute the values of the sample monotone dependence function $\hat{\mu}_{X,Y}(p_1), \ldots, \hat{\mu}_{X,Y}(p_k)$ (corresponding to an empirical distribution based on a raw sample).

— *data format*

A raw sample is given in the form of a sequence of points (x_i, y_i), $i = 1, \ldots, n$. Vector PP is defined as in Section A.2.

— *numerical method*

The sample quantiles are defined as

$$\tilde{x}_p = xo_{[np]+1},$$

$$\tilde{y}_p = yo_{[np]+1},$$

where n is the sample size; p specifies the quantiles ($p \in (0, 1)$); xo, yo denote the vectors x, y ordered nondecreasingly, with components denoted by xo_i, yo_i; $[np]$ denotes the greatest integer not greater than np.

The values of the sample monotone dependence function $\hat{\mu}_{X,Y}(p)$ are computed from the sample formula as in Section A.2 with consecutive components replaced by the respective estimators

$$\hat{E}X = \frac{1}{n}\sum_{i=1}^{n} x_i,$$

$$\hat{E}(X; Y < y_p) = \frac{1}{n}\sum_{i=1}^{n} x_i I(y_i < \tilde{y}_p),$$

$$\hat{E}(X \mid Y = y_p) = \frac{\sum_{i=1}^{n} x_i I(y_i = \tilde{y}_p)}{\sum_{i=1}^{n} I(y_i = \tilde{y}_p)},$$

$$\hat{P}(Y < y_p) = \frac{1}{n}\sum_{i=1}^{n} I(y_i < \tilde{y}_p),$$

$$\hat{E}(X; X < x_p) = \frac{1}{n}\sum_{i=1}^{n} x_i I(x_i < \tilde{x}_p),$$

$$\hat{P}(X < x_p) = \frac{1}{n}\sum_{i=1}^{n} I(x_i < \tilde{x}_p),$$

$$\hat{E}(X; X > x_{1-p}) = \frac{1}{n}\sum_{i=1}^{n} x_i I(x_i > \tilde{x}_{1-p}),$$

$$\hat{P}(X > x_{1-p}) = \frac{1}{n}\sum_{i=1}^{n} I(x_i > \tilde{x}_{1-p}),$$

where

$$I(r) = \begin{cases} 1 & \text{if relation } r \text{ holds,} \\ 0 & \text{otherwise.} \end{cases}$$

If the sequence $(x_1, \ldots, x_n)$ contains no ties, then the calculations can be simplified by the formula

$$\hat{\mu}_{X,Y}(p) = \begin{cases} \dfrac{\hat{L}}{M_1} & \text{for } \hat{L} \geqslant 0, \\[2ex] \dfrac{\hat{L}}{M_2} & \text{for } \hat{L} < 0, \end{cases}$$

where $\hat{L}$, M_1, M_2 are defined as follows:

$$\hat{L} = \hat{E}X - \frac{1}{np}\left\{\sum_{i=1}^{[np]} xyo_i + (n-[np])xyo_{[np]+1}\right\},$$

$$M_1 = \hat{E}X - \frac{1}{np}\left\{\sum_{i=1}^{[np]} xo_i + (n-[np])xo_{[np]+1}\right\},$$

$$M_2 = \frac{1}{np}\left\{\sum_{i=n-[np]}^{n} xo_i + (n-[np])xo_{[n(1-p)]}\right\} - \hat{E}X,$$

where xyo_i denotes the concomitant of the i-th order statistic xo_i in the sequence $(x_1, y_1), \ldots, (x_n, y_n)$.

— *program structure*

SUBROUTINE MDFP (PP, FMI, K, X, Y, N, PX, PY, IFAULT)

Input parameters:

PP	real array(K)	—vector containing values $p_1, \ldots, p_k$;
K	integer	—dimension of PP and FMI;
X	real array(N)	—vector containing $x_1, \ldots, x_n$;
Y	real array(N)	—vector containing $y_1, \ldots, y_n$;
N	integer	—dimension of vectors X, Y, PX, PY.

Output parameters:

FMI	real array(K)	—vector containing function values $\hat{\mu}_{X,Y}(p_1), \ldots, \hat{\mu}_{X.Y}(p_k)$;
IFAULT	integer	—error indicator; = 0 no error found; = 1 at least one component of PP is not from the interval (0, 1).

Work-space arrays:

PX	real array(N)
PY	real array(N)

— *comments*

Subroutine MDFP calls the subroutine FSORT, which sorts the vector X in nondecreasing order and, possibly, also the vector Y according to the ordering of X.

SUBROUTINE FSORT (X, Y, N, B)

Input parameters:

X	real array(N)	—vector X to be ordered nondecreasingly;
Y	real array(N)	—vector Y to be ordered according to the ordering of X;
N	integer	—dimension of X and Y;
B	logical	—if B = .TRUE. only vector X is to be

		ordered, if B = .FALSE. vector Y is to be ordered as well (see above).

Output parameters:

X	real array(N)	—vector X ordered nondecreasingly;
Y	real array(N)	—vector Y ordered according to the ordering of X.

```
      SUBROUTINE MDFP(PP,FMI,K,X,Y,N,PX,PY,IFAULT)
      DIMENSION PP(K),FMI(K),X(N),Y(N),PX(N),PY(N)
      REAL SRX,PN,S1,S2,S3,S4,C,RM,RL,PNMNP
      DO 1 I=1,K
      PN=PP(I)
      IF(PN.LE.0.0.OR.PN.GE.1.0) GOTO 501
    1 CONTINUE
      SRX=0.0
      DO 2 I=1,N
      PY(I)=Y(I)
      SRX=SRX+X(I)
    2 CONTINUE
      SRX=SRX/FLOAT(N)
      CALL FSORT(PY,PY,N,.TRUE.)
      DO 3 I=2,N
      J=I-1
      IF(PY(I).EQ.PY(J)) GOTO 50
    3 CONTINUE
C
C     CALCULATE IF NO TIES
C
      DO 11 I=1,N
      PX(I)=Y(I)
      PY(I)=X(I)
   11 CONTINUE
      CALL FSORT(PX,PY,N,.FALSE.)
      DO 12 I=1,N
   12 PX(I)=X(I)
      CALL FSORT(PX,PX,N,.TRUE.)
      DO 30 NUM=1,K
      PN=PP(NUM)*FLOAT(N)
      NP=IFIX(PN)
      IF(NP.GT.0.AND.NP.LT.N) GOTO 14
      PN=1.0
      NP=1
```

```
   14 NP1=NP+1
      PNMNP=PN-FLOAT(NP)
      RL=0.0
      RM=0.0
      DO 15 I=1,NP
   15 RL=RL+PY(I)
      RL=SRX-(RL+PNMNP*PY(NP1))/PN
      IF(RL) 18,21,16
   16 DO 17 I=1,NP
   17 RM=RM+PX(I)
      RM=SRX-(RM+PNMNP*PX(NP1))/PN
      GOTO 20
   18 J=N-NP+1
      DO 19 I=J,N
   19 RM=RM+PX(I)
      J=N-NP
      RM=(RM+PNMNP*PX(J))/PN-SRX
   20 FMI(NUM)=RL/RM
      GOTO 30
   21 FMI(NUM)=0.0
   30 CONTINUE
      IFAULT=0
      RETURN
C
C     CALCULATE IF TIES
C
   50 DO 51 I=1,N
   51 PX(I)=X(I)
      CALL FSORT(PX,PX,N,.TRUE.)
      DO 70 NUM=1,K
      PN=PP(NUM)*FLOAT(N)
      NP=IFIX(PN)
      IF(NP.GT.0.AND.NP.LT.N) GOTO 52
      PN=1.0
      NP=1
   52 NP1=NP+1
      S1=0.0
      S2=0.0
      S3=0.0
      S4=0.0
      C=PY(NP1)
      DO 55 I=1,N
      IF(Y(I)-C) 53,54,55
   53 S1=S1+X(I)
      S2=S2+1.0
      GOTO 55
   54 S3=S3+X(I)
      S4=S4+1.0
```

```
 55 CONTINUE
    RL=SRX-(S1+(PN-S2)/S4*S3)/PN
    S1=0.0
    S2=0.0
    IF(RL) 56,61,58
 56 J=N-NP
    C=PX(J)
    DO 57 I=1,N
    IF(X(I).LE.C) GOTO 57
    S1=S1+X(I)
    S2=S2+1.0
 57 CONTINUE
    RM=(S1+(PN-S2)*C)/PN-SRX
    GOTO 60
 58 C=PX(NP1)
    DO 59 I=1,N
    IF(X(I).GE.C) GOTO 59
    S1=S1+X(I)
    S2=S2+1.0
 59 CONTINUE
    RM=SRX-(S1+(PN-S2)*C)/PN
 60 FMI(NUM)=RL/RM
    GOTO 70
 61 FMI(NUM)=0.0
 70 CONTINUE
    IFAULT=0
    RETURN
501 IFAULT=1
    RETURN
    END


    SUBROUTINE FSORT(X,Y,N,B)
    DIMENSION X(N),Y(N)
    LOGICAL B
    M=-N
  3 M=M/2
    IF(M.GE.0) RETURN
    DO 1 J=1,N+M
    I=J
  2 IF(X(I).LE.X(I-M)) GOTO 1
    A=X(I)
    X(I)=X(I-M)
    X(I-M)=A
    IF(B) GOTO 4
    A=Y(I)
    Y(I)=Y(I-M)
    Y(I-M)=A
```

```
4 I=I+M
  IF(I.GT.0) GOTO 2
1 CONTINUE
  GOTO 3
  END
```

A.4. Algorithm for evaluating screening threshold

— *description*

For a raw sample from a bivariate distribution, an empirical screening threshold is computed according to formula (4.4.1).

— *data format*

A sample is given as a sequence (x_i, y_i), $i = 1, \ldots, n$ (in Section 4.4 this sequence was denoted by $(W^{(i)}, Z^{(i)})$). Constants θ_1^* and L are fixed.

— *numerical method*

Denote by *yo* the vector y ordered nondecreasingly. Let

$$u_j = \frac{1}{n-j+1} \sum_{i=1}^{n} I(x_i > L \wedge y_i \geqslant yo_j) \quad \text{for} \quad j = 1, \ldots, n,$$

where I is defined in Section A.3 and let $\{u_j^\circ\}_{j=1}^n$ denote the isotonic regression for the sequence $\{u_j\}_{j=1}^n$ (the definition of isotonic regression was given in Section 4.4).

The empirical screening threshold is computed according to the formula

$$t_0 = \frac{1}{l+1} \sum_{s=r}^{r+l} yo_s,$$

where

$$r = \begin{cases} 1 & \text{if } u_1^\circ \geqslant \theta_1^*, \\ j & \text{if } u_{j-1}^\circ < \theta_1^* \leqslant u_j^\circ, \\ n & \text{if } u_n^\circ < \theta_1^*, \end{cases}$$

$$l = \begin{cases} j & \text{if } u_{r+j}^\circ = u_r^\circ \text{ and } u_{r+j+1}^\circ > u_r^\circ, \\ n-r & \text{if } u_r^\circ = u_n^\circ < 1, \\ 0 & \text{if } u_r^\circ = u_n^\circ = 1. \end{cases}$$

— ***program structure***

SUBROUTINE SCREEN (N, X, Y, YO, U, UO, R1, R2, W, THETA, L, T, IFAULT)

Input parameters:

N	integer	—dimension of vectors X, Y, Z, U, UO;
X	real array(N)	—values $x_1, \ldots, x_n$;
Y	real array(N)	—values $y_1, \ldots, y_n$;
THETA	real	—value θ_1^*;
L	real	—value L.

Output parameters:

T	real	—threshold value t_0;
IFAULT	integer	—error indicator:

= 0 no error found;
= 1 if N < 2;
= 2 if $\theta_1^* < 0$ or $\theta_1^* > 1$.

Work-space arrays:

YO	real array(N)
U	real array(N)
UO	real array(N)
R1	real array(N)
R2	real array(N)
W	real array(N)

— *comments*

To compute the sequence $\{u_j^\circ\}_{j=1}^n$ the algorithm AMALGM (Cran, 1980) was used.

The subroutine SCREEN calls the sorting routine SORT. The text of this routine is also included.

```
      SUBROUTINE SCREEN(N,X,Y,YO,U,UO,R1,R2,W,THETA,L,T,IFAULT)
      DIMENSION X(N),Y(N),YO(N),U(N),UO(N),R1(N),R2(N),W(N)
      REAL L
      IF(N.LT.2) GOTO 21
      IF(THETA.LT.0.0.OR.THETA.GT.1.0) GOTO 22
      DO 1 I=1,N
    1 YO(I)=Y(I)
      CALL SORT(YO,N)
      DO 3 J=1,N
      I1=0
      I2=0
      P=YO(J)
      DO 2 I=1,N
      IF(Y(I).LT.P) GOTO 2
      I2=I2+1
      IF(X(I).GT.L) I1=I1+1
    2 CONTINUE
    3 U(J)=FLOAT(I1)/FLOAT(I2)
      DO 4 I=1,N
    4 W(I)=1.0
      CALL AMALGM(N,U,W,R1,R2,UO,IFAULT2)
      J=0
    5 J=J+1
      IF(UO(J).GE.THETA) GOTO 6
      IF(J.LT.N) GOTO 5
    6 IR=J
      IL=0
      UOR=UO(IR)
      IF(UOR.EQ.1.0) GOTO 8
      IL=N-IR
      IF(UOR.EQ.UO(N)) GOTO 8
      IL=-1
```

```
  7 IL=IL+1
    K=IR+IL
    IF(UO(K).EQ.UOR.AND.UO(K+1).GT.UOR) GOTO 8
    IF(K.GE.N-1) GOTO 8
    GOTO 7
  8 CONTINUE
    P=0.0
    DO 9 I=IR,IR+IL
  9 P=P+YO(I)
    T=P/FLOAT(IL+1)
    IFAULT=0
    RETURN
 21 IFAULT=1
    RETURN
 22 IFAULT=2
    RETURN
    END


    SUBROUTINE SORT(XY,N)
    DIMENSION XY(N)
    M=-N
  1 M=M/2
    IF(M.GE.0) RETURN
    L=N+M
    DO 4 J=1,L
    I=J
  2 IMM=I-M
    IF(XY(I).LE.XY(IMM)) GOTO 4
    A=XY(I)
    XY(I)=XY(IMM)
    XY(IMM)=A
    I=I+M
    IF(I.GT.0) GOTO 2
  4 CONTINUE
    GOTO 1
    END
```

References

Aalen, O. Q.: (1976), 'Nonparametric inference in connection with multiple decrement models', *Scand. J. Statist.* **3**, 15–27.

Aalen, O. Q.: (1978a), 'Nonparametric estimation of partial transition probabilities in multiple decrement models', *Ann. Statist.* **6**, 534–545.

Aalen, O. Q.: (1978b), 'Nonparametric inference for a family of counting processes', *Ann. Statist.* **6**, 701–726.

Aitchison, J., Habbema, J. D. F., Kay, J. W.: (1977), 'A critical comparison of two methods of statistical discrimination', *Appl. Statist.* **26**, 15–25.

Alam, K., Wallenius, K. T.: (1976), 'Positive dependence monotonicity in conditional distributions', *Comm. Statist. A—Theor. Meth.* **56**, 525–534.

Andersen, E. B.: (1970), 'Asymptotic properties of conditional maximum likelihood estimators', *J. Roy. Statist. Soc. Ser. B* **32**, 282–301.

Andersen, E. B., (1972), 'The numerical solution of a set conditional estimation equations', *J. Roy. Statist. Soc. Ser. B* **34**, 42–54.

Andersen, E. B.: (1973a), *Conditional Inference and Models for Measuring*, Mentalhygienijsk Forlag, Copenhagen.

Andersen, E. B.: (1973b), 'A goodness of fit test for the Rash model', *Psychometrika* **38**, 123–140.

Andersen, E. B.: (1973c), 'Conditional inference for multiple choice questionnaires', *British J. Math. Statist. Psych.* **26**, 31–44.

Andersen, E. B.: (1980), *Discrete Statistical Models with Social Science Applications*, North-Holland, Amsterdam, New York, Oxford.

Anderson, J. A.: (1969), 'Constrained discrimination between *K* populations', *J. Roy. Statist. Soc. Ser. B* **31**, 123–139.

Anderson, J. A.: (1982), 'Logistic discrimination', in: *Handbook of Statistics*, vol. 2, *Classification Pattern Recognition and Reduction of Dimensionality*, eds. P. R. Krishnaiah, L. N. Kanal, North-Holland, Amsterdam, New York, Oxford, 169–191.

Anderson, T. W.: (1958), *An Introduction to Multivariate Statistical Analysis*, John Wiley and Sons, New York.

Bahadur, R. R.: (1954), 'Sufficiency and statistical decision functions', *Ann. Math. Statist.* **25**, 423–462.

Barlow, R. E., Bartholomew, D. J., Bremmer, J. M., Brunk, H. D.: (1972), *Statistical Inference Under Order Restrictions*, John Wiley and Sons, New York.

Barra, J. R.: (1971), *Notions fundamentales de statistique mathematique*, Dunod, Paris; English translation: *Mathematical Basis of Statistics*, John Wiley and Sons, New York 1981.

Baserga, R.: (1978), 'Resting cells and G_1 phase of the cell cycle', *J. Cel. Physiol.* **95**, 377–386.

Basu, D.: (1958), 'On statistics independent of a complete sufficient statistic', *Sankhyâ* **20,** 223–226.

Beckman, R. J., Johnson, M. E.: (1981), 'A ranking procedure for partial discriminant analysis', *J. Amer. Statist. Assoc.* **76**, 671–675.

Bednarski, T.: (1981), 'On optimal properties of the chi-square test of fit testing ε-validity of parametric models', *Math. Operationsforsch. Statist. Ser. Statist.* **12**, 31–41.

Bednarski, T., Filipowski, H., Jagielski, J.: (1978), 'Identification of human chromosomes, a decision theoretic approach', Preprint 130, Inst. Math. Pol. Acad. Sci.

Bednarski, T., Gnot, S., Ledwina, T.: (1982), 'Testing approximate validity of Hardy–Weinberg law in population genetics', in: *Probability and Statistical Inference*, ed. W. Grossman, Reidel, Dordrecht, 35–46.

Berkson, J., Gage, R.: (1950), 'Calculation of survival rates for cancer', *Proc. May. Clinic* **25,** 270–286.

Bickel, P. J., Doksum, K.: (1977), *Mathematical Statistics: Basis Ideas and Selected Topics*, Holden-Day Inc., San Francisco.

Bickel, P. J., Lehmann, E. L.: (1975a), 'Descriptive statistics for nonparametric models, I. Introduction', *Ann. Statist.* **3**, 1038–1044.

Bickel, P. J., Lehmann, E. L.: (1975b), 'Descriptive statistics for nonparametric models, II. Location', *Ann. Statist.* **3**, 1045–1069.

Bickel, P. J., Lehmann, E. L.: (1976), 'Descriptive statistics for nonparametric models, III. Dispersion', *Ann. Statist.* **4**, 1139–1158.

Birch, M.: (1964), 'A new proof of Pearson–Fisher theorem', *Ann. Math. Statist.* **35**, 818–824.

Birnbaum, A.: (1968), 'Some latent trait models and their use in inferring on examines ability', in: *Statistical Theories of Mental Test Scores*, eds. F. M. Lord, M. R. Novick, Addison-Wesley, Reading, Massachusetts.

Birnbaum, Z. W., Chapman, D. G.: (1950), 'On optimum selections from multinormal populations', *Ann. Math. Statist.* **21**, 443–447.

Bishop, Y., Fienberg, S., Holland, P.: (1975), *Discrete Multivariate Analysis; Theory and Practice*, MIT Press, Cambridge, Massachusetts.

Bjørnstadt, J. E.: (1975), 'Inference theory in contingency tables', Statistical Research Report 2, Institute of Mathematics, University of Oslo.

Blackwell, D.: (1951), 'Comparison of experiments', in: *Proceedings of the Second Berkeley Symposium on Mathematical Statistics and Probability*, University of California Press, 93–102.

Blackwell. D.: (1953), 'Equivalent comparison of experiments', *Ann. Math. Statist.* **24**, 265–272.

Bock, R. D.: (1972), 'Estimating item parameters and latent ability when responses are scored in two or more nominal categories', *Psychometrika* **37**, 29–51.

Bock, R. D., Lieberman, M.: (1970), 'Fitting a response model for N dichotomously scored items', *Psychometrika* **35**, 179–197.

Breslow, N., Crowley, J.: (1974), 'A large sample study of the life table and product limit estimates under censorship', *Ann. Statist.* **2,** 437–453.

Broffitt, J. D.: (1982), 'Nonparametric classification', in: *Handbook of Statistics*, vol. 2, *Classification, Pattern Recognition and Reduction of Dimensionality*, eds. P. R. Krishnaiah, L. M. Kanal, North-Holland, Amsterdam, New York, Oxford, 139–168.

Broffitt, J. D., Randles, R. H., Hogg, R. V.: (1976), 'Distribution-free partial discriminant analysis', *J. Amer. Statist. Assoc.* **71**, 934–939.

Bromek, T.: (1988), 'Sufficiency and standard classes of statistical problems', in: *Transactions of the Tenth Prague Conference on Information Theory, Statistical Decision Functions and Random Processes, Prague, July 1986*, vol. A, Academia, 217–224.

Bromek, T., Kowalczyk, T.: (1986), 'Statistical inference in the case of unordered pairs', *Statistics* **17**, 357–364.

Bromek, T., Kowalczyk, T., Pleszczyńska, E.: (1988), 'Measurement scales in evaluation of stochastic dependence', in: *Proceedings of the International Conference on Advances in Multivariate Statistical Analysis*, eds. S. Das Gupta, J. K. Ghosh, Indian Statistical Institute, Calcutta, 83–96.

Bromek, T., Moszyńska, M.: (1983), 'Various categorical approaches to statistical spaces', *Dissertationes Math.* **224**, 1–29.

Bromek, T., Moszyńska, M., Prażmowski, K.: (1984), 'Concerning basic notions of the measurement theory', *Czechoslovak Math. J.* **34**, 570–587.

Bromek, T., Niewiadomska-Bugaj, M.: (1987), 'Threshold rules in two-class discrimination problems', *Probab. Math. Statist.* **8**, 11–16.

Brown, B. W., Hollander, H., Korwar, R. M.: (1974), 'Nonparametric tests for censored data, with application to heart transplant studies', in: *Reliability and Biometry*, eds. F. Proschan, R. J. Serfling, Society for Industrial and Applied Mathematics, Philadelphia, 327–354.

Brzeziński, J.: (1984), *Elements of Methodology of Psychological Research* (in Polish), PWN, Warszawa.

Cacoullos, T., (ed.): (1973), *Discriminant Analysis and Applications*, Academic Press, New York.

Castor, L. N.: (1980), 'A G_1 rate model accounts for cell-cycle kinetics attributed to »transition probability«', *Nature* **287**, 857–859.

Christoffersson, A.: (1975), 'Factor analysis of dichotomized variables', *Psychometrika* **40**, 3–32.

Conover, W. J., Iman, R. L.: (1980), 'The rank transformation as a method of discrimination with some examples', *Comm. Statist. A–Theor. Meth.* **9**, 465–489.

Cran, G. W.: (1980), 'Algorithms AS 149. Amalgamation of means in the case of simple ordering', *Appl. Statist.* **29**, 209–211.

Cuzik, J.: (1982), 'Rank test for association with right censored data', *Biometrika* **69**, 351–364.

Ćwik, J.: (1985), 'Statistical methods in measurement of statistical dependence' (in Polish), (Ph. D. Thesis, UAM, Poznań).

Ćwik, J., Gołembiewska, M., Kowalczyk, T., Pleszczyńska, E.: (1982), 'Conceptual and statistical problems of sister dependence', *Biometrika* **69**, 513–520.

Dąbrowska, D.: (1986), 'Rank test for independence for bivariate censored data', *Ann. Statist.* **14**, 250–264.

Dąbrowska, D.: (1985), 'Descriptive parameters of location, dispersion and stochastic dependence', *Statistics* **16**, 63–88.

Dąbrowska, D., Pleszczyńska, E.: (1980), 'On partial observability in statistical models', *Math. Operationsforsch. Statist. Ser. Statist.* **11**, 49–59.

Devijver, P. A., Kittler, J.: (1982), *Pattern Recognition: a Statistical Approach*, Prentice Hall, London.

Ducamp, A., Falmagne, J. C.: (1969), 'Composite measurement', *J. Math. Psych.* **6**, 359–390.

Dunn, O. J., Clark, V. A.: (1972), *Applied Statistics: Analysis of Variance and Regression*, John Wiley and Sons, New York.

Edwards, A. W. F.: (1963), 'The measure of association in a 2×2-table', *J. Roy. Statist. Soc. Ser. A* **126**, 109–114.

Efron, B.: (1967), 'The two sample problem with censored data', in: *Proceedings of the Fifth Berkeley Symposium*, Berkeley, 831–853.

Efron, B.: (1981a), 'Censored data and the bootstrap', *J. Amer. Statist. Assoc.* **76**, 312–319.

Efron, B.: (1981b), 'Nonparametric standard errors and confidence intervals', Technical Report 67, Division of Biostatistics, Stanford University.

Elandt-Johnson, R. C.: (1971), *Probability Models and Statistical Methods in Genetics*, John Wiley and Sons, New York.

Elandt-Johnson, R. C., Johnson, N. L.: (1980), *Survival Models and Data Analysis*, John Wiley and Sons, New York.

Elston, R. C., Forthofer, R. C.: (1977), 'Testing for Hardy–Weinberg equilibrium in small sample', *Biometrics* **33**, 536–541.

Esary, I. D., Proschan, F.: (1972), 'Relationships among some concepts of bivariate dependence', *Ann. Math. Statist.* **43**, 651–655.

Essen-Möller, E.: (1938), 'Die Beweiskraft der Ähnlichkeit im Vaterschaftsnachweis. Theoretische Grundlagen', *Mitt. Arthropol. Ges. Wien.* **68**, 2–53.

Feràk, V., Spevárová, E.: (1972), 'Discrimination between »true« and »false« parents based on the number of dermatoglyphic concordances', *Acta Facul. Rev. Nat. Univ. Comen. Anthrop.* **20**, 65–73.

Ferrari, D., Serazzi, G., Zeigner, A.: (1983), *Measurement and Tuning of Computer Systems*, Prentice-Hall, Englewood Cliffs.

Fleming, T. R., O'Fallen, J. R., O'Brien, P. C.: (1980), 'Modified Kolmogorov–Smirnov test procedures with application to arbitrarily right censored data', *Biometrics* **36**, 607–625.

Főldes, A., Rejtő, L.: (1981), 'A LIL type result for the product limit estimator', *Z. Wahrsch. Verw. Gebiete* **56**, 75–86.

Frankel, J.: (1965), 'The effect of nucleic acid antagonists on cell division and oral organelle development in *Tetrahymena Pyriformis*', *J. Exp. Zool.* **159**, 113–148.

Fritz, J. M.: (1975), 'Archaeological systems for indirect observation of the past', in: *Contemporary Archaeology, A Guide to Theory and Contributions*, ed. M. P. Leone, London, 135–157, 413–450.

Fukunaga, K.: (1972), *Introduction to Statistical Pattern Recognition*, Academic Press, London.

Gáfriková, V.: (1987), 'On nonparametric latent traits models', *Quality and Quantity* **21**, 71–79.

Gehan, E. A.: (1965), 'A generalized Wilcoxon test for comparing arbitrarily single-censored samples', *Biometrika* **52**, 203–223.

Geyer, E.: (1938), 'Die Beweiskraft der Ähnlichkeit im Vaterschaftsnachweis. Praktische Anwendung', *Mitt. Anthrop. Ges. Wien*, **68**, 54–87.

Gill, K. S., Hanson, E. D.: (1968), 'Analysis of preffission morphogenesis in *Paramecium Aurelia*' *J. Exp. Zool.* **167**, 219–236.

Gill, R. D.: (1980), 'Censoring and stochastic integrals', Mathematical Centre Tracts 124, Mathematisch Centrum, Amsterdam.

Gill, R. D.: (1981), 'Testing with replacement and the product limit estimator', *Ann. Statist.* **9**, 853–860.

Gillespie, M., Fisher, L.: (1979), 'Confidence bounds for the Kaplan–Meier survival curve estimate', *Ann. Statist.* **7**, 920–924.

Glick, N.: (1973), 'Sample-based multinomial classification', *Biometrics* **29**, 241–256.

Gnot, S.: (1977), 'The essentially complete class of rules in multinomial identification', *Math. Operationsforsch. Statist. Ser. Statist.* **8**, 381–386.

Gnot, S.: (1978), 'The problem of two-group identification', *Math. Operationsforsch. Statist. Ser. Statist.* **9**, 343–349.

Gnot, S.: (1979), 'Statistical methods in population genetics' (in Polish), (Unpublished paper).

Gnot, S., Ledwina, T.: (1980), 'Testing for Hardy–Weinberg equilibrium', *Biometrics* **36**, 161–165.

Goldstein, M., Dillon, W. R.: (1978), *Discrete Discriminant Analysis*, John Wiley and Sons, New York.

Goodman, L. A., Kruskal, W. H.: (1954), 'Measures of association for cross classifications, I', *J. Amer. Statist. Assoc.* **49**, 732–764.

Goodman, L. A., Kruskal, W. H.: (1959), 'Measures of association for cross classifications, II. Further discussion and references', *J. Amer. Statist. Assoc.* **54**, 123–163.

Goodman, L. A., Kruskal, W. H.: (1963), 'Measures of association for cross classifications, III. Approximate sampling theory', *J. Amer. Statist. Assoc.* **58**, 310–364.

Goodman, L. A., Kruskal, W. H.: (1972), 'Measures of association for cross classifications, IV. Simplification of asymptotic variances', *J. Amer. Statist. Assoc.* **67**, 415–421.

Gourlay, N.: (1951), 'Difficulty factors arising from the use of tetrachoric correlation in factor analysis', *British J. Psychol. Statist. Sec.* **4**, 65–76.

Green, P. J.: (1980), 'A »random transition« in the cell cycle?', *Nature* **285**, 116.

De Groot, M. M.: (1970), *Optimal Statistical Decisions*, McGraw-Hill, New York.

Guttman, L.: (1944), 'A basis for scaling qualitative data', *Amer. Sociol. Rev.* **9**, 139–150.

Habbema, J.: (1979), 'A discriminant analysis approach to the identification of human chromosomes', *Biometrics* **35**, 919–928.

Haber, M.: (1982), 'Testing for independence in intraclass contingency tables', *Biometrics* **38**, 93–103.

Haberman, S. J.: (1977), 'Maximum likelihood estimates in exponential response models', *Ann. Statist.* **5**, 815–841.

Haldane, J.: (1954), 'An exact test for randomness of mating', *J. Genetics* **52**, 631–635.

Hall, W., Wellner, J.: (1980), 'Confidence bands for survival curve from censored data', *Biometrika* **67**, 234–250.

Halmos, P. R., Savage, L. J.: (1949), 'Application of Radon–Nikodym theorem to the theory of sufficient statistics', *Ann. Math. Statist.* **20**, 225–241.

Hand, D. J.: (1981), *Discrimination and Classification*, John Wiley and Sons, New York.

Hellermann, H., Conroy, T. F.: (1975), *Computer System Performance*, McGraw-Hill, New York.

Hinkley, D. V.: (1973), 'Two-sample tests with unordered pairs', *J. Roy. Statist. Soc. Ser. B* **35**, 337–346.

Hirszfeld, L.: (1948). *Father Recognition in the Light of Blood Groups Science* (in Polish), Lekarski Instytut Naukowo-Wydawniczy, Wrocław.

Hirszfeld, L., Łukaszewicz, J.: (1958), *Father Recognition in the Light of Blood Groups Science* (in Polish), PZWL, Wrocław.

Holland, P. W., Rosenbaum, P. R.: (1986), *'Conditional association and unidimensionality in monotone latent variable models'*, Ann. Statist. **14**, 1523–1543.

Hulanicka. B.: (1973), 'Anthroposcopic features as a measure of similarity' (in Polish), *Materiały i Prace Antrop.* **86**, 115–156.

Hulanicka, B.: (1981), 'Using an anthroposcopic features in genetic expertises' (in Polish), *Materiały i Prace Antrop.* **100**, 103–120.

Hummel, K.: (1961), *Die Medizinische Vaterschaftsbegutachtung mit Biostatischem Beweis*, Gustav Fischer Verlag, Stuttgart.

Hummel, K., Ihm, P., Schmidt, V.: (1971), *Biostatische Abstammungsbegutachtung mit Blutgruppenbefunden*, Gustav Fischer Verlag, Stuttgart.

Indow, T., Sano, K., Namik, K., Makita, H.: (1962), 'A mathematical model for interpretation in projective tests: an application to seiken SCT', *Japan. Psych. Res.* **4**, 163–172.

Jacquard, A.: (1974), *The Genetic Structure of Populations*, Springer-Verlag, Heidelberg, New York.

Kaczanowska, J., Hyvert, N., de Haller, G.: (1976), 'Effects of actinomycin D on generation time and morphogenesis in *Paramecium*', *J. Protozool.* **23**, 341–349.

Kalbfleisch, J., Prentice, R.: (1980), *The Statistical Analysis of Failure Time Data*, John Wiley and Sons, New York.

Kaplan, E., Meier, P.: (1958), 'Nonparametric estimation from incomplete observations', *J. Amer. Statist. Assoc.* **53**, 457–481.

Keiter, F.: (1956), 'Zur Methodendiskussion in der Vaterschaftsdiagnostik', *Homo* **7**, 219–223.

Keiter, F.: (1960), 'Prolegomena zu Einem Wahrscheinlichkeitsbrevier für Vaterschaftsbegutachtung', *Homo* **11**, 220–227.

Kendall, D. G.: (1971), 'A mathematical approach to seriation', *Philos. Trans. Soc. London* **269**, 125.

Kendall, M. G., Stuart, A.: (1961), *The Advanced Theory of Statistics*, vol. 2, *Inference and Relationship*, Griffin and Co., London.

Kimball, R. F., Caspersson, T. O., Svensson, G., Carlson, L.: (1959), 'Quantitative cytochemical studies on *Paramecium Aurelia*, I, Growth in total dry weight measured by the scanning interference microscope and X-rays absorption methods', *Exp. Cell Res.* **17**, 160–172.

Kimeldorf, G., Sampson, A.: (1975a), 'One-parameter families of bivariate distributions with fixed marginals', *Comm. Statist. A—Theory Methods* **4**, 293–301.

Kimeldorf, G., Sampson, A.: (1975b), 'Univariate representations of bivariate distributions', *Comm. Statist. A—Theory Methods* **4**, 617–627.

Kimeldorf, G., Sampson, A.: (1978), 'Monotone dependence', *Ann. Statist.* **6**, 895–903.

Klecka, W. R.: (1981), 'Discriminant analysis', Sage University Papers, Series on Quantitative Applications in the Social Sciences, 07–019, Beverly Hills and London: Sage Publications.

Kleinrock, L.: (1975), *Queueing Systems*, vol. 1, *Theory*, John Wiley and Sons, New York.

Koch, A. L.: (1980), 'Does the variability of the cell cycle result from one or many chance events?', *Nature* **286**, 80–82.

Kolakowski, D., Bock, R. D.: (1970), 'A Fortran IV program for maximum likelihood item analysis and test scoring; normal ogive model', Education Statistical Laboratory Research Memo 12, University of Chicago.

Kowalczyk, T.: (1977), 'General definition and sample counterparts of monotonic dependence functions of bivariate distributions', *Math. Operationsforsch. Statist. Ser. Statist.* **8**, 351–369.

Kowalczyk, T.: (1982), 'On the shape of the monotonic dependence functions of bivariate distributions', *Math. Operationsforsch. Statist. Ser. Statist.* **13**, 183–192.

Kowalczyk, T., Kowalski, A., Matuszewski, A., Pleszczyńska, E.: (1979), 'Screening and monotonic dependence functions', *Ann. Statist.* **7**, 607–614.

Kowalczyk, T., Kowalski, A., Matuszewski, A., Pleszczyńska, E.: (1980), 'Applications of monotonic dependence functions: descriptive aspects', *Zastos. Mat.* **16**, 579–591.

Kowalczyk, T., Ledwina, T.: (1982), 'Some properties of chosen grade parameters and their rank counterparts', *Math. Operationsforsch. Statist. Ser. Statist.* **13**, 547–553.

Kowalczyk, T., Mielniczuk, J.: (1987), 'A screening method for a nonparametric model', *Statist. Probab. Letters* **5**, 163–167.

Kowalczyk, T., Pleszczyńska, E.: (1977), 'Monotonic dependence functions of bivariate distributions', *Ann. Statist.* **5**, 1221–1227.

Koziol, J., Green, S.: (1976), 'A Cramer–von Mises statistics for randomly censored data', *Biometrika* **63**, 465–474.

Krantz, P. H., Luce, R. D., Suppes, P., Tversky, A.: (1971), *Foundation of Measurement*, vol. 1, Academic Press, New York.

Krishnaiah, P. R., Kanal, L. N. (eds.): (1982), *Handbook of Statistics*, vol. 2, *Classification, Pattern Recognition and Reduction of Dimensionality*, North-Holland, Amsterdam, New York, Oxford.

Lachenbruch, P. A.: (1975), *Discriminant Analysis*, Hafner Press, New York.

Lachenbruch, P. A., Mickey, M. R.: (1968), 'Estimation of error rates in discriminant analysis', *Technometrics* **10**, 1–11.

Lawley, D. N.: (1943), 'On problems connected with item selection and test construction', *Proc. Roy. Soc. Edinburgh Sect. A* **61**, 273–287.

Lavenberg, S. S., (ed.): (1983), *Computer Performance Modeling Handbook*, Academic Press, New York.

Lazarsfeld, P. F.: (1950), 'The logical and mathematical foundation of latent structure analysis', in: *Measurement and Prediction*, eds. S. A. Stouffer, L. Guttman, E. A. Suchman, P. F. Lazarsfeld, S. A. Star, J. A. Clausen, Princeton University Press, New Jersey.

Lazarsfeld, P. F.: (1959), 'Latent Structure Analysis', in: *Psychology: A Study of Science*, vol. 3, ed. S. Koch, McGraw-Hill, New York.

Lazarsfeld, P. F.: (1967), 'Des concepts aux indices empiriques', in: *Le vocabulaire, des sciences sociales concepts et indices*, Paris, 3–36.

Lebioda, H.: (1983), 'Inheritance of descriptive traits of the nose' (in Polish), *Materiały i Prace Antrop.* **103**, 93–130.

Lebioda, H., Orczykowska-Świątkowska, Z.: (1984), 'Two methods of determining paternity probability on the basis of structure and pigmentation of the iris', *Stud. Phys. Anthrop.* **8**, 71–83.

Lehmann, E. L.: (1959), *Testing Statistical Hypothesis*, John Wiley and Sons, New York.

Lehmann, E. L.: (1966), 'Some concepts of dependence', *Ann. Math. Statist.* **37** 1137–1153.

Lininger, L., Gail, M., Green, S., Byar, D.: (1979), 'Comparison of four tests for equality of survival curves in the presence of stratification and censoring', *Biometrika* **66**, 419–428.

Lissowski, G.: (1978), *Statistical Association and Prediction, Problems of Formalization of Social Sciences*, Ossolineum, Wrocław.

Lord, F. M.: (1952), 'A theory of test scores', *Psychometric Monograph* 19.

Lord, F. M.: (1953), 'An application of confidence intervals and of maximum likelihood to the estimation of an examines ability', *Psychometrika* **18**, 57–76.

Łukaszewicz, J.: (1956), 'On proving paternity' (in Polish), *Zastos. Mat.* 2, 349–379.

Mantel, N.: (1966), 'Evaluation of survival data and two new order statistics arising in its considerations', *Cancer Chemoterapy Reports* **50**, 163–170.

Marshall, A. W., Olkin, I.: (1968), 'A general approach to some screening and classification problems', *J. Roy. Statist. Soc. Ser. B* **30**, 407–444.

Meier, P.: (1975), 'Estimation of distribution function from incomplete observations', in: *Perspectives in Probability and Statistic*, ed. J. Gani, Academic Press, New York, 67–87.

Mielniczuk, J.: (1986), 'Some asymptotic properties of kernel-type estimators of a density function in case of censored data', *Ann. Statist.* **14**, 766–773.

Mielniczuk, J.: (1985), 'Note on estimation of a number of errors in case of repetitive quality control', *Probab. Math. Statist.* **6**, 131–137.

Mokken, R. J.: (1971), *A Theory and Procedure of Scale Analysis*, Mouton, Paris, Hague.

Morse, N., Sacksteder, R.: (1966), 'Statistical isomorphism', *Ann. Math. Statist.* **37**, 203–213.

Moszyńska, M., Pleszczyńska, E.: (1983), 'Equivalence relations for statistical spaces', *Math. Operationsforsch. Statist. Ser. Statist.* **13**, 193–218.

Muthén, B.: (1978), 'Contributions to factor analysis of dichotomous variables', *Psychometrika* **43**, 551–560.

Nachtwey, D. S., Cameron, I. L.: (1968), 'Cell cycle analysis', in: *Methods in Cell Physiology*, ed. D. M. Prescott, Academic Press, New York, London, 213–259.

Nelson, W.: (1972), 'Theory and application of hazard plotting for censored failure data', *Technometrics* **14**, 945–965.

Neyman, J., Scott, E.: (1971), 'Outlier processes of phenomena and of related distributions', in: *Optimizing Methods in Statistics*, ed. J. S. Rustagi, Academic Press, New York, 413–430.

Niewiadomska-Bugaj, M.: (1987), 'Data transformation in two-class discriminant analysis' (in Polish), (Ph. D. Thesis, UAM, Poznań).

Nowak, S.: (1977), *Methodology of Sociological Research. General Problems*, PWN Reidel, Warszawa Dordrecht.

Nowakowska, M.: (1975), *Quantitative Psychology with Elements of Scientometrics* (in Polish), PWN, Warsawa.

Oakes, D.: (1982), 'A concordance test for independence in the presence of censoring', *Biometrics* **38**, 451–455.

Orczykowska-Świątkowska, Z.: (1972), 'Differentiation and inheritance of ridge counts in the soks' (in Polish), *Materiały i Prace Antrop.* **83**, 291–308.

Orczykowska-Świątkowska, Z.: (1977), 'Obiectivization of methods of paternity probability estimation' (in Polish), *Materiały i Prace Antrop.* **94**, 3–36.

Orczykowska-Świątkowska, Z., Krajewska, A.: (1984), 'The probability of paternity on the basis of 70 dermatoglyphic features', *Stud. Phys. Anthrop.* **8**, 53–69.

Orczykowska-Świątkowska, Z., Lebioda, H., Szczotka, H., Szczotkowa, Z.: (1985), 'The biometric method of paternity probability determination on the basis of morphological features, dermatoglyphics and blood groups', *Stud. Phys. Anthrop.* **8**, 85–104.

Orczykowska-Świątkowska, Z., Świątkowski, W.: (1970), 'Computation of the probability of paternity on the basis of blood group analysis' (in Polish), *Materiały i Prace Antrop.* **79**, 125–152.

Owen, D. B., Boddie, J. W.: (1976), 'A screening method for increasing acceptable product with some parameters unknown', *Technometrics* **18**, 195–199.

Owen, D. B., Hass, R. W.: (1978), 'Tables of the normal conditioned on *T*-distribution', in: *Contributions to Survey Sampling and Applied Statistics*, ed. H. A. David, Academic Press, New York, 295–318.

Owen, D. B., McIntire, D. M., Seymour, E.: (1975), 'Tables using one or two screening variables to increase acceptable product under one-sided specifications', *J. Quality Technology* **7**, 127–138.

Owen, D. B., Li, L.: (1979), 'Two-sided screening procedures in the bivariate case', *Technometrics* **21**, 79–85.

Owen, D. B., Su, Y. M.: (1977), 'Screening based on normal variables', *Technometrics* **19**, 65–68.

Parson, J. A., Rustad, R. C.: (1968), 'The distribution of DNA among dividing mitochondria of *Tetrahymena Pyriformis*', *J. Cell. Biol.* **37**, 683–693.

Peterson Jr., A.: (1977), 'Expressing the Kaplan–Meier estimator as a function of empirical subsurvival functions', *J. Amer. Statist. Assoc.* **72**, 854–858.

Pfanzagl, J.: (1968), *Theory of Measurement*, John Wiley and Sons, New York.

Poljakoff, N. L.: (1929), 'Die Vererbung und Ihre Praktische Anwendung (Übersicht Einiger Forschungen)', *Dtsch. Zeit. Gerichtl. Med.* **13**, 407–427.

Polya, G.: (1976), 'Probabilities in proof reading', *Amer. Math. Monthly* **83**, 42.

Prentice, R. L., Marek, P.: (1979), 'A qualitative discrepancy between censored data rank tests', *Biometrics* **35**, 861–867.

Randles, R. H., Broffit, J. D., Ramberg, J. S., Hogg, R. V.: (1978), 'Discriminant analysis based on ranks', *J. Amer. Statist. Assoc.* **79**, 379–385.

Rao, C. Radhakrishna: (1973), *Linear Statistical Inference and Its Applications*, John Wiley and Sons, New York.

Rash, G.: (1961), 'On general laws and the meaning of measurement in psychology', in: *Proceedings of the Fourth Berkeley Symposium on Mathematical Statistics and Probability*, vol. IV, ed. J. Neyman, Berkeley, 321–333.

Reche, O.: (1926), 'Anthropologische Beweisführung in Vaterschaftsprozessen', *Österr. Richterztg.* **19**, 157–159.

Reid, N.: (1981a), 'Estimating the median survival time', *Biometrika* **68**, 601–608.

Reid, N.: (1981b), 'Influence functions for censored data', *Ann. Statist.* **9**, 78–92.

Rejtő, L., Revesz, P.: (1973), 'Density estimation and pattern classification', *Problems Control Inform. Theory* **2**, 67–80.

Rényi, A.: (1959), 'On measures of dependence', *Acta Math. Acad. Sci. Hungar.* **10**, 441–451.

Roberts, F. S.: (1979), *Measurement Theory*, Addison-Wesley, Reading, Massachusetts.

Rogucka, E.: (1968), 'The differentiation of fingerprints in the Polish population' (in Polish), *Materiały i Prace Antrop.* **76**, 127–142.

Sacksteder, R.: (1967), 'A note on statistical equivalence', *Ann. Math. Statist.* **38**, 787–794.

Samejima, F.: (1973), 'A comment on Birnbaum's three-parameter logistic model in the latent trait theory', *Psychometrika* **38**, 221–233.

Samejima, F.: (1974), 'Normal ogive model on the continuous response level in the multidimensional latent space', *Psychometrika* **39**, 111–121.

Schade, H.: (1954), *Vaterschaftsbegutachtung*, Schweitzerbart Verlag, Stuttgart.

Schaeuble, J.: (1961), 'Die Haut', in: *De Genetica Medica*, ed. L. Gedda, Istituto Mendel, Roma, 157–183.

Scherr, A. L.: (1967), *An Analysis of Time-Shared Computer Systems*, MIT Press, Cambridge, Massachusetts.

Schoenfeld. J. R.: (1967), *Mathematical Logic*, Addison-Wesly, Reading, Massachusetts.

Schweitzer, B., Wolf, E. F.: (1981), 'On nonparametric measures of dependence for random variables', *Ann. Statist.* **9**, 879–885.

Schwidetzky, I.: (1956), 'Vaterschaftsdiagnose bei Unfraglichen Vätern, I. Mitteilung: Das Schauverfahren, II. Mitteilung: Essen-Möller-Methode und Vaterschaftslogarithmus', *Homo* **7**, 13–27, 205–214.

Shepard, D. C.: (1965), 'Production and elimination of excess DNA in ultraviolet-irradiated *Tetrahymena*', *Exp. Cell Res.* **38**, 570–579.

Shields, R.: (1977), 'Transition probability and the origin of variation in the cell cycle', *Nature* **267**, 704–707.

Shields, R.: (1978), 'Further evidence for a random transition in the cell cycle', *Nature* **273**, 755–758.
Shirahata, S.: (1981), 'Intraclass rank tests for independence', *Biometrika* **68**, 451–456.
Sisken, J. E.: (1963), 'Analyses of variation in intermitotic time', in: *Cinemicrography in Cell Biology*, ed. G. G. Rose, Academic Press, New York, London, 143–168.
Skibinsky, M.: (1967), 'Adequate subfields and sufficiency', *Ann. Math. Statist.* **38**,155–161.
Smith J. A., Martin, L.: (1973), 'Do cells cycle', *Proc. Nat. Acad. Sci. U.S.A.* **70**, 1263–1267.
Steinhaus, H.: (1954), 'The establishment of paternity' (in Polish), *Zastos. Mat.* **1**, 67–82.
Stevens, S. S.: (1951). 'Mathematics, measurement and psychophysics', in: *Handbook of Experimental Psychology*, ed. S. S. Stevens, John Wiley and Sons, New York.
Stidham, S.: (1972), '$L = \lambda W$. A discounted analogue and a new proof', *Oper. Res.* **20**, 1115–1126.
Stokman, F. N., Van Schuur, W. H.: (1980), 'Basic scaling', *Quality and Quantity* **14**, 5–30.
Stone, G. E., Miller, O. L.: (1965), 'A stable mitochondrial DNA in *Tetrahymena Pyriformis*', *J. Exp. Zool.* **159**, 33–37.
Stouffer, S. A., Guttman, L., Suchman, E. A., Lazarsfeld, P. P., Star, S. A., Clausen, J. A.: (1950), *Measurement and Prediction*, Princeton, New Jersey.
Susarla, V., van Ryzin, J.: (1978), 'Large sample theory for a Bayesian nonparametric survival curve estimator based on censored samples', *Ann. Statist.* **6**, 755–768.
Szczotka, H., Schlesinger, D.: (1980), 'Tables for calculation of paternity probability in the Polish population' (in Polish), *Materiały i Prace Antrop.* **98**, 3–52.
Szczotka, H., Szczotkowa, Z.: (1981), 'Interrelations of cephalometric traits between parents and offspring' (in Polish), *Materiały i Prace Antrop.* **101**, 56–132.
Szczotkowa, Z.: (1977), 'Morphological studies of the auride. Part I. Orthogenetic variability and sexual dimorphism' (in Polish), *Materiały i Prace Antrop.* **93**, 51–88.
Szczotkowa, Z.: (1979), 'Morphological studies of the auride, Part II. Heredity' (in Polish), *Materiały i Prace Antrop.* **96**, 77–128.
Szczotkowa, Z.: (1985), *Anthropology in Paternity Testing* (in Polish), PWN, Warszawa.
Szczotkowa, Z., Szczotka, H.: (1976), 'Attempt at an estimation of the utility of paternity probabilities. Calculated from blood groups' (in Polish), *Przegl. Antrop.* **42**, 3–18.
Takeuchi, K., Akahira, M.: (1975), 'Characterizations of prediction sufficiency (adequacy) in terms of risk functions', *Ann. Statist.* **3**, 1018–1024.
Torgersen, E. N.: (1977), 'Prediction sufficiency when the loss function does not depend on the unknown parameter', *Ann. Statist.* **5**, 155–163.
Torgerson, W. S.: (1958), *Theory and Methods of Scaling*, John Wiley and Sons, New York.
Tou, P., Gonzalez, C. G.: (1972), *Pattern Recognition Principles*, Addison-Wesley, London.
Tsiatis, A.: (1975), 'A nonidentifiability aspect of problem of competing risks', *Proc. Nat. Acad. Sci. U.S.A.* **72**, 20–22.
Tucker, L. R.: (1951), 'Academic ability test', Research Memorandum 51–17, Educational Testing Service, Princeton, New Jersey.
Turnbull, B., Brown, B., Hu, M.: (1974), 'Survivorship analysis of heart transplant', *J. Amer. Statist. Assoc.* **69**, 74–80.
Vithayasai, C.: (1975), 'Exact critical values of Hardy–Weinberg test statistics for two alleles', *Comm. Statist. A—Theory Metods* **1**, 229–242.
Wagner, S., Altman, S.: (1973), 'What time do the babooms come from the trees? (An estimation problem)', *Biometrics* **29**, 623–635.
Weninger, J.: (1935), 'Der Naturwissenschaftliche Vaterschaftsbeweis', *Wiener Klin. Wochensch.*, 1–10.
Wiese, B. N.: (1965), 'Effects of ultraviolet microbeam irradiation on morphogenesis in euplotes', *Exp. Cell Res.* **159**, 241–269.
Yahav, J.: (1979), 'On a screening problem', *Ann. Statist.* **7**, 1140–1143.
Yanagimoto, T.: (1972), 'Families of positively dependent random variables', *Ann. Inst. Statist. Math.* **24**, 559–573.
Yang, G.: (1977), 'Life expectancy under random censorship', *Stochastic Process. Appl.* **6**, 33–39.

Index

THEORY AND DECISION LIBRARY

SERIES B: MATHEMATICAL AND STATISTICAL METHODS

Editor: H. J. Skala, *University of Paderborn, Germany*

1. D. Rasch and M.L. Tiku (eds.): *Robustness of Statistical Methods and Nonparametric Statistics*. 1984 ISBN 90-277-2076-2
2. J.K. Sengupta: *Stochastic Optimization and Economic Models*. 1986 ISBN 90-277-2301-X
3. J. Aczél: *A Short Course on Functional Equations*. Based upon Recent Applications of the Social Behavioral Sciences. 1987 ISBN Hb 90-277-2376-1; Pb 90-277-2377-X
4. J. Kacprzyk and S.A. Orlovski (eds.): *Optimization Models Using Fuzzy Sets and Possibility Theory*. 1987 ISBN 90-277-2492-X
5. A.K. Gupta (ed.): *Advances in Multivariate Statistical Analysis*. Pillai Memorial Volume. 1987 ISBN 90-277-2531-4
6. R. Kruse and K.D. Meyer: *Statistics with Vague Data*. 1987 ISBN 90-277-2562-4
7. J.K. Sengupta: *Applied Mathematics for Economics*. 1987 ISBN 90-277-2588-8
8. H. Bozdogan and A.K. Gupta (eds.): *Multivariate Statistical Modeling and Data Analysis*. 1987 ISBN 90-277-2592-6
9. B.R. Munier (ed.): *Risk, Decision and Rationality*. 1988 ISBN 90-277-2624-8
10. F. Seo and M. Sakawa: *Multiple Criteria Decision Analysis in Regional Planning*. Concepts, Methods and Applications. 1988 ISBN 90-277-2641-8
11. I. Vajda: *Theory of Statistical Inference and Information*. 1989 ISBN 90-277-2781-3
12. J.K. Sengupta: *Efficiency Analysis by Production Frontiers*. The Nonparametric Approach. 1989 ISBN 0-7923-0028-9
15. A. Rapoport: *Decision Theory and Decision Behaviour*. Normative and Descriptive Approaches. 1989 ISBN 0-7923-0297-4
16. A. Chikán (ed.): *Inventory Models*. 1990 ISBN 0-7923-0494-2
17. T. Bromek and E. Pleszczyńska (eds.): *Statistical Inference*. Theory and Practice. 1991 ISBN 0-7923-0718-6
18. J. Kacprzyk and M. Fedrizzi (eds.): *Multiperson Decision Making Models Using Fuzzy Sets and Possibility Theory*. 1990 ISBN 0-7923-0884-0

KLUWER ACADEMIC PUBLISHERS — DORDRECHT/BOSTON/LONDON